Curtiss SB2C Helldiver

CROWOOD

AVIATION SERIES

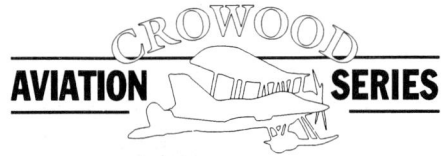

Curtiss SB2C HELLDIVER

Peter C. Smith

The Crowood Press

First published in 1998 by
The Crowood Press Ltd
Ramsbury, Marlborough
Wiltshire SN8 2HR

British Library Cataloguing-in-Publication Data
A catalogue record for this book is available from
the British Library.

ISBN 1 86126 141 1

Photograph previous page: Early test model in a
dive. Courtesy of Curtiss via Stanley I. Vaughn
Archive

Dediction
To the memory of Robert Olds,
my mentor and my friend.

Typefaces used: Goudy (*text*),
Cheltenham (*headings*)

Typeset and designed by
D & N Publishing
Membury Business Park, Lambourn Woodlands
Hungerford, Berkshire.

Printed and bound by Butler & Tanner, Frome,
Somerset.

HELLDIVERS' DRIVERS' SONG

O mother, dear mother, take down that blue star,
Replace it with one that is gold,
Your son is a Helldiver driver,
He'll never be thirty years old.

The people who work for Curtiss
Are frequently seen good and drunk,
One day with an awful hangover,
They mustered and designed that-there clunk.

Now the wings are built with precision –
The fuselage so strong it won't fail,
But who were the half-witted people,
Who designed the cockpit and tail.

The wings they fold and the safety belt snaps
I can open the bomb bay and split all the flaps,
But I've looked quite in vain (and I'm sure its not there),
For the gadget that retracts that big landing gear.

The skipper *hates* Helldiver drivers,
And he doesn't think much of that clunk,
Each time we slap down on his flight-deck –
He prays that this ship won't be sunk!

High, low, fast, slow,
'Wave him off' he yells to the l.s.o.
Slow, fast, low, high –
My God won't they ever learn how to fly?

My body lies under the ocean
My body lies under the sea
My body lies under the ocean
Wrapped up in a SB2C.

(Courtesy of Charles R. Shuford, John Olson and George Centre)

Contents

Introduction and Acknowledgements

The full story of the Curtiss SB2C Helldiver and its Army Air Force counterpart, the A-25 Shrike, has long needed recording in its entirety. Too often just the American Navy viewpoint has been recorded, and the US Army, the Royal Navy's involvement, the RAAF's indecision, and the post-war service of other nations – France, Greece, Italy, Portugal and Thailand – have been ignored completely in many accounts. From my decades' long research into dive-bombers and dive-bombing much of this fascinating story has been unearthed and is herewith included. Many of these facets of the Helldiver story, and much information that I located about this aircraft, were omitted from other dive-bomber volumes because of lack of space, so I am doubly happy to incorporate it here. So I again give thanks to Crowood for their great inspiration of the Aviation series, which surely deserves the enthusiastic support of both historian and model-maker alike.

A great many people were very kind and helpful to me during my researches in Columbus, Ohio, including my old friend the late Mark A. Savage; Ed Gillespie, Aviation legend and businessman, whose renovations of the old Columbus airport have preserved, and enhanced a great part of the original Curtiss buildings; Robert C. Tourt; Captain Jack M. Kennedy, USN (Rtd.), National Vice-President of the Navy League of the United States; Jay Eichenlaub, Vice-President of Nu-Tex; John B. Cornett, Chief, Division of Aviation and Andrew A. Doll, both of the Ohio Department of Transportation; Mike Smith of Lane Aviation; Hal Thorley; Chuck Williams of 288 WTTE, Columbus; E. A. Waller; R. W. Land, of Kniland Aerospace Industries; and E. Kenneth Hadden of the United States Aerobic Foundation. They were all most informative and knowledgeable. I would also like to acknowledge all the help and assistance I received during my visits to the US Air Force Museum at Wright-Patterson Field, Ohio.

I have been able to trace and make much use of a great deal of the original Curtiss documentation from copies held in various Government agencies; more exists but for some unknown reason is jealously guarded from most researchers. Fortunately the official papers of the US Navy, Marine Corps and Army Air Force were obtained from various centres in the USA. In a similar manner the official documents relating to the Royal Navy, Royal Australian Air Force, Royal Canadian Air Force and Defence Ministry, and the French Navy operations have all been made available to me, and between them tell the full story. In addition I have been greatly fortunate in locating former Curtiss-Wright employees and many former SB2C test pilots, combat aircrew and operational staff who knew many hitherto unrelated parts of the story and who were happy to talk to me, share their notes, diaries, scrapbooks and photographs to complete the picture.

I have again been fortunate in that I have been able to make several research trips to centres of American aviation history excellence in Ohio, California and Florida, where I have once more been afforded every facility and help by their historians, archivists and curators; these have, without exception, made me welcome and given me valuable direction and advice. As always I have also received widespread help and encouragement from friends and fellow historians in the UK, France and the USA who have again been generous with time, assistance and valuable criticism. The search for good quality and original photographs owes much to these kind persons and from the huge number examined, I sincerely hope that my final selection does the subject due justice. As is my usual practice, I have tried to limit as much as possible the use of any previously published sources, not all of which are reliable; but I have to pay tribute to the very first book on the Helldiver, written by my old friend and fellow writer, Robert Olds: *Helldiver Squadron*. Bob

wrote this book while working for Curtiss-Wright at the end of the war, and his widow, Peggy, has kindly made available to me all his papers and photographs. Also to Charles Crump's *History of VB-85* and Harold L. Buell's *Dauntless Helldivers*, both excellent sources.

Besides the above, I very much wish to thank the following individuals and organizations for their help and kindness to me in tracking down the true story of this much-maligned aircraft:

Hill Goodspeed, Emil Buehler Naval Aviation Library, National Museum of Naval Aviation, Pensacola, Florida, for his enthusiastic backing and his patience in researching photographs and archives during my stay at Pensacola making it such a pleasant and rewarding one; Vernon R. Smith, Textual Reference Division, National Archives at College Park, Maryland, for his outstanding dedication in tracing for me the many combat reports and war diaries of the various US units involved; Lawrence D. Webster for his help in tracking down survivors and wrecks; and George D. Bokos, PhD, Deputy Director, National Library of Greece, Athens; Captain N. Kostakis HN, Hellenic Embassy, London; Simon Watson, Aviation Bookshop, London; Danny J. Crawford, Head, reference section, History and Museums Division, Marine Corps Historical Center, Washington DC; Michael E. Starn, aviation specialist, Museums Branch, Marine Corps History and Museums Division, Quantico, Virginia; Dan Hagedorn, reference team leader, Archives Division, Smithsonian Institution, National Air and Space Museum, Washington, DC; M. Vincent Mollet, conservateur des archives centrales, le chef du Service Historique de la Marine, Marine Nationale, Vincennes, France; M. Jean Cuny, historian, Soignolles-en-Brie, France, for invaluable help in unearthing little-known documents on the French purchasing commission in the USA; Lieutenant Duarte Monteiro, Portugal.

Arthur Pearcy, historian and friend, Sharnbrook, Bedford with his unique knowledge of US Coastguard operations; Thomas M. Alison, curator, Aeronautics Department, National Air & Space Museum, Washington DC; Edwin R. Walker, historian, Great Bookham, Surrey, my old friend, for his support and company while I was researching at the Public Record Office, Kew, London.

Teruaki Kawano, Military History Department, National Institute for Defence Studies, Tokyo, again an invaluable source of strength in the Japanese archives and in assessing the accuracy of various battle claims and reports. Lieutenant Colonel Edward A. Miller, USAF, Department of the Air Force, Washington DC; David T. Selzam, senior editor of *The Bugler*, USMC; Jim Shaw, 'Last of the Great Stoof Drivers' Columbus, Ohio; Robert J. Cressman, Department of the Navy, Contemporary History Branch, and William T. Baker, Aviation History Branch, Washington DC; Bill Johnston, Directorate of History and Heritage, National Defence Headquarters, Ottawa, Canada; Jerry R. Shore and Debbie

Holland of the Fleet Air Arm Museum, RNAS Yeovilton; Sean Paul Milligan, author; Martha L. Fruth, Registrar of Collections, Ohio Historical Society, Columbus; G. A. Smith, Office of the Second Sea Lord & C-in-C Naval Home Command, Portsmouth: R. S. 'Dickie' Rolph, BEM; Joe Pruden, archives reference team, Smithsonian Institution, Washington DC; my friend and fellow author; Mel L. Shettle; Colonel Ted Short, Confederate Air Force, Texas; Ray C. Sturtivant, author; my very good friends Stanley I. Vaughn III and his charming wife Sandi Vaughn and to his father and Columbus factory manager, Stanley Vaughn II for his memoirs, photographs and notes. Colonel Tom Barnes, Confederate Air Force, Marietta, Georgia; Les Schnyder, restoration manager, National Museum of Naval Aviation, Pensacola, Florida; Group Captain Apisit Chulamokha, Royal Thai Air Force Directorate of Intelligence.

Special thanks to those Helldiver men whose personal memoirs were given so freely: to Commander James Alton Chinn for permission to quote from several remarkable letters he received from SB2C men; to Arthur N. Avery; Bill Adam; James T. Booker; Charles M. Crump; George Center; Chuck Downey; Peg Gumb, widow of Commander I. T. Gumb, USN; Frank M. Gallagher, USMC; Commander Richard W. Mann, USN Rtd; Ron McMasters; William S. Palmer; John P. Piercy; Okey C. Roush; Colin Shadell; D. F. Sidnell; Lawrence D. Webster; Ron Hinrichs; Benjamin G. Preston; Captain Robert B. Wood, USN Rtd.; Commander Charles R. Shuford, USN Rtd; ARM2c William C. Pinkerton; Roger H. Went; and Captain M. H. de Lestapis.

To all these fine people my sincere thanks for helping me tell the Helldiver story.

Peter C. Smith,
March, 1998

A Note on Ranks

British readers are probably unfamiliar with many of the ranks of the personnel serving in the US Navy, the Marine Corps and Army Air Force and the French Navy and Air Force as recorded in this book, although the ranks of the Royal Australian Air Force, being based on the RAF, will be familiar. Often no direct comparisons with British ranks are possible and the following is for general guidance only. In the U S Navy the rank of Lieutenant is sub-divided further as Lieutenant and Lieutenant (j.g.). The 'JayGees' – standing for Lieutenant (Junior Grade) – form the bulk of the young naval pilots who flew the SBD in combat, with full Lieutenants above them and Ensigns (which has no British equivalent) below them. Roughly Royal Navy ranks would be Midshipman, Sub-Lieutenant and Lieutenant. In the US Marine Corps and USAAF the same seniority ranks of the Dauntless or Banshee flyers would be Colonel, 1 Lieutenant, 2 Lieutenant. In the French Navy these ranks would be *Enseigne de Vaisseau 2me Classe*, *1re Classe* and *Lieutenant de Vaisseau*. Higher ranks stepping up were *Capitaine de Corvette*, (Lieutenant Commander), *Capitaine de Fregate* (Commander) and *Capitaine de Vaisseau* (Captain). Rear-seat men were aviation radiomen (ARs) of various classes in the USN; Sergeants or Staff Sergeants in the USAAF and wireless operator/rear gunner (WOAGs) in the RAAF.

A Note on Aircraft Names

It was not until 1 October, 1941, that U S Navy aircraft were 'officially' allocated names although Curtiss had used the name 'Helldiver' twice before, for the F8C/O2C biplane of the early 1930s, and the SBC (known in Britain as the Cleveland) dive-bomber of the late 1930s. The USAAF followed this practice even later, in April 1942, when the name 'Shrike' was allocated to the A-25, again this had been used before for the Curtiss A-12 of 1931 and the twin-engined A-18 of later years. As before, in neither service did these names really find much use among the aircrew themselves. They rejected the 'official' names, and also the name by which Curtiss themselves desperately tried to promote their product, 'Fist of the Fleet'. This was certainly to be an accurate descriptive name, once the Helldiver got into action, but before then it earned itself a far more widespread appellation, as 'The Big-Tailed Beast', or simply, 'The Beast', along with an unsavoury (and undeserved) reputation for killing its pilots. The many veterans who contributed to this book still use that name – some affectionately, some with venom – but she is mainly known as the SB2C: a unique dive bomber. The official designations therefore remained predominant, SB2C being the composite of the aircraft's two main functions, namely S = Scout and B = Bomber; 2 indicates that she is the second series of dedicated dive bombers for the Navy, and C is the manufacturer's designation for Curtiss, followed by various mark numbers. On the Army side they allocated all their dive-bombing aircraft to the A for 'Attack' range, plus their number, while the RAAF allocated their own designations to all aircraft and the Shrike became the A-69 in RAAF service.

CHAPTER ONE

The Development of the Helldiver

The United States Navy had 're-invented' dive-bombing in 1928, utilizing fleet fighters to prove to their own satisfaction that delivery of ordnance in a near-vertical dive was by far and away the most accurate method of attacking both fixed and, more importantly, swift-moving targets like warships.[1] Several years of experiments by both West Coast and East Coast-based squadrons, plus work done by the United States Marine Corps earlier that decade, had shown that the potential of the dive delivery of bombs was outstanding, although it was found that the stresses on the aircraft of the day were great. The Bureau of Aeronautics therefore came to the logical conclusions that first, dive-bombing was a necessary requirement for the Navy and that second, specialized aircraft were required to carry it out.

They wasted no time in putting their findings into practice, and starting with the Martin BM-1 and proceeding through the various experimental models such as the Chance Vought SBU-1 and the Northrop BT-1, a whole range of biplane and monoplane dive-bombers were produced by a selection of companies which led to rapid advances in technique, speed, strength and power. With the development of the Douglas SBD Dauntless from the Northrop design the all-round experimentation had resulted in a fine, sturdy and reliable platform, easy to fly and maintain and able to deliver its externally carried payload of 1,000lb (450kg) bombs with great precision to the target. The Dauntless went on to carve itself a memorable place in American naval history and was one of the most successful dive-bombers of all time, ranking with the German Ju 87 Stuka and the Japanese Aichi D3A1 Val as the premier anti-warship weapons of World War II.[2] The fact remained, however, that the basic design of the Dauntless dated back to 1934 and, towards the end of the decade, and with war clearly more and more likely, the US Navy could not be content with things as they then stood. The SBD was scheduled for early replacement and what was needed was a much more powerful dive-bomber with which to equip the brand-new aircraft carriers of the *Essex* class which were then being laid down on the stocks as part of a hugely expanded naval re-armament programme, the urgency of which grew with each year that passed as the threat of war quickly became reality.

Aviation technology was moving on apace at this period, with new engine designs appearing in quick succession, each providing greater power than its predecessor and thus enabling higher speeds and great bomb loads to be carried over further distances. The Bu-Aer engineers had to specify a design which combined what then seemed the very last word in such engines (which in 1938 meant the Wright R-2600 engine developing 1,500hp to give a maximum speed of 319mph (513km/h), compared to the R-1535 engine of 700hp and maximum speed of 223mph (359km/h) that had equipped the XBT-1 only four years earlier), but combine the extra size, weight and power available to fit the new concept of aircraft operations for the new fleet of aircraft carriers being constructed.[3]

BuAer VSB Competition 1938

As was customary in those days, early in 1938 the BuAer announced a competition open to all reputable builders of service aircraft. The new VSB specification, SD-110-25, of 29 June 1938,[4] when presented by Admiral Cook, was a gruelling and demanding one even then and was to impose limitations and requirements which scuppered one promising design, the Brewster SB2A Buccaneer, and almost aborted the Curtiss SB2C Helldiver on several occasions. To take advantage of the newly developed air-cooled radial engines the Navy wanted to squeeze the absolute maximum out of the designers. The new dive-bomber was to carry internally a 1,000lb bomb for a range of 1,000 miles (1,609km).

A monoplane was required, '… a single-engined, two-place scout and dive-bomber land plane for use aboard aircraft carriers.'[5] It had to be capable of, '… taking off from the deck of an aircraft carrier with or without the aid of a catapult and landing on deck in an arresting gear or on an ordinary landing field.' Retractable landing gear and armour protection for the aircrew was to be built in. New to Curtiss also was the fitting of double-split flaps and high-capacity dive brakes. Innovations also included rubber de-icing equipment, a rebreather oxygen system and alternate fitting retrospectively of foreign-built 20mm cannon. This caused retro-design problems when the choice of cannon was widened and different types had to be accommodated in the wings.

One of the most telling of the requirements was the decision by the BuAer that *two* of the new dive-bombers must be able to fit on the 41ft (12.5m) by 48ft (14.6m) lifts (elevators) of the new aircraft carriers. Thus the specification stated:

> The limiting overall dimensions shall be those which will permit handling the airplane with wings folded on an elevator 41 feet by 48 feet with a clearance of not less than 12 inches all around. However, minimum dimensions are of great importance. Attention shall be given to the possibility of handling two airplanes on an elevator of these dimensions.

The thinking behind this was for a faster turn-around of operations, which was sound enough. A carrier's air complement is always limited, and even the big new carriers would only carry a finite number of aircraft. Their air groups had to be divided into types for different missions – scouts, dive-bombers, torpedo bombers and fighters – and the balance of each was always changing according to the needs of offensive action (to strike hard at the enemy before he hit you) and defence (to provide, with a combination of defending fighters, warning patrols of scouts and later, efficient radar, plus the maximum number of long-range

THE DEVELOPMENT OF THE HELLDIVER

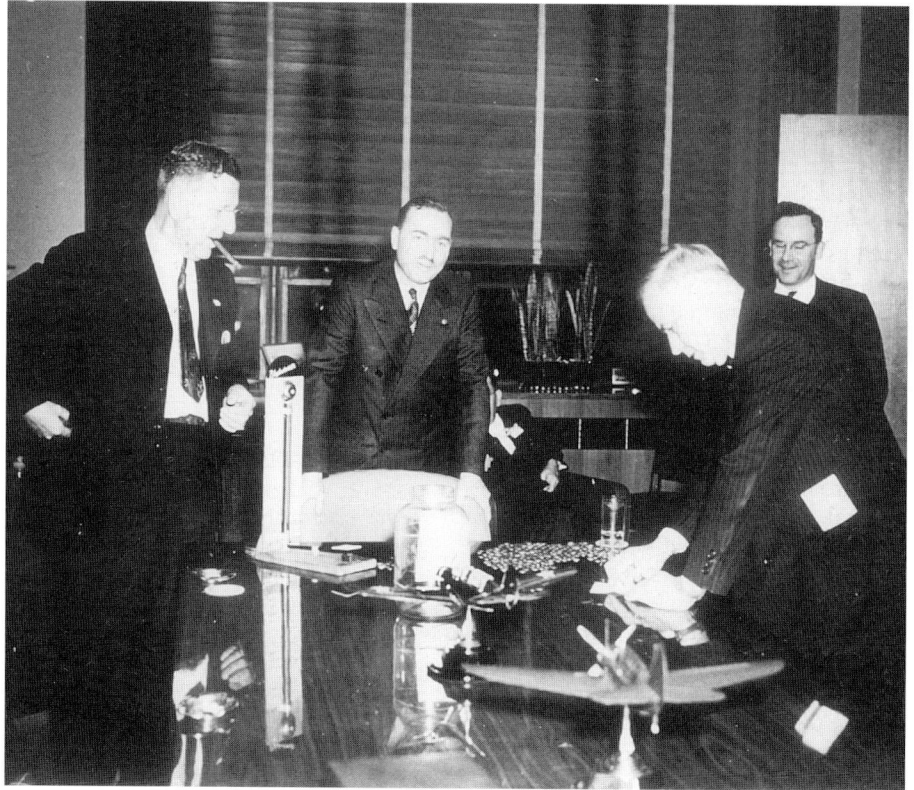

Raymond C. Blaylock (left) **and Stanley I. Vaughn** (front right). Stanley I. Vaughn Archive

and close-range anti-aircraft guns, a ring of invulnerability around your own ships). Obviously the fleet which could turn around its aircraft the fastest stood a better chance of survival on all counts, but especially on the offensive side, if it could get its dive-bombers refuelled and re-armed and then back up on deck to fly off again quickly. That was the theory, anyway (proven by the fate of the four Japanese carriers at Midway caught and sunk in the middle of this classic dilemma).

The new carriers themselves were designed and under way, so the new aircraft had to fit the carriers. This priority was necessary because huge and complex warships take a long time to design and build, so it was felt it would be easier to fit the planes to the ship than the other way round.

This left a tricky problem for chief engineer Don Berlin's team of designers of the new dive-bombers who, quite naturally, had they been left to design them with only aeronautical limitations, would have produced a very different aircraft. It would have been bigger, thus maintaining aerodynamic ratios, but as this was ruled out right at the start, the new dive-bombers had to be designed much broader in relation to their length than was normal, and

in compensation for the resulting lack of directional stability, the tail area had to be increased enormously, giving the Helldiver its distinctive feature. But this was to prove insufficient, and even with the fitting of the upward hydraulically-folding outer wing panels, the fit was tight. The SBD had fixed wings, which the Navy liked for its inherent strength, so important for dive-bombers, and they agreed only reluctantly to the adoption of folding wings for the new aircraft. The Navy was quite specific: 'Overall height with wings folded shall not exceed seventeen feet'.

And so it was to prove, in the long run, that when the new carriers were ready for combat, the new dive-bombers to operate from them were not – and unfortunately this unhappy result came in the middle of the largest sea/air war in history! Much of the subsequent trials and tribulations of the SB2C came about by trying to make two round pegs fit into an oblong hole. Ironically, when the Helldiver did finally appear, the actual need for two of them to fit on one elevator had passed and new techniques, equipment and operating methods had rendered it unnecessary ...[6]

On weights the specification demanded the following:

Minimum weight and size are desirable, but consideration will be given to designs of 9,000 to 9,500 pounds (in the 500lb bomb condition with sufficient fuel for 1,000 miles range at economical speed), if increased performance sufficient to warrant such weight increase proves attainable.

[Also specified was that] The first airplane under Item 3(a) shall be delivered, set up, serviced with 200 gallons of gasoline and 12 gallons of oil, and ready for flight at or near the contractor's plant ...

Innovations, desirable enough, turned the SB2C into the most complicated Navy aircraft designed up to that time. The hydraulic system was comprehensive and worked not only the folding wings actuator and locking pins, but also the landing flaps and dive brakes; the internal bomb-bay doors and the bomb release crutches themselves; the main landing gear which was fully retractable, the tail wheel, also initially fully retractable, as well as the Hayes expander tube brakes on the wheels themselves; the power turret and the collapsible turtle-deck mechanism; and the forward fixed armament chargers.

Another of the original major requirements which caused endless problems was that of a power gun turret for defensive fire. The fitting of such a heavy unit into what was already a tight design gave both Brewster and Curtiss enormous headaches, and in the end had to be abandoned – but only after causing considerable delay and complication. Again, in 1938, the power operated machine-gun turret was considered essential to provide sufficient defensive fire against fighters. By the time the Helldiver had entered service it had been shown by four years of war that no defensive weaponry could fully protect any bomber, and that what was required was escorting fighter protection. Thus the eventual abandonment of this feature, like the two on a lift concept, caused no operational problems when it was finally decided upon; but that was many years in the future.

The original specification was built around the new experimental Wright R-2600-8 Cyclone 14 engine which was rated at 1,700hp. This was expected to give a top speed of 313mph (504km/h), and a minimum speed of 71mph (114km/h). To assist aerodynamic performance an internal bomb-bay was specified for the first time, and a much larger fuel capacity was featured to increase the range of the striking power of the fleet. Tight limits were put on other performance aspects, with minimum requirements specified for the maximum stalling speed; the service ceiling; take-off into a 25-knot headwind within a distance of 265ft (80.7m); maximum braked diving speed with the new flaps and range.

Following invitations to tender for a Class VSB aircraft on the 6 August 1938, at the Curtiss-Wright Buffalo plant Ray Blaylock was given the assignment as chief engineer and set up a team of 300 engineers to tackle it. In all, six different aircraft companies responded to the challenge in December, but only two, the Brewster and the Curtiss, featured the new engines and the rest were soon discarded.

Curtiss-Wright Proposal: Initial Figures

Curtiss-Wright at Buffalo based their preliminary weight and balance estimate, Proposal 'A' on that specification, and took the useful loads directly from it, except for the fuel and oil capacities, which were calculated; the optical sight type N-2A used in place of telescopic sight; and flexible gun

equipment, altered to conform with the new Curtiss mount and Radio Equipment weights, which were in accordance with Specification SR-79A.[7] Their calculations came up with a set of figures compared with the current P-36A fighter aircraft under production, thus:

Wing Group	P-36A	VSB
Wing area	237	385
Design gross weight	5,535lb	10,000lb
Aspect ratio	5.93	6.5
Load factors	12,12,7	14,6,8
Thickness	15.9%	18.9%
Panel weight	745	–
Panel unit weight	4.05	–
(184.4sq ft)		

Wing loading increase = $\dfrac{26.0}{23.4}$ = 1.11 or 11%

or

Weight increase = 0.5 × 11 = 5.5%

AA load factor increase = $\dfrac{14}{12}$ = 1.17 or 17%

or

Weight increase = 0.5 × 17 = 8.5%

$\qquad$ 4.5 × 1.055 × 1.085 = 4.64lb/sq ft

$\qquad$ 4.64 × 313sq ft = 1,452lb

Additions

Folding outer panels: 8 hinge forgings req'd		
	est. steel	108lb
	est. dural	87.0lb
Retracting mechanism (hydraulic)		
Cylinders (2)		24.0lb
Links and bolts (4)		8.0
Hinge-pin mechanism		
Hinge pins (steel)		
4 + screw assembly	9.0	
Cable 70ft ¹⁄₁₆ flax	3.6	
Crank unit	1.5	
AN bolts 1in (4)	3.6	
Total	17.7	
Tubing 500in of ¼in × .035 Everdue		4.0
Oil (est 3.5# in 2 cylinders)		5.0
Double split flaps		
(41.6sq ft proj. area @ 2.23lb/sq ft)		93lb
Ailerons		
(30.5 sq ft total @ 1.44lb/sq ft)		44
Aspect ratio increase		15
Bomb rack supports and		
reinfor. (1,000lb bomb)		6
Root thickness increased 20%		–100
Torque-box type construction		
vs. multi-cellular		+30
Integral oil tank (32 gal vol.)		
0.2in/gal + filter cap and fittings		+13
Tip slots and hydraulic cyls (2), cables (int), oil, tubings, etc.		
2#/ft × 14' = 28 slots, 3 × 2 = 6# cyls.		+40
Integral fuel tanks (2) 90 gal,		
each @ 0.25#/gal		+45

Main beams bent to provide clearance
for 1,000lb bomb $\qquad$ +50

Total wing group: $\qquad$ 1,775lb

$\qquad \dfrac{1775}{385}$ = 4.61lb/sq ft

Tail Group

Construction similar to SBC-3 tail group, except stabilizer will be full cantilever design and higher aspect ratio on vertical tail.

	VSB	SBC-3		
	lb/sq ft	lb/sq ft	Area	Weight
Stabilizer	1.64	1.80	36.0	65
Elevator	1.55	1.56	46.9	73
Fin	1.37	1.61	12.4	20
Rudder	1.65	1.65	17.8	29
Total		1.65	113.1	187

Body Group

All metal (24ST Dural), all monocoque construction similar to P-36 fuselage except for the following additions and modifications:

Y1P-36 fuselage weighted 1.61lb per sq ft skin area,	
therefore 258 sq ft × 1.61 =	415lb
Plus:	
Arresting gear fittings	
and reinforcements	20
Catapult hooks and reinforcements	15
Bombing window	5
Hoisting sling compartment and	
reinforcements	10
Additional cockpit	15
Total fuselage	48lb

Windshield, hoods, cabin between cockpits and gunner's turret will be equipped with 'Plexiglas'. The weights for the remaining items of this group are based on corresponding parts of the SBC-3.

The weight of 20lb for the bomb-door hydraulic controls includes only the operating cylinders, liens, oil and control valve. The main oil-supply tank, engine pump, accumulator tank etc. are included under the landing gear hydraulic system.

Submission Made

The Curtiss submission of 6 December, the Model 84, showed all the outward hallmarks of the company's contemporary aircraft, including its immediate predecessor, the SBC-3 biplane dive-bomber, as can be seen from its fuselage shape aft of the cockpit with its fold-away turtleback. The name 'Helldiver' itself dated back to the company's F8C fighter of the early thirties

Specification – Model 84 XSB2C-1

Engine type:	XR-2600-8
Armament:	Two 30mm forward-firing fixed machine guns; two 50mm rear-cover flexible machine guns; maximum bomb load 1,000lb (454kg)
Dimensions:	Length 35ft 5in (10.79m); height 10ft 4in (3.14m); wingspan 50ft (15.24m); wing area 385 sq ft (422), 35.8 sq m (39.2)
Weights:	Maximum weight 10,189lb (4,622kg); empty weight 7,030lb (3,189kg)
Performance:	Maximum speed 325mph (523km/h)
	Take-off power 1,500hp; (1,094kW)
	Range as dive-bomber 1,330 miles (2,140km)
	Climb height 15,000ft/5.6min (4,572m/5.6min)
	Service ceiling 30,000ft (9,144m)
Number completed:	1
Serial number:	1758 (Bu. No.-Navy – Serial Army)
Contract:	Order date; 15 May 39
	Delivery date; 18 Dec 40

which was utilized for early dive-bombing experimental work, and was attributed to Charles E. Hathorn, that aircraft's designer. It had become famous world-wide and a company tradition for its dive-bombers, later formally adopted by the US Navy itself.

The preliminary design was examined in detail by BuAer, and although not completely satisfied, they decided that the Model 84 came closest to the required configuration, with a few modifications being required. Accordingly, on 15 May 1939 the Navy placed an order for a single prototype, which became the XSB2C-1. (Brewster were also awarded a similar order for their XSB2A-1, presumably as insurance, but in the event, the Buccaneer was to face equally as large a series of problems as the Helldiver and never saw combat service.) The contract (No. (a) 66638) was signed, and the BuAer number 1758 was allocated to the prototype. (The Curtiss-Wright number was 13032, and the engine government number 4559.)

Nor did the original design stay immutable (save in the dimensional respect), because as the war in Europe progressed, new lessons and ideas had to be continually absorbed and assimilated. This led to many changes, and inevitably those changes led to increases in weight – which had increased from a fully laden weight of 10,131lb (4,595kg) (empty weight of 7,988lb) (3,623kg) in November 1940, up to 11,900lb (5,398kg) by 1939 and to 16,800lb (7,620kg) six years later) – which in turn resulted in a reduction in performance, with top speed down to 296mph (476km/h) at 15,000ft (4,570m) with a single 500lb bomb), a higher stalling speed of 74mph (119km/h) and an extended take-off distance, now 382ft (116m), as well as more delays. New inventions, airborne radar, and the operational needs of war – such as

G. B. Clark and Stanley I. Vaughn.
Stanley I. Vaughn Archive

Charles E. Hawthorn, George Page and Stanley I. Vaughn. Stanley I. Vaughn Archive

ever-increasing bomb loads (from the original 1,000lb, through 1,600lb to 2,000lb), designing and fitting the new 20mm cannon and replacing the original .50 calibre machine guns all added to the problems.

The mock-up which was built at the Curtiss-Wright company's Buffalo plant incorporated the enclosed turret for the rear gunner, and this was inspected at the end of May 1939. This was the first time such an installation had appeared on carrier-born aircraft, and it caused endless problems: a major redesign was called for. Two further mock-ups followed in an effort to resolve this problem, and final approval was given in February 1940. The installation might have been all right but the mounting itself continued to cause much heartache. However, it was not until the spring of 1942 that BuAer finally relented and authorized Curtiss to replace this turret with a twin .30 calibre flexible mounting, and contracted with three more firms to develop independently a turret for the Helldiver. This later had to be abandoned, as further modifications had already impinged on the marginal stability to an alarming extent.[8]

Good starboard broadside view of the XSB2C-1, Model 84, No. 1758 Helldiver on the ground. Note the old-type telescopic bombing sight, retractable tail-wheel and small tail. Mrs E. J. Cooper via National Aviation Museum of Canada

An in-flight view of the XSB2C-1 Helldiver. Mrs E. J. Cooper via National Aviation Museum of Canada

In July 1939 wind-tunnel testing took place, and continued through February 1940, along with full-scale experimental flight tests, and these revealed a major problem by showing that the wing maximum lift coefficient would turn out to be lower than planned, which would result in a high stalling speed. Extending the nose by one foot was the recommended solution to the instability in the fore-and-aft axis. Other problems meant that solutions had to be sought quickly. Blaylock tried increasing the size of the leading edge slats until they encompassed the full span – but to no improvement. The ailerons were drooped – still no improvement. Finally the wing area was increased from 385 to 422sq ft, which meant a complete redesign of the whole wing. This resulted in increased weight and loss of speed. In compensation the widespread use of magnesium alloy was adopted for parts: but this, too, proved in most cases to be a failure and they had to be reworked in aluminium. Another attempt to lose some excess weight was the incorporation of aluminium

forgings for the wing carry-through area and a redesign of structural members. Curtiss-Wright turned to the experienced firm of Alcoa in this, but that company was experiencing expansion difficulties of its own, and its problems of inaccurate machined forgings and late deliveries, added another layer of complication to the SB2C's tale of woe! And of course the weight still remorselessly increased.

Another difficulty encountered was with the new (to Curtiss) dive brakes, which, when extended, failed to slow the aircraft down as much as predicted. Varying the angles of opening of both the split dive flaps and the landing flaps were tried in tests to overcome this problem.

Then the new powerplant, the experimental Wright Aeronautical R-2600-8 engine of their incorporated company being built out at Lockland, Ohio, ran into quality problems and fell further and further behind schedule throughout 1939 and on into 1940. Rated at 1,500hp at 5,800ft, it had a military rating of 1,700hp at 3,000ft. It was not finally delivered until August

1940 – which, due to the many other design and testing problems encountered, proved just as well! Similarly the propeller was also experimental, being of the new Curtiss-Electric 12ft, constant speed, three-bladed type which also had teething problems.

Meanwhile events on the world's stage had moved on apace, with the Stuka dive-bombers of the Luftwaffe spearheading the swathe of the Blitzkrieg that quickly over-ran Poland, Norway, the Netherlands, Belgium, Luxembourg and France. At sea the same Junkers Ju 87 aircraft had sunk and severely damaged a large number of Allied warships, while the Royal Navy's Blackburn Skua dive-bombers had sunk the German cruiser *Königsberg*. There could be little doubt in the minds of the US Navy chiefs that their faith in the dive-bomber was justified by these events, which were to continue at a fast pace throughout 1941 when both the British aircraft carriers *Illustrious* and *Formidable* were heavily damaged, and several cruisers and a large number of destroyers were sunk by dive-bombing in the Mediterranean.

CHAPTER TWO

Rapid Expansion

The great expansion bill for the Navy, signed by the President on 14 June 1940 under the National Defense Program, with its accompanying vast increase in the number of naval aircraft required, was America's answer to this worsening situation. One facet of this mighty planned growth was to turn American aviation production from a virtual 'cottage industry' producing handfuls of machines, into a vast, automated powerhouse turning out thousands to meet the demands of its own forces and that of Great Britain. This America proved able to do in the long run, and it really did become 'The Arsenal of Democracy'; but it could not be done overnight!

Existing production facilities were simply inadequate to meet this new demand, and further, the scheme called for a dispersal of production both for security reasons, and also to take advantage of fresh, untapped labour forces in areas hitherto not noted for heavy industries. The recommendation of the National Advisory Committee (more commonly known as the Knudsen Committee, after its chairman), was that the Government subsidize the building of brand-new aircraft plants away from existing centres, and that it use a fresh labour force and brand-new techniques from scratch. Moreover, this all had to be done within one year!

This was to prove difficult enough, but unskilled labour had to be trained, vast quantities of specialized machinery had to be obtained and installed, production lines for a thousand and more parts had to be established, and sub-contractors had to be located, all under enormous pressure with the clock running out. Curtiss was affected no less than other specialist dive-bomber builders – Vultee, Brewster, even to some extent Douglas. All had to face these enormous expansion and relocation problems while simultaneously designing, testing and amending brand-new aircraft with untried powerplants and constantly increasing changes dictated by war experience. For some, like Brewster, it eventually

Publicity shot of 'Rosie the riveteer', a Curtiss-Wright wartime employee, with a model of the SB2C Helldiver, in front of Old Glory, to stir the American heart and stimulate production. Front cover of Curtiss Fly Leaf, **Vol. XXV, No. 6, January–February 1943 edition.** Curtiss-Wright via Stanley I. Vaughn Archive

proved insurmountable; for others, like Vultee and Curtiss, it was to involve costly delays which ultimately affected their products, good though they were.

Curtiss was an old-established company with an track record of supplying the Navy with the aircraft it needed, and as such it deserved, and earned, the trust of the Navy Department that it would come through.

Moreover it was in a vice of the Navy's own making that the company strove to perfect the XSB2C against increasing odds, as we have seen. That faith was ultimately to be justified, but only after years of frustration.

The first move that the Curtiss-Wright Corporation made in compliance with the National Defence Program directives was the urgent seeking out of new plant sites, as existing manufacturing facilities at Buffalo and St Louis could not be expanded enough, due to shortage of labour. At the end of July 1940, therefore, they sent the vice president, Charles W. Loos, along with the airplane division factory manager, Peter N. Jansen, to look at some suitable cities in the mid-west, and sum up their potential. A former Curtiss chief test pilot, John Corrodi, had his own flying company based at Port Columbus airfield and he drew their attention to the potential of the area. With the enthusiastic backing of the Mayor of Columbus, Floyd F. Green, and with the aid of his chief engineer, Paul Maetzel, a suitable site to the south-west of the airfield was selected and a report recommending it was submitted to the national government within three days!

Following study and inspection by the Civil Aeronautics Administration and the Reconstruction Finance Corporation, a government agency founded to provide the funding for the Navy of such new factories, approval was given. It was estimated that a pool of 40,000 existing workers could be tapped – but only at the expense of existing businesses who could not match the new pay rates. By the end of the war Curtiss-Wright had trained an extra 84,000 workers, while sub-contractors had trained another 16,000 on top of that! Thus at a special meeting of Columbus City Council convened on 1 November 1940, the leasing of the area was agreed. Less than a month later, on 28 November 1940, Guy Vaughan, the president of Curtiss-Wright, Burdette S. Wright, vice-president, and Major Green attended the ground-breaking ceremony, along with the new factory manager designate Vice President J. A. Williams. Things were moving at breathtaking speed – but the actual work on the XSB2C was moving ahead of even these events.

The construction of a virgin factory on a regal scale, along with the various ancillary buildings, offices and railroad sidings was going to take time, even at the speed of the Americans; but work on the new dive-bomber could not wait. Accordingly the company took out a lease on the summer Cattle Exhibition Barns at the Ohio State Fairground in the city for the winter months as temporary accommodation, and by 10 December 1940 the new employment, training, plant layout and other officers were being set up there and machinery

Specification – Model 84 SB2C-1	
Engine type:	R-2600-8
Armament:	Four 50mm forward-firing fixed machine guns; two 30mm rear-cover flexible machine guns; maximum bomb load 2,000lb (907kg); 1-Mk-13 Torpedo
Dimensions:	Length 36ft 8in (11.17m); height 13ft 1⅝in (4.01m); wingspan 49ft 8⅝in (15.15m); wing area 422 sq ft (39.2 sq m)
Weights:	Maximum weight 16,607lb (7,533kg); empty weight 10,114lb (4,588kg)
Performance:	Maximum speed 273mph (439km/h) Take-off power 1,700hp (1,268kW) Range as dive-bomber 1,375 miles (2,213km) Climb height 10,000ft/11.4min (3,048m/11.4min) Service ceiling 21,200ft (6,462m)
Number completed:	200
Serial numbers:	00001/00200 (Bu. No.-Navy – Serial Army)
Contract:	Order date; 19 Nov 40 Delivery date; Sep 42–Aug 43

was being moved into the barns. From here the Development Planning Committee, with Don Berlin, Ray Blaylock, J. A. Williams and John Corrodi, started planning the new factory in every detail. The actual fabrication of parts for both the SB2C and the equally new SO3C seaplane, to be built on the same site, was commenced under the direction of E. K. Fry, sheet-metal general foreman, with materials being obtained from the Buffalo plant.

It was on 19 November 1940, that Contract No(s). 79082,[9] with amendment, was finally obtained for the production of 370 SB2C-1 aircraft for the Navy. This was duly signed, despite the fact that the prototype had still not flown! This was unprecedented, but it was a reflection of the faith that the US Navy had in Curtiss, a company which had provided them with many great aircraft in the past. It also reflected the increasing anxiety of the Navy, with the war in Europe threatening to extend itself further and involve America. The 'Two-Ocean' fleet was authorized and under construction, and the need for an up-to-date dive-bomber was all too evident, so chances were taken. Further orders followed in January 1942, which increased the total the US Navy wanted to 578; moreover USAAC and the British were also showing great interest in the Helldiver, until no fewer than 7,000 were being asked for.

The A-25A Shrike

On the last day of 1940, the USAAC placed an official order with Curtiss-Wright for 100 Curtiss SB2C-1As, which they gave the attack designation A-25A, having no experience or use of dive-bombers or dive-bombing themselves since the early 1920s. In fact the USAA, like their counterparts in the RAF, had firmly rejected the concept and were only brought to their senses by the spectacular successes of the Junkers Ju. 87 Stukas in Poland, Norway, Belgium and France. Even then not everyone was convinced – until, that is, the revolutionary effect of the combination of Panzer and Stuka in crushing old-type armies in weeks rather than years, could not be ignored. Originally the Shrike was to be identical to the Helldiver in most respects, to make for smooth production, and no great problems were foreseen in fulfilling the Army requirements. However, when the Stukas continued their remarkable achievements with the crushing of Yugoslavia, Greece and Crete in the spring of 1941, followed by the spectacular advances into the Soviet Union that summer, the Army quickly increased its requirements ten-fold, and by the spring of 1942 no fewer than 3,000 A-25s were on order!

In addition to the new Army requirements swamping the Navy's own mushrooming production requirements, it soon became apparent that the needs of the two

Specification – Model S84 SB2C-1A	
Engine type:	R-2600-8
Armament:	Four 50mm forward-firing fixed machine guns; one 50mm rear-cover flexible machine gun; maximum bomb load 1,600lb (725kg)
Dimensions:	Length 36ft 8in (11.17m); wingspan 49ft 8⅝in (15.15m); wing area 453 sq ft (42.08 sq m); height 16ft 11in (4.91m)
Weights:	Maximum weight 17,162lb (7,791kg); empty weight 10,363lb (4,704kg)
Performance:	Maximum speed 272mph (437km/h)
	Take-off power 1,700hp (1,268kW)
	Range as dive-bomber 1,030 miles (1,657km)
	Climb height 1,940ft/min (591m/min)
	Service ceiling 21,200 ft (6,462m)
Number completed: 410	
Serial numbers:	75218/75588 (Bu. No.-Navy – Serial Army); 76780/76818 (Marine ex Army)
Contract:	Order date; 31 Dec 40
	Delivery date; Jul 43–Feb 44

services were so different that the Shrike quickly began to take on its own separate identity which made for further complications of supply. It was clear that more drastic measures would have to be taken. The solution was that the Material Command of the USAAC ordered Curtiss-Wright to convert its sprawling St Louis, Missouri, plant (nowadays located at Airport Road and McDonnell Boulevard) to the almost exclusive construction of the A-25, and set up its own production lines. This allowed the Army to start introducing its own specialized equipment in the form of radios, extra crew protection with armour plate forward and underneath, and so on.

The first A-25 prototype was Bu No. 00006, which kept the Helldiver characteristics right down to the power turret and DF loop in the after cockpit, and in the first ten production models, wing slats and fold-

The XSB2C-1 (Bu. No. 1758) with wings folded. Official US Navy Photo

ing wings! But even in this batch distinctive features included earlier adoption of the ring-and-bead gunsight rather than the telescopic sight, an extended exhaust and smaller, fixed rather than fold-away wings, larger main wheels, and a pneumatic, rather than solid, tail-wheel. The wing slats were fixed in the retracted position and later done away with completely, while arresting gear and catapult launching spools and even radar went the same way. The Navy compromised by adopting the larger main wheel in an attempt at standardization, but the differences of requirements were too many and kept escalating.

Eventually these differences between the Helldiver and the Shrike were so numerous that, even though St Louis and Columbus were both part of the same company, and most of the sub-contractors for the Navy-type aircraft were retained for the Army-type, standardization had to go by the board as attempts to enforce it merely delayed the production of both types. Curtiss-Wright at St Louis assigned its own model number to the Shrike, S84 (which later became the 84B), and began producing their own sets of drawings for it.

Further complications arrived when, in 1942, the Royal Australian Air Force placed orders under Lend-Lease for 150 Shrikes for their own needs, which required some further modifications. These Shrikes were formally assigned to the RAAF during 1943 under the USA presidential commitment to Dr Evatt, in the combined chiefs of staff minute CCS 277/2 and the munitions assignment committee (Air) Case No. 200.[10] This model was known by the RAAF as the A69 and assigned serial numbers A69-1 to A69-150.

The Army requirements were transferred from Navy contracts to Army contracts, and a whole new series of inspection bodies, sub-contractors and government-supplied equipment had to be established, which caused yet further hold-ups. It was thus not until 20 September 1942, that the first A-25A had its maiden flight, with the last of the remaining nine of the initial batch completing as late as March 1943.

U S Navy Production Requirements

The US Navy's production schedule required Curtiss to have the first SB2C-1 ready by December 1941 (a prophetic, if unattainable date!) and by April 1942 they were to be producing the Helldiver at a rate of eighty-five aircraft per month. This was pretty ambitious, because the Navy was calling for a whole list of changes to be incorporated over the XSB2C-1 as well. These modifications were to take into account mass-production methods, including the use of large forgings and die castings to speed the assembly lines.

Unfortunately for Curtiss-Wright the power hammers necessary to forge the main fuselage members were in short supply and enough could not be obtained: and those they had were running to maximum capacity. The parts had therefore to be redesigned, and the enormous surge in demand far outstripped even the brand-new plant's capacity to meet its commitments.

Specification – Model S84 A-25A/RA25A	
Engine type:	R-2600-8
Armament:	Two 50mm forward-firing fixed machine guns; two 30mm rear-cover flexible machine guns; maximum bomb load 1,600lb (725kg)
Dimensions:	Length 36ft 8in (11.17m); height 16ft 11in (4.91m); wingspan 49ft 8⅝in (15.15m); wing area 453 sq ft (42.08 sq m)
Weights:	Maximum weight 17,162lb (7,791kg); empty weight 10,363lb (4,704kg)
Performance:	Maximum speed 272mph (437km/h) Take-off power 1,700hp (1,268kW) Range as dive-bomber 1,030 miles (1,657km) Climb height 1,940ft/min (591m/min) Service ceiling 21,200ft (6,462m)
Number completed: 480	
Serial numbers:	41-18774/18783 (1-Cs); 41-18784/19923 (5-Cs); 41-19924/19973 (10-Cs); 42-79663/79672 (10-Cs); 42-79673/79682 (15-Cs); 42-79693/79732 (15-Cs); 42-79733/79972 (20-Cs); 42-79973/80132 (25-Cs); 42-80133/80462 (30-Cs) (Bu. No.-Navy – Serial Army)
Contract:	Order date; 31 Dec 40 Delivery date; Dec 42–Mar 44

Specification – RAAF A-69	
Engine type:	R-2600-8
Armament:	Four 50mm forward-firing fixed machine guns; one 50mm rear-cover flexible machine gun; maximum bomb load 1,600lb (725kg)
Dimensions:	Length 36ft 8in (11.17m); height 16ft 11in (4.91m); wingspan 49ft 8⅝in (15.15m); wing area 453 sq ft (42.08 sq m)
Weights:	Maximum weight 17,162lb (7,791kg); empty weight 10,363lb (4,704kg)
Performance:	Maximum speed 272mph (437km/h) Take-off power 1,700hp (1,268kW) Range as dive-bomber 1,030 miles (1,657km) Climb height 1,940ft/min (591m/min) Service ceiling 21,200ft (6,462m)
Number completed: 10	
Serial numbers:	(ex A25S) 4583/4592 (ex-42-79683/79692) (Bu. No.-Navy – Serial Army)
Contract:	Order date; Mar 43 Delivery date; 23/25 Nov 43 (130 cancelled)

As a result of European combat experience, extra armour-plated protection was worked in, and leak-proof wing fuel tanks were another requirement. Changes in the gun armament saw the dropping of the 20mm cannon as foreign deliveries were not considered to be reliable, and American licence-built supplies were as yet inadequate to meet the demand. There followed the substitution of the .50 calibre machine guns again, and the doubling of these weapons from one to two in each wing for the production model.

Meanwhile work on the prototype continued, the various sections being 'handmade' in the traditional manner and assembled at Buffalo where the completed aircraft was finally rolled out on 10 December 1940. The Japanese attack on Pearl Harbor intensified feeling against the enemy considerably, and along with it, the pressure on Curtiss-Wright. The continuing modifications had hindered the setting up of the production lines so it was decided to give urgent priority to the initial four SB2C-1s, which were to be hand-built in advance in order that several test-bed machines could be quickly available to expedite flight testing.

First Flights, First Problems

The XSB2C-1 (Navy No. 1758), built on contract 66638, was originally weighed on 2 December 1940. It was not until 18 December 1940, however that its first flight was made from Buffalo, with the chief of the flight test section, Lloyd Childs, at the controls. Testing continued through January 1941, and quickly showed up yet more problems. These were listed as difficulties with the engine cooling system, buffeting in the dive condition, complications in operating the trim tabs correctly, and weakness of some of the hatches. Lack of stability was the biggest concern.

This all culminated in an engine failure during the landing approach on the twenty-sixth test flight, which took place on 9 February 1941. The prototype made a very heavy landing, the fuselage snapping at the turtle-deck and the aircraft ending up on its nose. Although the pilot survived this crash the aircraft was severely damaged. The whole machine had to be rebuilt incorporating further modifications, and it did not fly again until 6 May 1941.

On 27 April 1941 a group weight statement was issued for the XSB2C-1 at Buffalo, witnessed by G. Hicks for the Navy and W. G. Pendergast and B. R. Machin of Curtiss. Their report listed all the changes made while the aircraft was undergoing repair.[11]

(a) Improvement of flexible gun installation with permanently installed 500-round capacity ammunition box instead of removable magazines.
(b) Detachable parts of flotation equipment removed.
(c) Flap controls improved at fold fitting.
(d) Oil cooler changed from 11in dia. × 12in (UAP) to 18in dia. × 9in core (Harrison).
(e) Turtleback and controls improved.
(f) Hydraulic controls for wheel cowl added.

The prototype XSB2C-1, which crashed on 9 February 1941 and had to be rebuilt. It featured gun ports on either wing and above the engine cowling. Its bulky body and stubby shape resulted from the need to get two aircraft side-by-side on the deck lifts of the new Essex-class carriers then being built for the fleet. Note the conventional tail and the Aldis telescopic bomb sight. US National Archives, College Park

Weight Problems

	Weight (lb)	Horiz arm to lew (in)	Balance % MAC	Vertical arm to T. L. (in)
Weight empty (wheels up)	8,030	+18.25	16.4	+1.68
Weight empty (wheels down)	8,030	+17.7	15.9	-1.09
Normal useful load, 1,500lb Bomb	3,152			
Normal gross weight, 1,500lb Bomb (wheels up)	11,182	+31.1	28.1	+0.78
Normal gross weight, 1,500lb Bomb (wheels down)	11,182	+30.6	27.7	-1.21
Bomb, 2,500lb, 180gal fuel	11,734	+32.0	29.0	-0.18
Bomb, 1,500lb, 270gal fuel	11,767	+32.2	29.2	+1.42
Bomb, 1,100lb, 180gal fuel	11,699	+31.9	28.9	-0.27
Bomb, 2,100lb, 180gal fuel	10,885	+30.2	27.3	+1.33
Scout, no bombs, 270gal fuel	11,215	+31.2	28.3	+2.45
Scout, no bombs, 400gal fuel (max)	12,178	+32.5	29.5	+0.75
Smoke-screen layer, Mark V Mod, I tank	11,890	+31.7	28.7	+1.03
Most aft C.G. including flares	12,227	+34.0	30.8	-0.44
Most forward C.G. (wheels down)	9,345	+24.5	22.1	+0.76

(g) Hydraulic accumulator changed from 5in dia. to 10in dia.
(h) Engine cowl flaps and controls added.
(i) Carburettor air scoop improved.
(j) Cabin construction revised, eliminating magnesium alloy material.
(k) Wheels and tyres replaced with new set.

After the repairs and changes were completed, but before engine test, the aircraft was reweighed: first: weight empty, thrust line horizontal; second: weight empty, tail down. The summary is as given in the box (*left*).

All the C.G. (Centre of Gravity) locations were for the wheels up (flying condition) except as noted. The C. G. moved forward approximately .5in MAC and down 2in when the wheels were lowered. The balance figures were based on MAC = 109.3in located .34in aft of the LEW. A detailed breakdown and explanation of the overweight since first weighing are given in the box (*below*).

Breakdown of Overweights

Centre Panel	lb
Reinforced front and rear beams	+10.0
Reinforced flap beams at fold	+2.0
Ribs and stiffeners	+4.0
Pin pull-out mechanism changes	+2.0
Added gussets and reinforcements	+4.0
Additional di-chromate paste for fuel tank	+2.0
Main landing-gear fitting	+4.0
Metal covering changes	+4.0
Paint	+2.0
Misc. and nuts, bolts, etc.	+2.0
Total centre panel	**+36.0**

Outer Panels	
Reinforced front and rear beams	+7.0
Reinforced flap beams at fold	+1.0
Ribs	+1.0
Added gussets and reinforcements	+2.0
Wing hinges	+2.0
Metal covering changes	+4.0
Paint	+2.0
Misc. and nuts., bolts, etc.	+2.0
Total outer panel	**+21.0**
Repairs to flaps	+1.0
Repairs to stabilizer	+2.0
Repairs to fin	+1.0

Fuselage Monocoque	
Reinforcements added to longerons	+1.0
Additional gussets and reinforcements	+2.0
Stringers	+1.0
Reinforced bulkheads	+2.0
Added heavier gauge skin	+8.0
Paint	+2.0
Misc. and nuts, bolts, etc.	+1.0
Total fuselage monocoque	**+17.0**

Turtleback	
Changed from magnesium to dural	+4.0
Changed collapsing mechanism	+5.0
Total turtleback	**+9.0**
Changed magnesium to dural on pilot's cabin	+3.0

Gunner's Cabin	
Changed magnesium to dural and revised cabin with quick release doors	+6.0
Revised controls	+1.0
Total gunner's cabin	**+7.0**
Reinforced pilot's and gunner's floors	+2.0
Gunner's canopy – estimated original weight	-4.0
Landing-gear hold-up mechanism changed from cable to rod type	+5.0
New wheels, tyres and tubes	+6.0
Added inboard doors to landing-gear cowl	+14.0
Repaired engine ring cowl	+2.0
Added two cowl flaps to engine cowl	+5.0

Engine Accessories	
Changes made to shroud for jet exhaust stacks	+1.0
Air intake system revised	+6.6
Changed oil coolers	+14.2
Revised engine breather tubes	+2.2
Total engine accessories	**+24.0**
Added controls for engine cowl flaps	+4.0
New propellers with cuffs	+4.0

Misc. changes to fuel and oil systems	+2.0
Revised flap controls, new housings at fold end, misc. items	+18.0

Flexible Gun	
Gun mount redesigned	+12.0
Ammunition box transferred from U.L. to W.E.	+27.0
Case, feed and link chutes transferred from U.L. to W.E.	+10.0
Revised sight mount	+1.0
Added locking mechanism	+1.0
Added gun guard	+1.0
Misc. supports, flooring etc.	+4.0
Added gun fire interrupter	+7.0
Total flexible gun	**+63.0**
Parachute flare cans changed from magnesium to dural	+1.0
Oxygen inst. supports changed – box for rebreather for rear man	+2.0

Hydraulic System Inst.	
Added landing gear cowl controls	+10.0
Changed accumulator from 5in to 10in dia.	+9.1
Added larger tank	+5.2
More oil used in system	+20.0
Misc. piping, etc.	+9.7
Total hydraulics	**+54.0**
Loaded check valve, etc, to automatic pilot	+1.0
Changed magnesium tubing and added boxes to electrical system	+4.0
Eliminated removable parts from flotation gear	-45.0
Total overweight increase	**+250.0**

HELLDIVER MEN: Stanley Irving Vaughn, Snr

Stanley I. Vaughn Snr. Stanley I. Vaughn Archive

born on 16 December 1886 in Rootstown, Ohio, that state so intricately linked with America's aircraft heritage and the Helldiver itself; after routine high school education he expressed an early interest in aviation, and after working as a mechanic he became a pilot as early as 1907.

It was with gliders and dirigibles rather than heavier-than-air flying machines that he first made his mark, building his own experimental gliders. Later, working for the Strobel Amusement Company between 1909 and 1915, he recorded many notable 'firsts' in that line of exploit, and learned to fly by risk, daring and intuition. He flew a one-man airship whose sole motive power was a motorcycle engine, and did the rounds of state carnivals and fairs.

He purchased his first Curtiss biplane, the 30hp powered serial no.1, and using this as a model, he built six more to his own improved design. In 1915 he joined the staff of the Curtiss Aeroplane & Motor Company, which later became Curtiss-Wright when the feud between Glen Curtiss and the Wright Brothers was finally ended in a merger. His first assignment there was as mechanic and instructor at Newport News, and in 1917 he was supervising the manufacture of mechanical parts for dirigibles at Buffalo. As the years went by he diligently applied himself, and was project engineer at the Garden City plant from 1917 onwards, for several new designs of aircraft, and was credited with the assembly of the first Curtiss JN-4 Jenny. He then became superintendent at Buffalo, before his appointment to Columbus with first the tough assignment of factory manager, then as director of the Experimental and Development Department.

After thirty-five years with Curtiss-Wright (which was absorbed by North American) he retired from the company on 27 November 1950. He kept his interest in all matters aeronautical, however, becoming vice-president of the 'Early Birds', that nationwide organization of men who flew before World War I, as well as a Mason.

The family tradition was maintained when his own son, Stanley Vaughn II, on graduation, also went to work at the Curtiss-Wright plant in 1941; and in turn, his son, Stanley Vaughn III, also graduated and started his aerospace career working for the North American division of Rockwell International in 1983, 'cutting his teeth' on the B-1B bomber, the MX missile and the space shuttle programmes. After North American vacated the Columbus facility in 1989, the Douglas division of McDonnell Douglas moved in and he worked on the C-17, MD-11, MD-80, the Titan missile and the FA-Hornet, until an ever-shrinking defence budget and a weak commercial market forced Douglas to vacate, too, after only about five or six years in occupancy. The Vaughn family and Ohio had therefore seen everything, from the birth of aviation to the conquest of the moon and beyond.

The Curtiss SB2C Helldiver fought many battles during its wartime operational career, but perhaps the most crucial that was ever experienced in connection with it was for it rather than with it! Had not the dedicated team at Curtiss-Wright persevered despite all the problems and setbacks, there would never have been a Helldiver to bring havoc to the Japanese fleet at the battles of the Philippines Sea, Leyte Gulf and Okinawa and finally to avenge Pearl Harbor at Kure itself. One of the many stalwart and unsung heroes behind the scenes at the Columbus plant, those who kept the aircraft lines rolling despite the many setbacks, alterations, changes in requirements and vitriolic attacks by politicians, was Stanley Irving Vaughn, Snr.

A veteran of early aviation pioneering long before he became manager of the new Columbus factory, Stanley I. Vaughn had long been associated with innovations in the aeronautical history of the United States. He was

Almost immediately she was involved in a further crash when, on 10 May the left landing gear failed on landing, giving way outwards and causing the machine to ground loop. Fortunately the damage this time was limited, and the flight tests had resumed by the end of the month. In June and July almost a flight a day was made as the intensity grew, and on 11 August a 'Weight and Balance Status' was issued on the SB2C-1.

The orders from the Navy kept piling up. In October 1941 orders had increased to 1,000, and in the spring of 1942 the Navy told Curtiss they would require 3,000 more Helldivers; this contract was finalized on 4 June 1942, taking the total on order to 3,685 aircraft.

The flight evaluations had shown many failings which were remedied, but the problems of stability and weight continued. To try and offset the latter's adverse effect on the Helldiver's speed, experiments were made with a jet exhaust, but to no avail. The cowling and its flaps were altered to try to overcome the engine cooling fault, and a chin scoop was fitted at the lower edge cowling.

To try and improve on the plane's stability there were several alterations to the horizontal and vertical surfaces of the tail which gradually grew into its distinctive size and shape, culminating in the fitting of a whole new tail assembly in September. In a major effort to improve the still poor stability in the yaw axis, along with tightness in the pitch stability – both inherent faults in the design due to the original specification – the Navy allowed the engine mounting to be moved forward by one foot; in order to accommodate this change the fuselage had to be lengthened, and this was done in August. Static tests had also revealed further weaknesses in the wings and the fuselage of the XSB2C-1, and these also had to be addressed.

Testing resumed, and in all, ninety-five test flights were made from Buffalo during the rest of the summer and autumn, including propeller stress tests, followed by preliminary demonstration flights and controlled spins which continued through October. By November 1941 the XSB2C-1 was considered ready for dive tests, and after some of these had been conducted, on 12 November 1941, she was transferred from Buffalo to the new Curtiss plant under construction at Columbus, Ohio, to continue these.

New Columbus Plant Initiated

Excavation work on the site of the new facility at the airport itself had commenced on 20 January 1941. Associated with the construction work was the building of new rail links and new roads, and the widening of old roads in the city to accommodate the vastly increased freight traffic anticipated. By June 1941 enough of the new factory was built to enable the transfer from the Fairground site to commence in earnest. It took just 147 days for this move, and full production and the new factory were officially under way in 318 days from the first spade being turned. In the event the Columbus area proved woefully unable to come up with a pool of experienced engineers, and the training programme had to be expanded far beyond that original envisaged. Already the original staff of thirteen people was mushrooming, and it was destined to reach 25,000 by the war's end. The training facility designed to turn farmhands and housewives into skilled aircraft production workers was established under Ray Watkins in downtown Columbus itself.

On 4 December 1941 was the formal plant dedication ceremony, conducted by Rear Admiral John Towers, chief of the Bureau of Aeronautics. Towers had first learned to fly with Glen Curtiss himself as long ago as 1911, he had been a Navy test pilot himself with Curtiss at San Diego, and had piloted the Curtiss NC-3 for the Atlantic crossing in 1919. Also present were, amongst others, Artemus Gates, assistant secretary of the Navy for Air; Merrill Meigs, chief of the aircraft section, Office of Production Management; Colonel Philip Schneeberger, industrial planning head of USAAC, Wright Field; John W. Snyder, Defense Plant Corporation; and Major Floyd Green, with Curtiss itself being represented by Guy Vaughan, Burdette Wright, Charles Loos and J. W. Williams.

Key man under J. A. Williams was Stanley I. Vaughn, who had been with the Curtiss organization since 1915 and who was a real veteran flyer. A vice-president of the 'Early Birds' – those Americans who had flown balloons and string-and-wire flying machines in the first decades of the century – Stanley Vaughn had built experimental gliders as early as 1907, had toured the States two years later exhibiting a dirigible for the Strobel Dirigible Company, and in 1910 had become the proud owner of a 30hp Curtiss biplane, Serial no.1. He had built six just like it before joining the Curtiss organization in 1915. After training early flyers at Newport News he had been appointed project engineer at Garden City in 1917. He told me:

> The plant itself was actually opened on 6 June 1941, and we started using the building right away, although the actual dedication ceremony did not occur until November. We initially trained people at the Fairground site. There was no flooring in half the workspace, but we set up tooling and sheet-metal fabrication in the half that had it, and worked right on while they were laying concrete over the rest.[12]

The chief engineer, Raymond C. Blaylock, had joined Curtiss back in 1929 from the University of Michigan, and had worked on many of their aircraft, including the early Helldiver, the XSBC. His project engineer for the SB2C project at Columbus was Roland W. Holmes, whose Curtiss credentials went back even further, to Garden City, Long Island in 1928. Others involved in the design and production of the XSB2C were George T. Bauman, quality control manager, William J. Flood, comptroller, and William's two assistants: W. P. Kennedy who joined Curtiss from Martin in 1939 and John T. Corrodi from Columbus itself, and a former USAAC aviator who had joined Curtiss originally at Port Columbus in 1930 and rejoined in 1940. When Curtiss disintegrated after the war, Blaylock went on to Chance-Vought at Dallas/Fort Worth, Texas.

The test pilots were headed by Lloyd Childs, and then Barton T. 'Red' Hulse, who became the chief test pilot in November 1941; the team eventually comprised test pilots Elwood Collins and Harry Hill, both from Ferry Command; Lynn Hurst and William Webster, both ex-Navy flyers, Hurst having served with the American volunteer group as well; and Lee Miller, a former civilian instructor. All these men were to be tested in the crucible of the SB2C furnace …

CHAPTER THREE

Disaster!

Testing of the prototype had recommenced with a series of dives that built up steadily in intensity; during these, tail buffeting was encountered which resulted in minor damage as the XSB2C-1 was pushed harder and harder, both with and without the new dive brakes being utilized. Shortly after the dedication ceremony, on 21 December 1941, when the XSB2C was being flown by 'Red' Hulse on its thirty-third demonstration flight, disaster struck again when it broke up during a terminal velocity dive from a height of 22,000ft. At the start of the pull-out from this dive the right wing and the tail both failed. Stanley Vaughn II told me:

At 8,000 feet the wing skin started peeling off in strips; Hulse himself managed to bail out at around 1,500 feet and parachuted safely down –

he noticed the tail had gone. He was unharmed save for a sprained ankle. Part of the problem, it was concluded, was that the problem was 'stick force reversals' – at some point forces would reverse, increasing the G-load.[13]

This was far from his only escape from death with the 'Beast': during one test of a Helldiver, Red found that his ailerons were crossed; on another the control stick of the Helldiver he was piloting came out of its socket – but he survived it all!

This disaster left the company without a test vehicle while the wreckage was assembled and an inquiry begun as to its causes – and the next aircraft was not scheduled to be ready for six months! It was ultimately concluded that the cause of the crash was the failure of the horizontal stabilizer.

In the meantime, and as the urgency grew, Curtiss was forced to sub-contract more and more. Inevitably this led to inexperienced firms being used, with a resulting loss of standards. The placing of orders for 1,000 Helldivers with the Canadian Car and Foundry Company at Fort William, Ontario, through the auspices of the Canadian government agency War Supplies Ltd in May 1942, was one such move designed to ease the situation, although almost half the planned production placed there was to be allocated to the needs of the Royal Navy (450 machines). However, to start with the machinery and plant in Canada was not as up to date as the new Columbus plant, making it hard to keep to the high standard required; also, this sub-contracting in some other ways

Curtiss XSB2C-1 on 11 January 1941. National Archives via Emil Buehler Naval Aviation Museum Library, Pensacola

increased the problem for the already over-stretched Columbus plant because they still had to supply the Canadians with every bit of supply engineering and design information, including the continuous stream of drawings with all the modifications coming from BuAer as the war progressed. And this problem was extended when a second Canadian company, Fairchild, was similarly sub-contracted for.

HELLDIVER MEN: Barton T. 'Red' Hulse, Test Pilot

Crucial to the development, and indeed the survival, of the SB2C was the team of dedicated Curtiss-Wright test pilots. Risk-taking was their business, but they had to be real professionals with an extra dash of derring-do and a lot of luck to survive long. Breaking in the Helldiver was certainly no sinecure, and it needed a big man for the job – and with Red Hulse, Curtiss-Wright had all these qualities wrapped up in one package.

Barton (*not* Baron as many would have it) Traver Hulse was born at Perth Amboy, New Jersey, on 21

He worked on both the engineering and production flights with the Helldiver, and also the SO3C programme during his time at Columbus; but it was the 'Beast' that took him to the limits of his own capabilities and beyond!

He survived the horrendous disintegration of the XSB2C on 21 December 1941, getting out just in time as the aircraft fell to pieces around him, and parachuting down to land intact some miles east of Reynoldsburg, Ohio, in a field at the Derr Farm along Refuge

Curtiss ace test pilot Barton T. 'Red' Hulse *(extreme right)* talking to Raymond C. Blaylock *(left)* and Stanley I. Vaughn *(centre)* at Columbus. Stanley I. Vaughn Archive

August 1910; he was educated at Barringer High School between 1923 and 1928, and at Raymond Riordon School, New York, between 1928 and 1929. He moved on to the Rensselaer Polytechnic Institute between 1929 and 1931, and finished his higher learning as a student at the College of William and Mary between 1931 and 1933, where he studied aviation and took his private pilot degree for his commercial pilot's rating (No. 25352). He then flew as a pilot for the Universal Credit Company of the General Electric Contracts Corporation.

He joined the US Navy in September 1935 as an aviation cadet, being promoted to ensign and reaching the rank of lieutenant (jg) USNR, and honing his skills in the service environment before leaving in December 1939. In this same month he joined the Curtiss aeroplane division at Buffalo, New York, where he was employed as the army and navy sales demonstration pilot, and demonstration pilot in foreign countries for new Curtiss products. In November 1941, he was switched to being their Chief Test Pilot for the SBC2C project at Columbus.

Road. Ironically, having got out and descended without a scratch, he twisted his ankle walking down the rutted farm track to report his landing! This experience did not stop him appearing in newspaper and magazine articles promoting Camel cigarettes in cartoons and full page spreads, which sang the praises of the Helldiver to the skies!

In December 1942, after a traumatic year with the Helldiver, Hulse expressed his distaste for the quality of the aircraft coming off the production lines and which he was expected to test, and he was moved to the research division at Buffalo as head of flight research. He remained there for just a year, then in January 1944 he took up a new appointment with General Motors Corporation's Fisher Cleveland aircraft division as chief test pilot again. In December 1944 he was transferred to their aircraft development section at Detroit, Michigan, still as chief test pilot. However, his name will always be associated with the Helldiver and its early tribulations.

The main structures for the SB2C were all designed and drawn at Buffalo, but it was on the overstretched Columbus team that the task of preparing the actual production drawings for all these plants rested. The acute shortage of skilled draughtsmen and engineers compounded the difficulties, and much work on the shop floor was seriously delayed without them. All changes – and eventually there were to be many hundreds – had to be redrawn in, approved by Buffalo, altered again and then sent to the factories, and each delay and change meant alterations to the planning of the production line or its disruption.

According to one source, already massive administration problems were compounded for the Columbus plant by the heavy-handed management style from Curtiss-Wright which '… tied the hands of the Columbus plant manager and required him to forward all important questions to Buffalo, New York, plant for decision.'[14] In the spring of 1942 an inspection team from the US Navy visited the plant and found that this red tape complicated an already complex situation still further; under pressure to change, Buffalo reluctantly released additional staff to Columbus to try and clear this bottleneck and skilled engineers were taken off the far less vital seaplane project, the SO3C, and assigned to the SB2C. This all helped a little.

Thus it was not until 30 June 1942 that the first production SB2C-1 (00001) took to the air, again with Hulse at the controls. He reported that this machine was even worse than the prototype had been! In fact this was not too surprising, because the prototype had passed its tests without such essential equipment as the armour plating and armoured glass protection for the crew, ASB radar equipment with underwing Yagi rotating antenna, and the direction-finding loop at the front of the rear cockpit; furthermore it had magnesium rather than aluminium forgings. All these things meant its all-up weight was considerably greater, and this reduced top speed by 40 knots, having been an originally impressive 320 knots, while the landing speed *increased* by 10 knots to 79 knots.

Other changes included the suppression of the top-cowl mounted machine guns which were relocated in the wings, while a single .50 calibre machine gun replaced the twin .30 calibre for a while. The obsolete Aldis telescopic gunsight was dropped and a reflector sight fitted instead. Modifications were made to the already capacious bomb-

19ft ½in (5.80m)

12ft (3.66m)

wing highest position
during folding plane
in three point position

49ft 8⅜in (15.15m)

22ft 6½in
(6.87m)

20ft 6½in
(6.26m)

16ft 10in
(5.13m)

12ft 2in (3.71m)

16ft (4.88m)

36ft 8in
(11.18m)

rudder hinge

elevator hinge

3½in (9cm)

23ft 2in
(7.06m)

10ft 8in
(3.25m)

9ft 7¹³⁄₁₆in
(2.94m)

16ft 11in
(5.16m)

13ft 1½in
(4.00m)

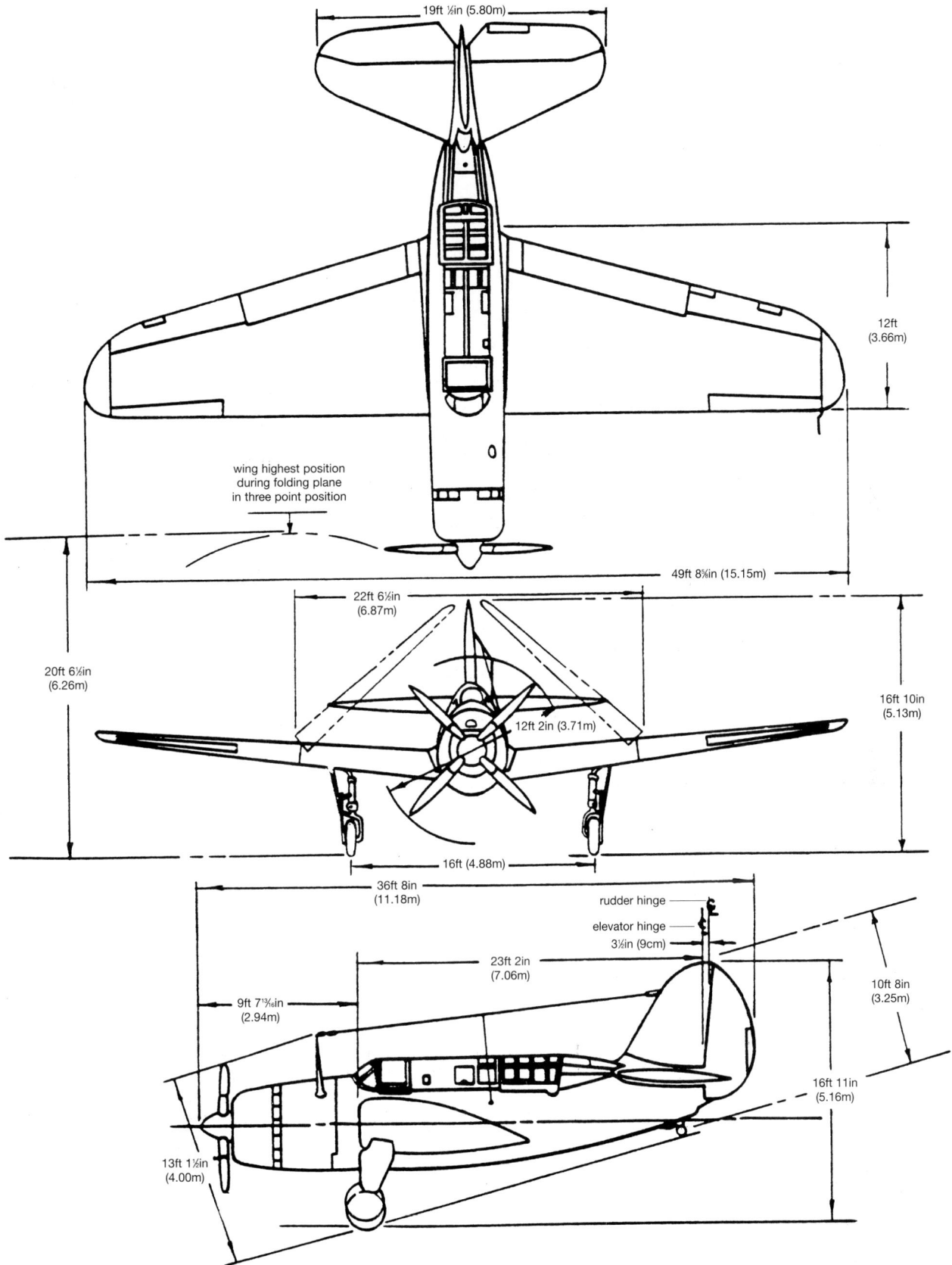

Overall dimensions.

bay, and underwing racks for light bombs were added – another increased drag factor, of course. The doors for the tail-wheel were done away with. An oil cooler outlet was located under the fuselage. No. 00001 thus remained a Curtiss test machine throughout its short life.

A preliminary demonstration of the SB2C-1 was conducted by Curtiss-Wright in the vicinity of Columbus, Ohio, during the period from July 1942 to January 1943.[15] The demonstration was conducted in accordance with the Bureau of Aeronautics specification SR-38B, with the exception of the armament items, which were omitted due to previous delivery of other SB2C-1s to Norfolk and Quonset Fields for separate armament trials.

BuAer Trials Procedures Initiated

On 24 June 1942 Captain Ralph Davison, assistant to the chief of the BuAer informed the President of the Board of Inspection and Survey that:

It is expected that the first SB2C-1 airplane will be delivered to the Navy Air Station, Anacostia, in July 1942 and that four (4) additional airplanes will be delivered shortly thereafter to the Naval Aircraft Factory, the Naval Air Station, Norfolk (Aircraft Armament Unit), the Naval Air Station, Quonset and the Naval Air Station, Anacostia, for use in connection with the subject trials.[16]

He then listed the criteria for these trials:

a. Performance characteristics as a scout (290 gal) and as a bomber 1,500lb, 210gal)
b. Comprehensive investigation of stability and controlability, and take-off characteristics. 'It is desired that this investigation cover gross weights up to, or near, the maximum emergency overload, which is obtained with the torpedo loading plus full tanks, all ammunition and complete radar equipment; this gross weight is estimated to be about 15,100lb.'
c. Routine arresting and catapulting test. 'The limited gross weights are 12,600 and 11,700lb respectively, including bomb, smoke tanks, auxiliary fuel tank or torpedo (Mk. 13-2) installations.'[17]
d. Routine night flying test.
e. Comprehensive armament installation tests, including the AN Armament Test

SB2C-1 in flight. Curtiss-Wright via Cdr. Alton Chinn Archives

Programme and tests of the fixed guns, the turret, camera guns and both flight and ground testing of bombs, torpedoes and fusing arrangements; pyrotechnics (flares and signal pistols) plus smoke equipment with the Mk V-3 smoke tank; and tow target fitting.
f. Routine vibration survey etc. etc.

Anacostia Trials

The second production model (00002) did not follow until August, and was not received at Anacostia until 14 September 1942;[18] there it underwent further engine cooling tests, as well as stability and control studies, which brought about yet more modifications. Although lateral and directional stability was considered satisfactory, these tests again showed that the longitudinal stability problem was not. There was also high friction in the aircraft's aileron controls, which had a single balance tab on the starboard aileron and a single trim tab on the port aileron.

This machine then took part in mock carrier trials, with catapult launches and arrested landings being tried out as part of accelerated service evaluation. This soon revealed further weaknesses in several vital areas: for instance, it was found that in the approach to the carrier, as well as in pullouts

after dives, heavy lateral and directional pressures built up on the controls. This led to the fitting of a second balance tab on the port aileron which helped control and high speed. Power and non-power stalling characteristics were reported as being within limits. These findings again required the strengthening of some aspects of the Helldiver, including the main fuel tank in the fuselage and supporting structure, and of course, the tail-wheel assembly. Serviceability was also found to be high.

Carrier Acceptability Tests with #00002

On 18 October this aircraft was sent directly to the Ship Experimental Unit, Philadelphia, to undergo carrier landing and take-offs under the experienced supervision of Commander Tobey Hogle. This officer had two test pilots under his command, Lieutenant Denbo and Lieutenant Heep, who were to put the aircraft through its paces.

According to Roland Holmes,[19] Hogle at once examined the Helldiver's arresting hook gear. This consisted of an aluminium carriage to which the hook was fixed, and which moved up and down in a channel on rollers to a fixed stop. It was actuated by the pilot control through cables on a geared drum. No lock was fitted to keep the carriage hard against the stop, the Curtiss thinking being that inertia and cable tension would suffice. In Hogle's opinion it would not, and he recommended that the trials be delayed until a modification had been made. He was overruled on this, and the designed programme of ninety-seven tests commenced.

These tests were conducted through to 22 December 1942, and they resumed again in March 1943.[20] On the third of these tests, however, Hogle's dire predictions proved only too accurate, and the concussion of the hook hitting the deck bucked the alloy channels. After hasty repairs the carriage was kept fully locked aft to prevent any repeat of this while the remaining trials were conducted – but this meant that yet another modification was required back at Columbus.

On the eighth arrested landing the arrestor hook came free, and on the ninth the fuselage fuel tank shifted forward and was punctured. Further bracing was required which solved the problem. On the twenty-seventh and twenty-ninth arrested landings the rear wheel collapsed completely, a common fault found in sea trials.

Otherwise everything proceeded very well, and contrary to the Anacostia Flight Test centre, lateral and directional control and manoeuvrability was said to be very good. The aircraft was judged easy to land under all loading conditions in both two and three-point landings, while the undercarriage proved sturdy and reliable. The solitary jarring note was that slight longitudinal difficulties with a high loading factor were noted. From a total of ninety-six arrested landings the following data was presented:

a. Maximum ground speed into arresting gear — 50 knots @ 13,211lb (5,993kg)

b. Maximum deceleration — 2.1g

c. Maximum tail rise — 46in (116.8cm)

d. Propeller clearance at maximum tail rise — 14¾in (38cm)

e. Maximum dome pressure — 800psi

f. Minimum run-out — 83ft (25.3m)

g. Minimum retrieving load — 60lb (27.2kg)

h. Recommended orifice size — 0465in (#56 drill)

Suitability Tests of the Pneumatic Tail-wheel

One result of the USAAC belated interest in the dive-bomber and the setting up of the A-25 Shrike programme was an urgent need to standardize both types to speed production. One offshoot of this was the adoption of the Army-type pneumatic tail-wheel in place of the normal, and smaller, solid tyre used by the US Navy.

Tests of this new-type tail-wheel to examine if they were suitable for carrier deck landings, also took place at the Ship Experimental Unit, Philadelphia, again utilizing No. 00002.[21] A total of twenty-three arrested landings were made, all by daylight, with the new wheel inflated to air pressures of from 75 to 90lb psi on the Philadelphia platform, with aircraft varying gross weightloads, as the table (below) shows.

Arrested Landing Trial Results, Philadelphia				
No. of landings	Condition of SB2C-1	Loading weight (lbs)	Type	Arrested
5	Light	11,900	3P	Yes
1	Light	11,900	2P	Yes
3	Bomber	13,211	3P	Yes
2	Bomber	13,211	2P	Yes
7	Heavy bomber	13,729	3P	Yes
5	Heavy bomber	14,322	3P	Yes

'There was no noticeable tendency of the tail-wheel to bounce or for it to cause the tail hook to bounce upon contact with the deck in landing', was the conclusion reached, although some beefing up of the frail centring device was called for.

A Hopeful Forecast Made

Thus it was that Commander J. N. Murphy, US Navy, made a favourable report on the status of the SB2C-1 in which he stated that it had shown good landing, take-off and overload characteristics. He was hopeful that full production would soon swing into action. By early 1943 Murphy enumerated the three remaining problems as being:

a. the Emerson after power turret was exposed so that an enemy could put it

out of action with only a few hits. There was also severe sight tracking vibration;

b. the dive flaps were slow in closing;

c. the vapour dilution system of the central fuel tank was inadequately protected from gunfire.

Production remained at a trickle for the remainder of the year – seven in October, ten in November – but despite this there seemed to be room for optimism that the Helldiver was coming right. This report was to prove altogether too sanguine, however, for although extensive flight testing of 00001 since July showed that improvement had been made to the engine cooling and performance, the old problem of longitudinal stability remained in doubt. Further changes were necessitated to the control system before dive testing was commenced.

With regard to the power turret problems, because no other turret had been thought suitable, this was of Curtiss design and they had hoped to sub-contract the production to Emerson Electric; but that firm, like so many others, was already at capacity. Curtiss then turned to a division of Borg-Warner, Norge, to construct the turret and on 23 May 1942 gave that company the go-ahead to proceed with the necessary tooling-up for this. A mock-up was examined, changes made, and then in November 1942 Norge was told to proceed with the first fifty turrets to the new specification. The first of these had arrived at Columbus in January 1943, where testing had shown up '... excessive sight vibration, inconsistent sight racking with erratic and excessive fire interruption.'

The third production machine (00003) was retained at Columbus and had her flap-operating system modified to try and rectify the tardy opening problem identified by Commander Murphy. Rather than have the two actuating hydraulic motors, located on the centre-line of the fuselage of the aircraft, working through torque shafts and universal joints to the gearboxes on the flap beam, which worked the actuated screw jacks, now six hydraulic cylinders fed directly to the flap actuators. This worked splendidly and was scheduled to be introduced on the production line at Columbus, beginning with Helldiver 00201. Here, Building 3A was being finished off and expansion was necessitated by the need to concentrate SB2C-1 production along with the other lines (SO3CV and SC) for the sub- and final assembly work, while the original Building 3 was to be re-equipped for the manufacture of the smaller parts for the Helldiver line.

Trials and Tests

The last of the hand-built batch (00004) was completed, assigned on 1 October 1942, and ferried to the Aircraft Armament Unit in Norfolk, Virginia, to undergo both bombing and gunnery evaluations.

Patuxent River Tests

Starting in that month, comprehensive tests were conducted at NAS Patuxent River, which lasted into 1944.[22] These were summarized as follows:

Fixed Guns

Although the actual .50 installation was of itself satisfactory, their accessibility remained poor. Another change in the wing gun mountings was also introduced at this time, with a single 20mm cannon replacing one of the .05 calibre machine guns in each wing subsequent to aircraft No. 00200, and this installation was reported as acceptable in a separate report. The recommendation was that the 20mm wing guns be retained, '… when and if two .50 calibre synchronized guns are added.' Also that, '… the synchronized guns be installed below the engine instead of above.'

Fixed Gunsight

The Mk-8 sight was used initially, pending the installation of a Mk-30 director. It was considered necessary to retain a filter for strafing and dive-bombing operations although it, '… would lose some of its importance with the use of the new type sight lamp of multiplied brilliance.' Also that, '… one or more adjustable mirrors be provided for the pilot just above the sight installation.'

Turret/Flexible Guns

The turret difficulties suffered excessive sight vibration, inconsistent sight tracking and erratic and excessive fire interruption. As an interim measure a twin .30 calibre installation, with continuous feed as used on the SBD, was installed. They concluded that this was '… satisfactory until a more satisfactory .50 calibre turret becomes available.'

They added the important rider that '… fire interruption on the turret installation, when and if it is replaced, must be improved to give less firing restriction. It is understood that the Barco interrupter cam is being re-cut to closer tolerances and to a more representative airplane contour.'

Fixed Gun Camera

A trial installation was made in the leading edge of the right wing, and air firing tests proved satisfactory; this was later made retrospective. However, subsequent tests brought to light certain deficiencies of the camera mount, originally installed on the fuselage ahead of the pilot's cockpit. The camera location was therefore changed to the right wing stub at the wing fold point, approximately 20in (50cm) outboard of the 20mm wing gun. This in turn was found to be satisfactory for taking photographs when the guns were not firing, but excessive vibration was encountered when the guns were being used, which would be the norm. The recommendation was for a new camera mount design.[23]

Bomb Crutch

New type, ratchet sway braces were made available at Supply, Norfolk for installation in all aircraft #00300 and under. Curtiss was to furnish identical braces in aircraft subsequent to #00300.

Displacing Gear Reversing Valve

The angle at which the hydraulic pressure was reversed from down pressure to up pressure on the displacing gear was changed from 85 degrees to 45 degrees. 'Although no trouble was experienced in any releases by this unit with the angle at 85 degrees, other reports were received of the bomb-bay doors closing up on the lowered gear. Tests were conducted with the 45-degree reversing angle change made, and releases in level flight and all dive angles were made with satisfactory displacing gear action.'

A Navy SB2C Curtiss Helldiver dive bomber in late wartime colour scheme. NASM, Washington, DC

Further trials were conducted in December 1943 and January 1944, with Helldiver #00201, to determine why the bomb displacement gear on SB2C-1 1000# failed to retract properly after bomb release. No trouble was experienced, '... without deliberately offsetting the normal adjustment.' The conclusion was that careful adjustment of the reversing levers on the two displacing ear arms to under 5 degree coincidence, would avoid trouble of that nature.'[24]

Floatplane Helldiver

Airplane No. 00005 was modified further but without the controversial rear turret, and became the XSB2C-2. The thinking behind this variant, a floatplane version of the Helldiver, originated with a BuAer concept for the Marine Corps, to enable its amphibious operations to take place with air cover readily available even if the Navy's limited number of carriers were not available. All types – fighters, scouts and bombers – were considered. The Japanese had already very successfully adapted some of their own aircraft for such a purpose, and these, especially the float-equipped amphibious fighter (the Nakajima A6M2-N 'Rufe'), were to prove very successful indeed in the South Pacific campaigns.

Accordingly, in January 1940, it had been proposed to convert the XSB2C-1 prototype into the prototype for a floatplane dive-bomber, the SB2C-2, once its Navy testing programme had been completed. The Eldo company was earmarked for construction of the necessary floats for some 294 of the these aircraft – which showed how seriously the project was considered – these machines to be converted from existing SB2C-1 contracts. With the tragic demise of XSBC-1, production model 00005 was selected to carry out this evaluation.

Accordingly this machine was trialled as a conventional dive-bomber and then ferried over to Anacostia Naval Air Station. Here the Eldo floats were fitted to her, and testing began in her new configuration. Taxiing and take-off trials were carried out in varying weight conditions, and although the distance taken by the aircraft to 'unstick' was more than had been hoped for, they were generally satisfactory. Similarly, on 9 March 1943 rough water trials were conducted in Hampton Roads, and these also went reasonably well.

The silver model of the SB2C-1 presented to Robert Olds by the Curtiss-Wright Corporation as a memento for his publicity work during World War II. Peter C. Smith

Specification – Model 84C XSB2C-2	
Engine type:	R-2600-8
Armament:	Two 20mm forward-firing fixed cannon; two 30mm rear-cover flexible machine guns; 1-Mk 13 torpedo
Dimensions:	Wingspan 49ft 8⅜in (15.15m); wing area 422 sq ft (39.2 sq m)
Weights:	Empty weight 14,960lb (6,785kg)
Performance:	Take-off power 1,700hp (1,268kW)
Number completed: 1	
Serial numbers:	(ex-00005) (Bu. No.-Navy – Serial Army)
Contract:	Order date; Mar 41 Delivery date; 28 Oct 42

As with the carrier and land-based versions, the problem of the power turret caused further delays to this programme; finally the twin .30 flexible rear gun as used by the Dauntless was installed, and further tests conducted to see what difference this made. It was decided to adopt this as an interim measure pending satisfactory completion of a workable turret arrangement – but, by this time events in the actual combat zone had moved on, and the floatplane dive-bomber no longer seemed the urgent requirement it had once been. After some further re-thinking the whole project was finally cancelled completely on 14 April 1944.

Trials Continue

The final demonstration of #00001 was commenced on 21 January 1943 at the Naval Air Station, Anacostia, with the aircraft in the 1,000lb bomber condition. Four series of manoeuvres were conducted at this time:

a. Without speed reduction devices in operation but with bomb doors open, one dive to an indicated speed of 413mph (665km/h), with pullout at about 13,000ft (3,962m) altitude in which an acceleration of 6.7g was reached.

b. Without speed reduction devices in operation but with bomb doors open, one dive to an indicated speed of 435mph (700km/h), with pullout at about 11,000ft (3,353m) altitude in which an acceleration of 6.3g was reached.

c. With dive flaps and bomb doors open, one dive to terminal velocity in which an indicated speed of 363mph (584km/h) was reached. In pullout from this dive an acceleration of 8.6g was attained.

d. Without speed reduction devices and with bomb doors closed, one dive at between 50 and 60 per cent of design maximum speed. In this dive and pull-out therefrom an indicated speed of 280mph (450km/h) and acceleration of 6.3g respectively, were attained.

These dives completed the requirements as to speed and wingload in the 1,000lb bomber condition. The next dive was to be made without speed reduction devices in operation and with bomb doors closed, and was scheduled for 25 January.

Expansion of the Trial Programme

The original contract had required that trials should be conducted on the first Helldiver '... upon its delivery and acceptance', in an effort to accelerate delivery of a fully tested and satisfactory dive-bomber to the fighting fleets; however, in a letter dated 12 August 1942, the Bureau of Supplies and Accounts authorized further trials on the second, or any airplane, of the Helldiver's armaments, *in advance* of demonstration and final acceptance of the first airplane for these items. Subsequent Helldivers utilized thus included serial nos. 00002, 00003, 00007, 00015, 00149 and 00201.

The first of the mass-produced SB2C-1s had began to arrive in October 1942, but these were soon found to be wanting: '... the careful construction which went into making the parts for the first airplanes was replaced by mass-produced parts which were defectively designed, carelessly manufactured, or casually inspected. Curtiss could not continue production and still

Curtiss XSB2C-2 seaplane on 29 September 1942. Emil Buehler Naval Aviation Library, Pensacola

take the corrective action necessary to make the airplanes combat-worthy.'[25]

Of these aircraft, 00006 was used as test machine for the US Army's A-25A Shrike programme at Wright Field, of which more later. No. 00008 went to RNAS Anacostia with full radio and radar installations in place to undergo electrical and radio testing, and this was followed by service trials. These revealed the shoddy workmanship very quickly. Meanwhile 00009–00012 were used for training programmes.

The Helldiver as Torpedo Bomber

As a result of the Navy's requirement for a torpedo-carrying version of the Helldiver 00013 was especially modified in this role and was despatched to NAS Quonset Point, Rhode Island, for torpedo-launching trials; these commenced in November. Tests were made to determine the effect of the torpedo, torpedo fairing and torpedo drag ring on the speed of the aircraft; the results were as follows:

Maximum speeds (mph, km/h in brackets following) at 6,100ft (1,860m) altitude at 1,500hp:

1. With torpedo fairing, no torpedo	272 (438)
2. With torpedo fairing, with torpedo	260 (418)
3. With torpedo faring, with torpedo and drag ring	236 (380)
4. No torpedo fairing, no torpedo	266 (428)
5. No torpedo fairing, with torpedo	257 (414)

These trials were resumed at the naval proving ground, Dahlgren, Virginia between 27 and 31 May 1943, and will be discussed in detail later. They eventually led to a fully satisfactory system that worked perfectly – but the Helldiver was never used as a torpedo dropper in combat.

Perforated Dive Flaps

Zero lift dives using perforated dive flaps were conducted. These had been successfully utilized by the Douglas SBD Dauntless to overcome buffeting problems, and it was thought that something similar might help with the SB2C-1. Aircraft No. 00007, at a weight of 14,450lb (6,555kg) – 1,000lb combat bomber minus ¼ fuel – was the selected test bed for these trials. Terminal velocities were found to be:

280 knots with flaps open 30 degrees;
265 knots with flaps open 35 degrees;
240 knots with flaps open 45 degrees.

Already perforated with a series of a large diameter holes (fifteen in the upper and forty-one in the lower flap), these were supplemented by a series of smaller-diameter holes which finally had the desired effect. This type of dive brake was eventually introduced on the later models of the SB2C-3E.

A-25A with torpedo in place. Emil Buehler Naval Aviation Library, Pensacola

Encouraging Results of Early Trials

The overall summary of the first trials was that although, '... the Model SB2C-1 airplane was found to conform substantially to the specifications ...' of the contract, and '... was acceptable for the purposes intended and free from defects in material and workmanship ...',[26] there were some serious exceptions. These were in regard to the aircraft's stability and controllability characteristics and were summarized as:

a. Longitudinal stability was found to be negative or neutral in all normal flight conditions except those of low power or power off glide.
b. Longitudinal controllability was found to be satisfactory.
c. Lateral stability was satisfactory in the cruising condition, but weakly positive in the landing condition.
d. Lateral controllability was found to be

unsatisfactory in the carrier approach condition due to low aileron effectiveness, and in high speed flight due to high aileron control forces.
e. Directional stability was found to be positive in all normal flight conditions except at low speed where it became neutral to negative.
f. Directional controllability was unsatisfactory at high speed due to the excessively high forces required for rudder operation, while at low speeds rudder control forces were subject to reversal.[27]

In addition, the SB2C-1 was found to be, under all load conditions, '... controllable and stable on the ground; in taking off from and landing on a carrier (with some minor exceptions), and operating from catapults, once the tail-hook assembly modifications had been made.'

Although the original contract had specified that the empty weight of the

Helldiver should not have exceeded 7,988lb (3,623kg), the actual empty weight of the first aircraft to be received at Anacostia was 9,630lb (4,368kg). The US Navy's Board of Inspection merely commented that,

'This figure is not significant since a large number of modifications and changes have been made. The gross weight of the airplane during trials may be taken as approximately 13,240lb.'

Again, although the original contract called for performance characteristics to be determined by both the scout (290 gallons/1,318 litres fuel) and the dive-bomber (one 500lb bomb, 210 gallons/955 litres of fuel) conditions, for which guarantees had been made by the company, in actual fact 'in view of the lack of interest in non-combat loadings, and as authorized verbally by the Bureau of Aeronautics, all performance data were obtained with the airplane in the combat bomber condition ...'

Helldiver 0007 airborne from the Naval Air Station, Anacostia in December 1942. This model was equipped with ABD IFF and ASB radar and was used to test these and other radio equipment and electrical fittings.
US National Archives, College Park

A comparison between the actual performance of the SB2C-1 as determined by the Patuxent River flight tests and the guarantees (as revised[28]) make interesting reading, as can be seen by the figures given (*see* box).

During comprehensive investigations into stability and control carried out by flight test at Anacostia and Patuxent River, it was recommended that the centre of gravity limits for the SB2C-1 be established at 28 to 33.7 per cent of the mean aerodynamic chord. With regard to stability, the Aircraft Experimental and Development Squadron (later Tactical Test) commented:

The inherent longitudinal instability of the SB2C-1 for this loading at high speeds demands caution and complete command of the controls on the part of the pilot when pulling out and levelling off at the low altitudes necessary for torpedo release. This instability presents a problem when radical evasive manoeuvres are employed flat on the water. Much care must be exercised in the pull out from a dive to prevent excessive accelerations being imposed on the structure. When entering a dive and when recovering from a dive, it is necessary to make large adjustments with the rudder trim tab.[29]

Routine night-flying tests were carried out by the Naval Aircraft Factory. They commented:

a. Exhaust glare did not interfere with vision during night approaches.
b. Reflections on the airplane structure from 'hookspot light' did not interfere with the pilot's vision during approach.
c. The windshield was satisfactory with respect to vision and light reflections at night.

d. Landing lights were satisfactory.
e. Identification and approach lights were adequate and properly located.
f. Lighting in the pilot's cockpit was acceptable but could be improved.

Had they known of the night operation ordeal that was coming their way a year later in the Philippine Sea battle, no doubt Navy Helldiver crews would have been reassured by this report!

As well as comprehensive armament tests, a whole host of miscellaneous tests were carried out; these included:

a. Attitude gyro – Application for dive-bombing.
b. Curtiss model 150CH-2 turret.
c. T14 20mm aircraft automatic gun feed mechanism.
d. Optical ring sight.
e. Bomb-rack release, AN-A2A.
f. 2,000lb bomb installation.
g. Curtiss model 150CH-2 turret and illuminated optical sight, Mark 8.
h. Clearance between bomb path and propeller disc.
i. Loading trials of special incendiary bomb.

Vibration Tests, Philadelphia

A routine vibration survey was conducted by the aeronautical materials section of the Naval Aircraft Factory, Philadelphia (later the Aeronautical Materials Laboratory of the Naval Air Experimental Station).[30] These tests took place with Helldiver No. 00002 between 21 and 26 December 1942, to determine the vibration characteristics of the SB2C-1 in the laboratory and in flight. The objectives were to study anything which might affect the service usefulness of the Helldiver, and to determine the effects upon pertinent flutter parameters of varying the amount of fuel in the wing tanks. The conclusions reached were:

a. The vibration characteristics were considered 'satisfactory'.
b. The effects upon pertinent flutter parameters of filling the wing fuel tanks:
  1. The structural damping coefficients in bending and in torsion were increased.
  2. The natural frequency in bending was not materially changed.
  3. The natural frequency in torsion was decreased by about 2 per cent.

No recommendations were made.

Patuxent River Flight Tests			
Condition	Actual	Guaranteed (revised)	Guaranteed (original)
Maximum speed at engine rated power and airplane critical altitude (mph (km/h))	286 (460)	not less than 309.1 (497.3)	not less than 313 (504)
Minimum speed without power at sea level (mph, km/h))	75.5 (121)	not more than 75.9 (122.1)	not more than 71.8 (115.5)
Service ceiling, starting with normal load (ft (m))	25,800 (7,860)	not less than 26,730 (8,147)	not less than 29,000 (8,839)
Take-off distance in a 25 knot wind (ft (m))	350 (107)	not more than 356 (109)	not more than 265 (81)

CHAPTER FIVE

The Helldiver Described

Restoration work being carried out on the split flaps of the CAF SB2C, which gives a good view of the hole configuration. Commander J. Alton Chinn, USN, Rtd

The SB2C-1 was a two-place, single-engined, low-wing monoplane designed as a scout bomber. Its gross weight was 14,000lb (6,350kg). The engine was a Wright-Cyclone R-2600-8 radial, and air-cooled; with a propeller gear ratio of 16:9; rated at 1,700hp at 2,800rpm at take-off; 1,500hp at 2,700rpm at sea level to 6,700ft (2,042m); and 1,350hp at 2,400rpm at 13,000ft (3,962m).

The engine was attached to the mounting ring at seven points by means of shear-type rubber bushings. The engine mount was rigidly attached to the fuselage. The propeller was a 12ft (3.7m) Curtiss Electric, constant speed type having three alloy blades, model no. 89324-12.

The fuselage was of stressed skin, semi-monocoque, all-metal construction. The wings were of full cantilever construction consisting of a centre section with two built-in fuel tanks (100 gallons/455 litres each wing), and two folding outer panels on which were incorporated slots and split flaps. The slots were operated automatically with the retractable landing gear which folded span-wise into the centre section of the wing.

The control system was of the stick-and-pedal type with cables connecting to all surfaces. The fixed surfaces were of full cantilever, stressed skin, all-metal construction. The movable surfaces were of metal construction and were fabric covered. All the surfaces were fitted with controllable tabs.

All flight instruments were located in the front cockpit and were installed on two panels, each of which was shock-mounted on shear-type rubber bushings.

Pilot's protection

Gunner's protection

Armour plate

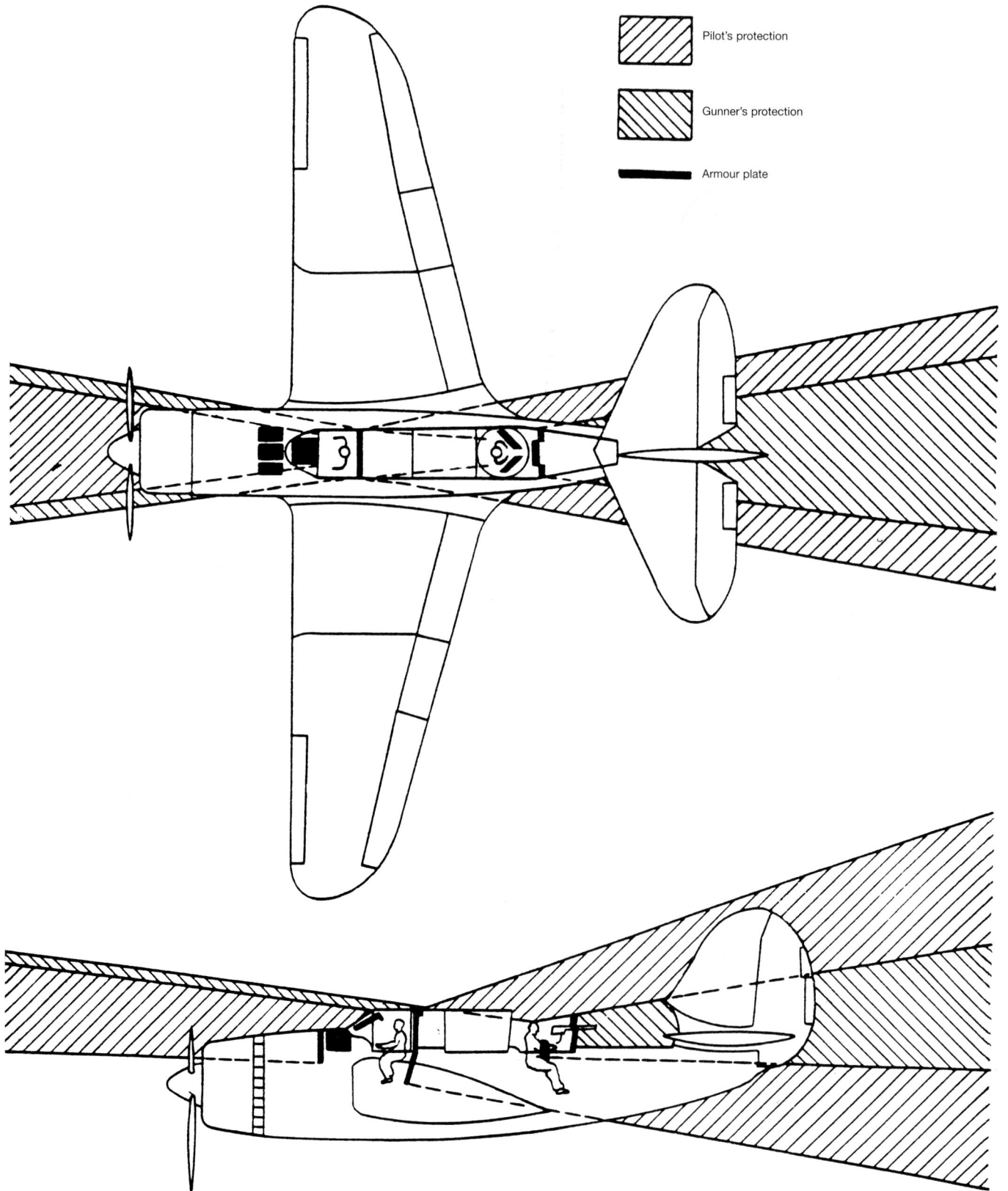

Gunfire protection.

Close up of the wingtips of the CAF SB2C showing the various lights and flaps to good effect.
Commander J. Alton Chinn, USN, Rtd

Radio and Electrical Tests

Routine radio and electrical inspections and tests were conducted by radio test, then located at the Naval Air Station, Anacostia, with additional tests by radio test at Patuxent River; these included:

a. Radio noise survey.
b. Laboratory model of AN/APS-4 radar.
c. Installation of AN/APS-4 equipment.
d. Communication antenna modification.
e. Relocation of tail lights.
f. IFF equipment.
g. Modified AN/APS-4 equipment.

The take-off characteristics of the SB2C-1 were determined at several gross weights, with results as listed (*see* box below).

It was recommended by Flight Test at Patuxent River that:

… with heavy loads, full flap (60 degrees) be used, and that the airplane be placed in a tail high position as early in the run as possible. The tail should then be pulled down smartly just before reaching the bow in order that minimum drop after leaving the deck would result. Wheels should be retracted immediately after take-off, and flaps retracted in steps with increase in speed to prevent too sudden changes in lift characteristics.[31]

A carbon-monoxide test was carried out with Helldiver No. 00002 with bomb-bay doors removed which found '…excessive concentrations of .019 and .014 per cents of CO were in front and rear cockpits

respectively, in rated power climb with cockpit hoods closed.'

The final summary is informative. The Board noted that:

The SB2C-1 airplane was one of a number of models which were, of necessity, obtained without full benefit which might have been derived from an experimental model completed and tested ahead of production. The experimental airplane crashed while undergoing contractor's preliminary tests, was rebuilt, suffered a landing gear failure, and was stated in September 1941 to be a true aerodynamic prototype of the production article. The contractor reported that after incorporation of a larger and redesigned vertical and horizontal tail, the airplane was stable in all conditions (tested). In December 1941, during the preliminary demonstration, the experimental model crashed to complete destruction, leaving the production model without a prototype. In spite of efforts to expedite deliveries, the first model SB2C-1 airplane was not delivered until September 1942, at which time nearly 7,000 airplanes were on order, of which 3,100 were for the Army Air Forces as the model A-25 airplane.

Early in the service operation of the model SB2C-1 airplane, many deficiencies in design, workmanship and inspection became apparent which required the setting up of modification programmes. Major changes included substitution of a Douglas flexible mount with twin .30 calibre guns in place of the turret, installation of a self-sealing fuselage fuel tank, and reinforcement of the rear portion of the fuselage. Modifications were put into the production line and in November 1943 the first airplane with all items incorporated was delivered.

With refreshing candour, the Bureau stated that the SB2C-1 was by that time, '… the only scout bomber available to replace the model SBD airplane which had borne the brunt of early dive-bomber actions, and, in spite of deficiencies, showed a number of superiorities over the latter model.[32] They gave the following comparison figures, which were not, however, all that convincing!

It was added that the SB2C-1 '… failed to fulfil the original guarantees by a large amount and was some 1,600 pounds above the original estimated weight. The longitudinal stability characteristics were unsatisfactory and other defects were found in the airplane. However, acceptance of the model SB2C-1 airplane for service use in spite of its deficiencies, is considered to be in the public interest.' The Board therefore

Take-off Data, Patuxent River Trials			
Aircraft No.	Gross weight (lb (kg))	Take-off distance in calm (ft (m))	Take-off distance in 25 knot wind (ft (m))
00002	13,239 (6,005)	890 (271)	350 (107)
00013	15,207 (6,898)	1,050 (320)	420 (128)
00170	14,361 (6,514)	1,030 (314)	440 (134)
00170	14,719 (6,677)	1,125 (343)	490 (149)
00003	13,289 (6,028)	900 (274)	356 (109)

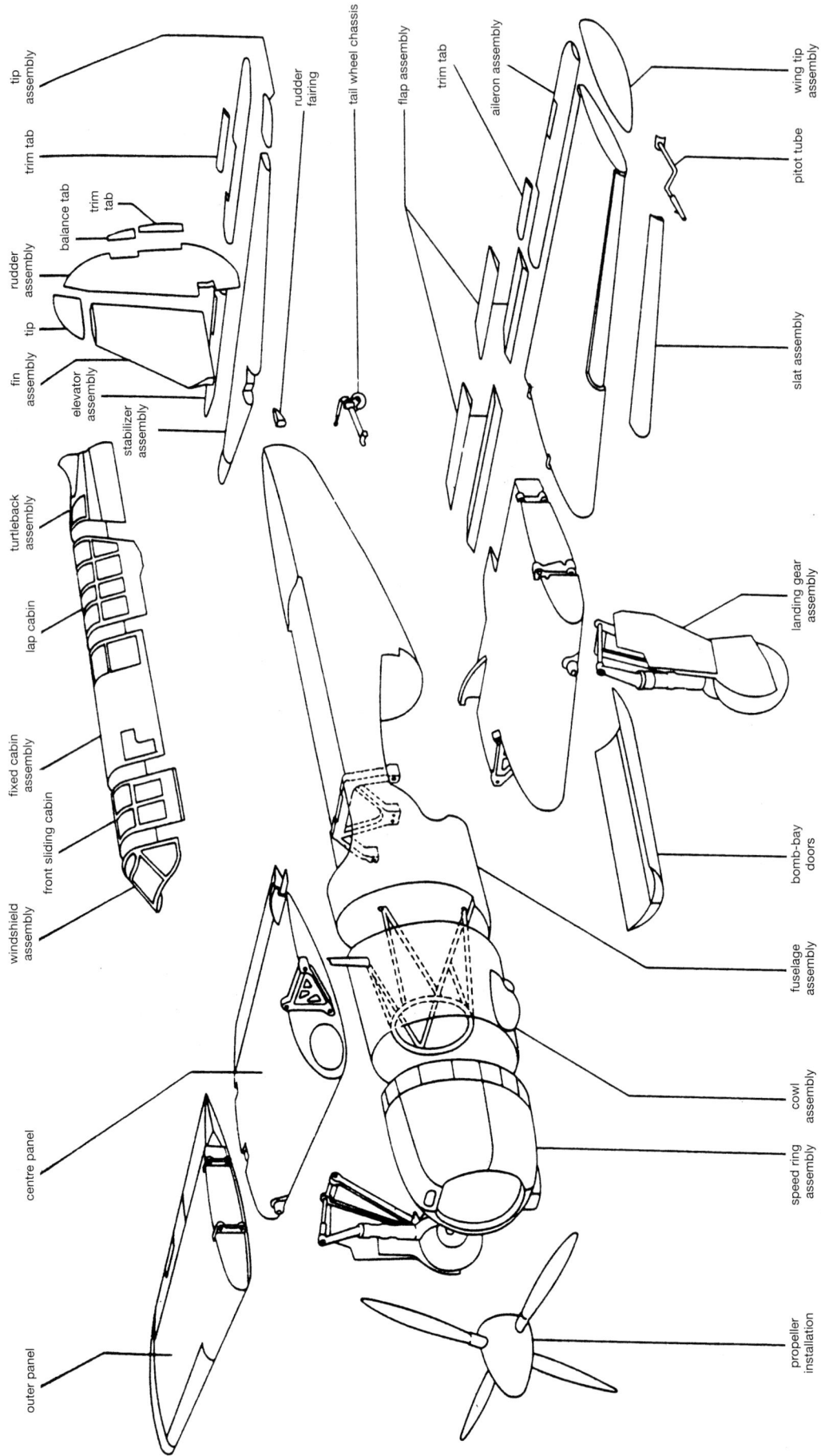

tip assembly

trim tab

rudder fairing

tail wheel chassis

flap assembly

trim tab

aileron assembly

wing tip assembly

balance tab

trim tab

pitot tube

rudder assembly

tip

fin assembly

elevator assembly

slat assembly

stabilizer assembly

turtleback assembly

lap cabin

landing gear assembly

fixed cabin assembly

front sliding cabin

bomb-bay doors

windshield assembly

fuselage assembly

cowl assembly

centre panel

speed ring assembly

outer panel

propeller installation

Major disassembly.

recommended that the model SB2C-1 '... with appropriate changes recommended ...', be accepted as a service type '... as a matter of expediency, no other airplane being available for the purpose intended.'[33]

The Initial NACA Tests

It is of interest to give some details also of the tests run by the National Advisory Committee for Aeronautics (NACA), tests for what was then the leading edge of aircraft technology. On 1 January 1943, the purpose of the NACA flight investigation of the SB2C-1 was defined as being to determine the cause of the tail failure during dive pullouts, and to obtain comprehensive data as to the magnitude (including anti-symmetrical distribution) of tail loads, and the span distribution of the wingloads. BuAer noted that it considered that the new design loading conditions must be obtained in this manner in order to insure the structural adequacy of the airplane.[34]

They took as their working hypothesis for this first phase, on the basis of available evidence, that wrinkling of the wing skin outboard of the wing fold had caused premature separation. Accordingly, the aircraft

Comparison Data between the SB2C-1 and the SBD-6		
Model	SB2C-1	SBD-6
Gross weight (lb (kg))	13,240 (6,006)	9,410 (4,268)
Rated power (bhp)	1,350	900
Rated altitude (ft (m))	13,000 (3,962)	18,500 (5,639)
Aeroplane critical altitude (ft (m))	13,400 (4,804)	17,800 (5,425)
Maximum speed (mph (km/h))	286 (460)	251 (404)
Stalling speed (mph (km/h))	75.5 (121.5)	75 (120.7)
Service ceiling (ft (m))	25,800 (7,864)	28,200 (8,595)
Take-off distance (ft (m))	350 (107)	360 (110)
Maximum fuselage bomb capacity (lb (kg))	2 × 1,000 (454) or 1 × 1,600 (726)	1 × 1,600 (726)
Fixed guns	4 × .50 or 2 × 20mm	2 × .50
Flexible guns	2 × .30	2 × .30
Armour (lb (kg))	195 (88.5)	167 (76)

required instruments to study separation and the resultant effects on the strain and deflection of the wing and horizontal tail. Again, BuAer noted that the data thus obtained would be compared with data obtained by static test and calculations which would then indicate discrepancies between the actual distributions of the air load and those assumed for design.

The instruments selected for both the first and second phases of the test were identical, except that during the second phase, extensive pressure distribution measurements and strain measurements on the wing and tail were to be made. The pressure measurements were expected to show the trends of pressure distribution at lower speeds at which the local effects of wrinkling will not affect the pressure determinations.

The instruments chosen for the first phase, and their status, are given below.

General Instruments

1. Accelerometer (at centre of gravity): ready for use.
2. Recording airspeed meter: ready, but a boom was to be prepared.

Curtiss SB2C-1 (note tail marking 'Ship 240', extended nose section and other modified areas) crash-landed on 20 November 1943 at North Island, San Diego, during experimental engine testing.
Emil Buehler Naval Aviation Library, Pensacola

3. V-G recorder: ready.
4. Angular velocity recorder: available, but would have to be taken from another aircraft.
5. Yaw meter: available, but would have to be taken from another aircraft.
6. Recording altimeter: prepared, but was being used on another aircraft.
7. Thermometer: available.
8. Timer: available
9. Mechanical control position recorders: available, and electrical CPRs available if needed.

Wing Installations

1. Optograph: was being adjusted and would be ready by 1 February.
2. Strain gauge equipment: ready, and eight strain gauges were to be placed on the wing spars by Mr LaVier.
3. Cine special motion picture camera, available to photograph wing wrinkles and tufts over the rear section of the wing from the wing fold outboard.
4. Three head rakes of three tubes each were being constructed and a fifteen-capsule manometer had been adjusted for use with these rakes at high speeds.

Tail Installations

1. Optograph: prepared to record motions of lights on lower surface at various locations. A mock-up had been made to check the sensitivity and range of these installations.
2. Strain gauges: the ready and remaining two channels would be used to two gauges located on or near the front spar of the tail on either side.
3. Pressure orifices: to be installed one on each side of the tail to record the angle of attack changes on two capsules of fifteen-capsule manometer used for the total head tubes.

Special Equipment

1. Radio-telemeter equipment to record acceleration, airspeed, and four channels of strain; to be supplied by the NAF. This equipment was undergoing development and its use would be to obtain data on the instrument and any information of value to the tests which could be obtained.

For the second phase of the tests a special tail in which orifices and strain gauges had been installed at the factory, was to be used. Forty-five capsules of manometer capacity were available, and ten channels of strain (neglecting telemeter). The tests for both phases of the programme were to consist of determination of deflections, strains and skin wrinkles as a function of airspeed and acceleration for comparison with static test data. Therefore measurements were to be made with all the instruments listed, in level flight and in pullups and pullouts at increasing accelerations and speeds up to maximum values of about 6g and 400mph (644km/h).

On 1 February 1943, a memo (Aer-E-2454-CG) revealed that a comparison of available data – Curtiss report no. 8151 - *Determination of Aerodynamic Coefficients for Structural Design*; NACA Preliminary Conference data, *Force Tests of Curtiss SB2C-1 Model (⅛- scale), Part II*, dated 7 December

Curtiss publicity photograph, 'Fist of the Fleet', which is how they wanted the SB2C to be known. Ultimately that is what the Helldiver became, but it is as 'The Beast' that she went down in history, rightly or wrongly! Curtiss-Wright Corporation via John F. Johnson, ADC, USNR

1942; and Curtiss report no. 8161, *Determination of Applied Loads, Shear, Bending Moments and Torques on Fuselage* – showed a '…considerable disagreement' with the tail-load coefficients between the first two. It stated that the horizontal tail-load developed in a pullout from a zero lift dive to a load factor of 9 at 400mph (644km/h) indicated airspeed with a gross weight of 12,600lb (5,715kg); this had been calculated using an assumed stick motion and a rational procedure based on the equations of dynamic stability. The results of these calculations gave an ultimate tail-load of 13,200lb (5,987kg) (up). (*See* box, right.)

Curtiss was told to strengthen the horizontal stabilizer to withstand a 15,000lb (6,804kg) static load, almost doubling the existing 8,100lb (3,674kg) static load on the production models, and to increase the thickness of the leading edge skins on both stabilizer and wings.

Air Group 9's Experiences

And still the orders were piling up, forcing Curtiss to sub-contract again, this time to the Canadian company Fairchild in December 1942, for what was to become the SBF-1. That same month also saw the first deliveries of SB2C-1s to the Navy's operational squadrons, VS-9 then working up for the new carrier *Essex* (CV-9) at NAS Fentress Field, Virginia, being the first unit assigned the new dive-bomber, quickly followed by VB-9 of the same air group, where they replaced the aged Vought SB2U Vindicators. Each squadron received fifteen of the new Helldivers apiece and eagerly commenced training with them.

Alas, they proved not to be the wonderful new dive-bomber they had been promised for so long. On 27 December 1942, VB-9 lost its first machine when No. 00023's engine cut out with fuel starvation and crash landed, an accident put down to the pilot's lack of familiarity with his new mount. The same pilot crashed again on 2 January 1943, writing off No. 00022; this time it was a failure of hydraulics which caused him to overshoot the runway, and again it was the pilot who was found to be at fault, and not the aircraft.

But these findings did not alleviate the problems encountered by Air Group 9 with these first production SB2Cs. Hydraulic leaks were a major problem for both pilots and mechanics alike, and

servicing the new machines proved a nightmare. There were also leaks in the self-sealing fuel tanks; problems with cracked Solar exhaust manifolds and bearings in the wing-folding joints; fractures in the carburettor air scoop; and the new propeller governor was unreliable.

Then came the tragic death of Lieutenant-Commander John Yoho, the executive officer of VS-9, when his machine, 00019, failed to come out of a terminal dive on 7 January 1943. A graphic recollection of the preliminaries to this tragedy was given by Bertram F. Rogers, then a young ARM3c with the squadron:

My pilot (it may have been Ensign Hampton) and I took off from NAS Norfolk at about 10:00 a.m. It was my first flight in an SB2C-1 which had a new ATC transmitter, and a hydraulically controlled power turret with one fixed .50 calibre machine gun in the rear. As you can imagine from my log, I was used to cruising at about 120 knots in SB2Us, and it was very exciting to cruise at about 170 knots in this 'giant' new plane which (I was told) weighed, with the turret, nearly 13,000lb.

When we were out about three-quarters of an hour, the pilot called me on the interphone, said he'd like to make a few shallow practice dives and would I mind closing the hatch because there was some vibration problem with the tail assembly. I agreed, closed my hatch, and we had made four or five dives when my pilot got on the intercom and said, 'Oh Christ, I forgot, the skipper is up next and I'm late!' We headed at top speed for Norfolk (I seem to remember nearly 240 knots), landed uneventfully, and taxied to the hangar.

There, in flight suits, waited Lieutenant Commander John R. Yoho, USN, and chief aviation radioman George T. Blalock, USN. The skipper was not in a bad mood, and took my pilot aside to ask about the plane's performance. George asked me what I thought about the radio. I remember saying 'The receivers are a little noisy, but everything's great.' George grinned at me and he and the skipper climbed in and taxied away.

It was about 11:30 or so. I ate lunch and headed for the radio room which our squadron had in the hangar – and by the time I arrived, at around 12:15, it was all over. The skipper and George had 'spun in' at Oceana in their first shallow practice dive at the target, the plane had exploded and both had been killed instantly. Subsequently the crash crew loaded all the debris from that plane into a 1½ ton stake body truck and returned to Norfolk. A memorial service was held at the NAS base chapel that Sunday, attended by the squadron members and the wives and children of Yoho and Blalock.

I have often wondered if this story was ever told – how an Ensign with about 500 hours and an ARM3c flew in that plane, made five dives and walked away, whilst Lieutenant Commander Yoho and Chief Blalock, with thousands of flying hours, were lost on their first dive. I have also often wondered if they were the Navy's first SB2C operational fatalities. They were both wonderful family men – and even today, I can literally see George's face as he grinned at me and walked toward the plane. In another tragic twist, Lieutenant Donaldson, who took over as Acting CO, was killed in one of our replacement SBDs within about a month or so after Lieutenant Commander Yoho. It is my recollection that both planes carried the serials VS9-1. I have (for some reason) always remembered that crashed SB2C's serial number as 00007, but unfortunately my log book does not support it.[35]

A lengthy investigation heard that as Yoho had commenced his pull out at the end of the dive, the right landing gear had fallen out of its well, causing the aircraft to roll at very low altitude, and the pilot had not had the chance to recover before the machine went into the ground at high speed. The cause was soon pinpointed to the failure of a locking pin to engage due to lack of pressure in the oleo strut; the cure was to relocate it from the moving part of the undercarriage leg to the fixed drag truss where hydraulic pressure would not affect its working.

Loading Test Figures				
Method	Condition	LFV mph (km/h)	Gross weight (lb (kg))	Tail-load (ultimate) (lb (kg))
Detailed spec.	9	210 (338)	10,982 (4,981)	14,500 (6,577) up
Detailed spec.	–4	509 (819)	10,982 (4,981)	13,900 (6,305) down
Stick motion	9	400 (644)	12,600 (5,715)	13,200 (5,987) up

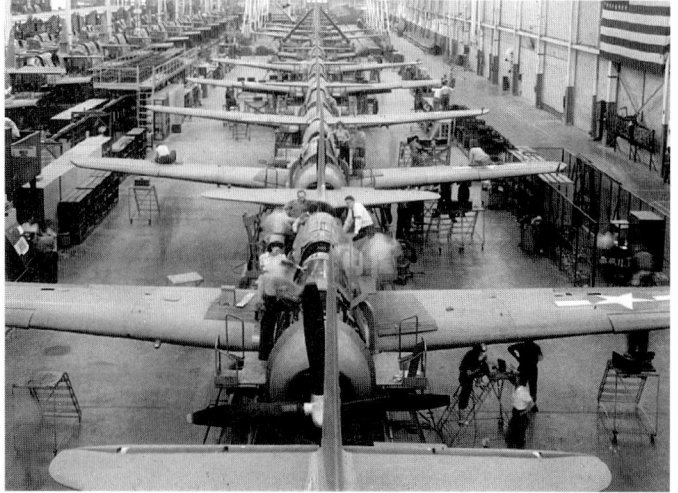

(Top left) **The SB2C Helldiver production line at Columbus, Ohio under way. The various lines can be seen with the centreline tail, fuselage and inner wing sections.**

(Above) **The Helldiver production line at Columbus. The various lines can be seen with, on the right-hand line, the tail, fuselage and outer wing sections in place.**

A pristine SB2C-4 fresh from the production line at the Columbus plant.

An immaculate line-up of SB2C Helldivers fresh from the production line at the Columbus plant.

All pictures Robert Olds Collection courtesy Peggy Olds

Hydraulics were not to blame for the next mishap to befall Air Group 9: on 25 January, following a routine take-off, the outboard flap of one aircraft did not retract. The pilot managed to land in one piece, and a detailed examination revealed that a pivot pin on one of the torque shaft universal joints had snapped; it was also discovered that no less than seventy-four of the other joints were equally badly worn already. All the Helldiver's outer flaps were disconnected, and the aircraft was restricted to level flying.

This was just shoddy workmanship and poor inspection, which reflected very badly on the Columbus plant – although it must be said that all the machines other than the first two hundred now had the new hydraulic actuating system so they were not affected. The early models had to have a stronger joint fitted as part of their rebuilding. But then, also on 25 January 1943, came news of another blow: the first production machine, 00001, suffered a structural failure during the pullout sequence from a high-speed dive similar to that of the XSBC-1. This time the test pilot, Byron A. Glover, was killed.

(Left) **A factory-fresh SB2C Helldiver in the new US Navy paint scheme, airborne in the summer of 1943.** Robert Olds Collection courtesy Peggy Olds

Underside view of Helldiver 0018. This aircraft served with VS-9, which was based at Norfolk NAS, Virginia. US Navy via National Archive, College Park

CHAPTER SIX

The Nadir

On the afternoon of 25 January 1943, Byron had taken off in SB2C-1 No. 00001 from NAS, Anacostia, on what he told the officers of the Flight Test Section, was to be a 'build-up' dive in preparation for the fifth dive of the final demonstration. This fifth dive was to have been a clean dive (with bomb-bay doors and dive flaps both closed) with the Helldiver in scout configuration. It was anticipated that 380 knots indicated speed would be attained in this dive, and that about 5g would be required. Glover said that he intended to make the speed but would 'ease' out on the 'build-up' dive at about 13,000ft (3,962m). It was later thought possible that he had changed his mind on this, and had gone for the full 'g' after all in order to avoid the necessity of a second dive in that condition. The machine was fitted with two V-g recorders, one camera directed at an auxiliary instrument board, and one audio accelerometer.

The SB2C-1 lifted off as planned and was followed by an observing SBD-3 Dauntless, piloted by Lieutenant Commander Crommelin, to watch the test. Crommelin himself was therefore in perfect position to observe exactly what occurred. He reported that after climbing to about 23,000ft (7,010m), Glover had called him on the radio and pushed over into the dive. The message sent was unintelligible, probably because of Glover's oxygen mask, and he was well into the dive itself before a repeat could be requested. At 15:55 Crommelin observed the start of the pull-out and estimated Glover's altitude at about 12,000ft (3,658m).

After some 10 degrees of the recovery Helldiver 00001 disintegrated. Of the cloud of wreckage which left the diving airplane, Crommelin could identify the left outer wing panel but nothing else – although closely observing it, he was not able to state which part left first. After making what was described as '… a few gyrations', the Helldiver settled into a spin, apparently striking the ground in an inverted attitude.

Glover himself bailed out of the aircraft at rather low altitude but in plenty of time to have effected a safe landing *if* his parachute had opened properly. However, it streamed, and would not 'Blossom out'; this was apparently due to flying metal which severed the lines and Glover plummeted to his death.[36]

The wreckage of the crash was scattered over a wide area from west to east, the direction of the dive had been north, with a strong wind blowing from the west. In sequence the parts were distributed over an area of 2½ to 3 miles across the rolling and wooded countryside, as follows:

1. The main wreckage, consisting of the forward part of the fuselage, the centre section, right wing and aileron, and both main landing gears.
2. The turret floor and the leading edge of the lower half of the rudder.
3. The rear part of the fuselage (with the break just forward of the stabilizer), fin, right stabilizer (with blue paint streaks on the underside) and part of the left stabilizer, indicating an up-failure. It was noted that no part of any movable surface, except the torque tube forging connecting the elevators at the fuselage, was found with the tail. The forging showed where both elevators had broken clear, but the elevator horn and centre hinge fitting were still attached to the rear of the fuselage.
4. The left outer panel with the aileron and slat attached. The fact that both ailerons were present was good, indicating that there was no aileron flutter. A piece of the left elevator, but this had not been present on the evening of the crash, and was probably found by a farmer and carried to this spot.
5. Pieces of the turtleback, a piece of the rear escape hatch, both right and left upper landing gear fairings, and the front part of the right bomb-bay door.

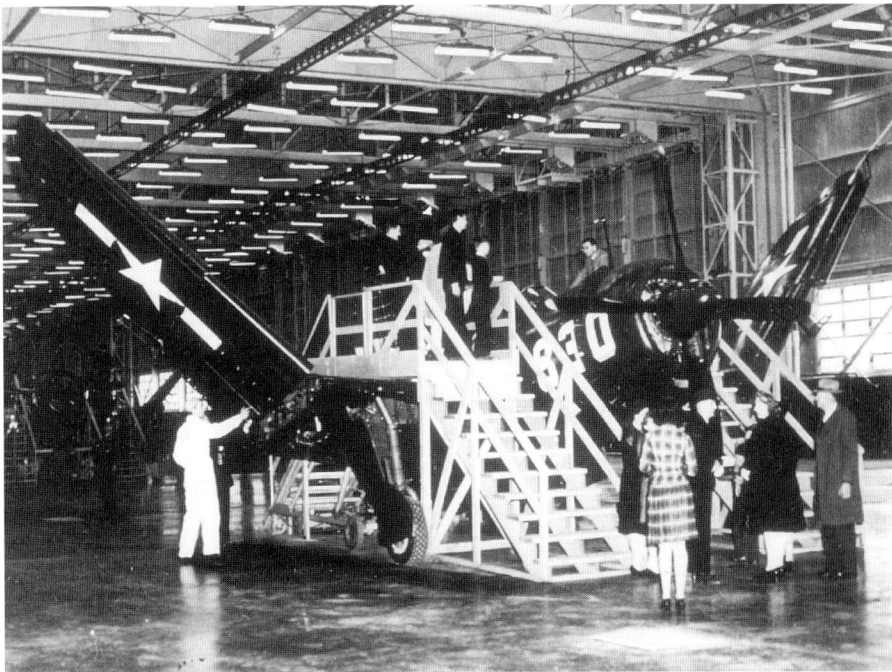

An SB2C at the Army-Navy 'E' Award and Family Day at the Columbus plant, 19 November 1944. Curtiss-Wright via Stanley I. Vaughn Archive

6. The inboard part of the left elevator, with the tab in good condition. A piece of wing spar cap strip, and miscellaneous small metal pieces.
7. The left outboard stabilizer (with dents in the leading edge, with blue paint marks).

Not found were the right elevator, most of the rudder, the right slat and the lower wheel fairing. Only one instrument was found that was of any value, and this was one V-g recorder which showed a speed of 455 (732km/h) mph indicated and 6.7g. This speed was almost 20 knots faster than was expected and the acceleration was 1.7g more than the test required.

That night a party of naval officers, with Mr Maxam of Curtiss-Wright, inspected the wreckage which was then placed under guard. Next day a thorough inspection was made by Commander Hatcher and Commander Murphy from BuAer; Lieutenant Lloyd of NAS Anacostia, and Mr Blaylock and Mr Maxam of Curtiss-Wright. The wreckage was then loaded on to trailers and taken to the NAS where it was laid out in the order found. On 27 January additional Curtiss people – Messrs Page, Jenkins, Rumpf and Clark – arrived, also Lieutenant Commander Brown and Mr Creel of BuAer, and a preliminary conference was held.

Commander Hatcher showed them curves plotted from wind-tunnel data which had been very recently obtained and which indicated that the tail-loads were considerably higher than those used in design. The model had only been tested in the clean condition, so it was agreed to repeat the tests at once with the flaps split. There followed a discussion as to what Curtiss should do next, and the following points were made:

1. To expedite the destruction tests of the unreinforced and the reinforced tails in the up-load condition. These had already been requested and would define the actual strength of the tail and would indicate the places where reinforcement should be immediately added, while maybe providing a firmer basis for the aircraft's restrictions.
2. Blaylock was to examine the fuselage and stabilizer to ascertain what additional strength, if any, was required to meet the additional loads found by the wind-tunnel tests.
3. The landing gear up-locks were to be redesigned as a highest priority.

Close up of crowd with SB2Cs at the Army-Navy 'E' Award and Family Day on 19 November 1944. Curtiss-Wright via Stanley I. Vaughn Archive

4. The allocation of a replacement Helldiver to NACA was recommended in order that the scheduled structural test programme could proceed simultaneously with the full-scale wind-tunnel tests.
5. The possibility of fuselage failure was discussed.
6. Since the condition of the only part of the rudder found was suspicious, the possibility of fin-rudder flutter was discussed. Calculations, however, showed ample margin of safety against this.
7. The possibility of extreme wrinkling of the wing leading edge under acceleration was then discussed, and it was pointed out that this might accelerate a compressibility breakdown of flow. Structure representatives said that by making some very sweeping assumptions, such as considering that all load was redistributed to the wing tips (this assumes a complete flow breakdown over the inboard portion of the wing),

a wing failure could be expected at about 7g.

The following day Blaylock told them that, after consultation with the plant and a further conference among themselves, the test of the un-reinforced stabilizer might be ready by 1 February, while that of the reinforced stabilizer would be available by the 3rd, as would the new loading calculations. The Curtiss wind-tunnel model would be taken straightaway to NACA and they would rush tests on it. The Buffalo plant's oscillograph would be sent to Columbus, and a group of flutter experts would be called in from the various Curtiss plants to remeasure the frequencies of all parts of the tail. Another conference would be set up at Columbus. NACA was to be supplied with another Helldiver (serial no. 00056) to enable them to proceed with the structural flight tests.

The matter of restrictions on the SB2C-1 as a result of the crash was then discussed in some detail. It was decided to proceed as follows with those Helldivers which had reinforced tails:

In the clean condition at 12,600lb (5,715kg)

0–10,000ft (0–3,048m)	6g
10,000–15,000ft (3,048m–4,572m)	5g
above 15,000ft (4,572m)	4g

In the arrested (with flaps split) condition, the above accelerations could be increased by 1g. For higher gross weights the allowable accelerations had to be reduced, keeping the product of the gross weight × the acceleration, a constant. The maximum permissible speed between sea level and 10,000ft (3,048m) was to be 350 knots indicated and 325 knots indicated, above 10,000ft. The Helldivers' serial nos. 00002 through 00016 (those with unreinforced tails) remained restricted to a maximum speed of 300 knots indicated and 4g at 12,600lb (5,715kg).

No conclusions as to the primary cause of the accident were possible at that time, but Commander Murphy raised the following pertinent questions:

1. Why were both landing gear up-locks undamaged, and the up-lock rollers on the landing gear struts in good shape, yet the wheels had extended? This was entirely *different* from what had occurred in the crash of the XSB2C-1.
2. Why did all control surfaces except the ailerons leave the airplane in the air? Could this have been primary?
3. Why did the fuselage fail, and was it primary?
4. How were the dents in that part of the left stabilizer which tore off to be explained? How about the blue paint on the under side of the right stabilizer?
5. How did one explain the up-failure of the left stabilizer? Was it primary?
6. When did the left outer wing panel fail in relation to the rest of the airplane?
7. Why were both landing gear fairings found a long way from the main wreckage? Only the tip of the left stabilizer was further away.

From subsequent studies of the speeds, accelerations and altitudes involved, demonstration requirements were modified and addendum no. 9E to the Bureau of Aeronautics Specification No. SR-38B was finally issued, as late as 4 July 1944.

The final assembly line of the Curtiss Ohio plant, where the Navy Helldivers were produced. Curtiss-Wright via Stanley I. Vaughn Archive

The final demonstrations themselves were completed in the vicinity of the Curtiss-Wright plant at Columbus on 7 September 1944. By this date this demonstration covered Helldiver models SB2C-1, SB2C-1C, SB2C-3 and SB2C-4.

US Navy top brass visited the Curtiss-Wright plant frequently. Here they view the new four-bladed Helldiver which was under production, on 8 April 1944. Curtiss-Wright via Stanley I. Vaughn Archive

One further SB2C-1C (Bu. No. 01159) was modified for experimental work and transferred to the Naval Air Modification Unit, Johnsville, Pennsylvania. Here it was tested by them and utilized in radio-controlled dive tests there. It was also used by NACA at the Langley Field experimental site.

Results of Accelerated Service Tests at Norfolk, Virginia on 17 February 1943

Hard on the heels of these disasters came yet a further damning report from the Accelerated Service tests conducted at Norfolk.[37] The total flying time on this aircraft was 309.3 hours, of which 121.1 hours were devoted to fuel consumption tests, while the remaining 188.2 hours (less 75 hours at 80 per cent power) was devoted to accelerated service tests at 90 per cent power. During that period the aircraft was loaded for take-off at various gross weights from 13,219lb (5,996kg) to 15,050lb (6,827kg). As the external torpedo installation was not available, fuel consumption tests were not made and take-off data was not taken with the external torpedo installed; a supplementary report was to be made covering this loading condition once the installation became available. Also during that period full take-off power was used for all take-offs and 90 per cent power in flight thereafter.

The following programme was carried out:

a. Flights between sea-level and service ceiling.
b. Operation of two-speed supercharger.
c. Operation of automatic and manual propeller control.
d. Operations on all fuel tanks.
e. Operation of emergency and booster fuel pumps.
f. Operation of hydraulic system (wing-folding, landing gear, slats, flaps, bomb-bay doors).
g. Use of oxygen equipment.
h. Dives made clean, and with dive flaps, bomb-bay doors, and approach doors open.
i. Take-off tests at various gross weights from 13,219lb (5,996kg) to 15,050lb (6,827kg).
j. Night flying.

Take-off Figures

Gross weight (lb (kg))	Wind (mph (km/h))	Density altitude (ft (m))	Flap (degrees)	Take-off distance (ft (m))
13,219 (5,996)	10 (16)	(-) 2,220 (677)	45	520 (158)
"	"	"	30	510 (155)
"	9 (14.5)	(-) 1,770 (539)	45	600 (183)
"	"	"	30	560 (171)
"	"	(-) 1,250 (381)	45	665 (231)
"	6 (9.7)	"	30	648 (198)
"	4 (6.4)	(-) 940 (287)	45	720 (219)
"	3 (4.8)	"	30	775 (236)
"	18 (29)	(-) 1,613 (492)	45	540 (165)
"	"	"	30	595 (181)
"	20 (32)	"	45	540 (165)
"	"	"	30	560 (171)
"	12 (19.3)	(-) 45 (14)	45	655 (200)
"	10 (16)	"	30	560 (171)
"	"	440 (134)	45	590 (180)
"	9 (14.5)	"	30	640 (195)
"	4 (6.4)	670 (204)	45	660 (201)
"	"	"	30	800 (244)
13,730 (6,228)	10 (16)	(-) 1,383 (422)	45	650 (198)
"	"	"	30	700 (213)
"	12 (19.3)	(-) 723 (220)	45	650 (198)
"	10 (16)	"	30	640 (195)
"	8 (12.9)	(-) 63 (22)	45	710 (216)
"	"	"	30	640 (195)
"	6 (9.7)	(-)1,428 (435)	45	665 (231)
"	9 (14.5)	"	30	645 (197)
"	14 (22.5)	(-) 1,354 (413)	45	660 (201)
"	"	"	30	675 (206)
"	9 (14.5)	(-) 1,319 (402)	45	657 (200)
"	10 (16)	"	30	645 (197)
"	13 (20.9)	(-) 1,630 (497)	45	625 (191)
"	11 (18)	"	30	600 (183)
14,330 (6,500)	6 (9.7)	(-) 1,540 (469)	30	705 (215)
"	7 (1.6)	(-) 1,460 (445)	30	720 (219)
"	5 (8)	(-) 110 (33.5)	30	650 (198)
"	10 (16)	(-) 2,040 (622)	30	800 (244)
"	12 (19.3)	(-) 1,720 (524)	30	660 (201)
"	14 (22.5)	(-) 1,360 (415)	30	710 (216)
"	14 (22.5)	(-) 1,360 (415)	30	750 (229)
15,050 (6,827)	8 (12.9)	(-) 760 (232)	30	1,020 (311)
"	15 (24)	(-) 850 (259)	30	1,012 (308)
"	"	(-) 1,360 (415)	30	850 (259)
"	8 (12.9)	(-) 1,980 (604)	30	875 (267)
"	15 (24)	(-) 1,100 (335)	30	645 (197)
"	12 (19.3)	(-) 700 (213)	30	840 (256)
"	29 (47)	(-) 3,080 (939)	30	705 (215)
"	13 (20.9)	(-) 300 (91)	30	750 (229)
"	10 (16)	(-) 2,650 (808)	30	695 (212)

Take-off data was recorded at various gross weights from 13,219lb (5,996kg) to 15,050lb (6,827kg) using both 30-degree and 45-degree flap. The take-off technique used was as follows: hold brakes with plane in three-point attitude; turn up engine to full take-off power; release brakes; allow plane to fly off from three-point attitude. The stalling speed was found to be 71 knots at 14,326lb (6,498kg) load. At a gross weight of 13,219lb the take-off data recorded was as shown (see box above).

Not surprisingly in view of this second disaster, and the trials undergone, many faults were shown up, and the recurring theme was the lack of longitudinal stability. The full list was quite a long one as the findings given clearly show (see box overleaf).

Problems and Recommended Solutions

Problem	Recommended Solution
(1) Right exhaust stack cracked. Exhaust manifold connection to No. 8 cylinder exhaust pipe cracked.	Redesign entire exhaust manifold assembly.
(2) Solid tail-wheel tyre loosened on wheel.	Tail wheel to have sufficient durability to withstand more than 150 hours' service operation.
(3) Hydraulic expander tube for right wheel brake leaking.	Redesign expander tube to withstand normal service operation.
(4) Servo cylinder piston of propeller governor control failed.	Design a more positive and secure means to hold spring and centre contact in place.
(5) Pilot's cockpit insufficiently ventilated.	Ventilate cockpit sufficiently to ensure pilot is reasonably comfortable in hot weather.
(6) Right exhaust manifold adjacent to stack failed.	Redesign exhaust manifold.
(7) Weld at outlet connection in exhaust manifold for vacuum line oil separator failed.	Redesign entire exhaust manifold assembly, and install longer tubing for vacuum line for greater flexibility.
(8) Monel metal kick pin installed in carburettor.	Exercise more care in inspection of carburettors prior to installation.
(9) Fracture of fuel pump cooling tube flange.	Strengthen flange to withstand engine vibration.
(10) Malfunction of propeller power unit brake discs.	Use better wearing material in manufacture of brake discs.
(11) Carburettor air scoop fractured on both sides adjacent to alternate air door hinge caused by failure of back fire door to take up force of the back fire.	Redesign this entire section of the carburettor air scoop.
(12) Exhaust valve of No. 11 cylinder burnt and warped.	Investigate material used in manufacture of these valves.
(13) Left exhaust stack cracked at lip.	Redesign entire exhaust manifold.
(14) Holes in universal joint for bottom engine cowl flaps elongated, causing looseness and vibration of flaps.	Redesign engine cowl flap-operating mechanism.
(15) Brake slip in propeller power unit.	Use better wearing material in manufacture of brakes.
(16) Looseness and slot motion in connection between propeller governor operating lever and propeller governor control lever due to wear in bushing installed on the bell crank at the centre of rotation.	Substitute a bearing for the bushing in bell crank.
(17) Exhaust valve of No. 13 cylinder burnt and warped.	Investigate materials used in manufacture of these valves.
(18) Inner spring for intake valve of No. 2 cylinder failed.	Investigate materials used in manufacture.
(19) Left exhaust collector cracked at collar of outlet part.	Redesign entire exhaust manifold.
(20) Right exhaust collector burned and cracked at junction of outlets from Nos 6, 7 and 8 cylinders.	Redesign entire exhaust manifold.
(21) Right exhaust outlet section cracked at connection for vacuum line oil separator.	Redesign entire exhaust manifold.
(22) Fuel entered after cockpit from left and right wing tank vents due to vent openings being flush with fuselage.	Extend vents about 2in outside the fuselage, and seal carefully at the point of passage through the fuselage both inside and out, or bring the vents out of the lower side of the wings.
(23) Malfunction of carburettor due to improper installation of accelerating check valve and accelerating check valve spring.	None.
(24) Right exhaust collector cracked around edge of weld of outlets of Nos. 7 and 8 cylinders.	Redesign entire exhaust manifold.
(25) Altimeter in rear cockpit not functioning properly due to static line being connected to airspeed meter pressure line.	That the contractor exercise more care during installation and inspection of these items.
(26) Severe wrinkles in left horizontal stabilizer.	Since this stabilizer did not have the latest reinforcement of the strength members incorporated by the contractor, no recommendation was made.
(27) Failure of turtleback diagonal brace, port side aft, at adjustable end.	Redesign braces for the turtleback to withstand loads imposed by normal dive pullouts and landings.
(28) Failure of electric primer lead at carburettor due to guide slot in amphenol plug being too wide and allowing plug to vibrate in receptacle.	Substitute an elbow-type Cannon plug.
(29) Lost motion in universal joint of flap actuating mechanism, due to fracture of short pin carried in ball section at hole for joining pin.	Make entire universal joint large enough and of sufficient strength to allow ample thickness of the walls of the balls and pins after drilling.
(30) Right exhaust stack failed at weld for connection of vacuum line oil separator.	Redesign entire exhaust manifold.
(31) Fuel leaked from right wing fuel tank due to cracks around flange of self-sealing liner for installation of fuel quantity gauge transmitter. Cracks appeared to be caused by strain being placed on this connection as a result of the forward upper half of the cell receding from the compartment beam.	Not known what caused deformation, so none made.
(32) Solid tail-wheel tyre torn and loose on wheel.	Give tail-wheel sufficient durability to withstand more than 150 hours' service operation.
(33) Fluorescent lights not placed for best lighting of pilot's cockpit.	Place light behind and on each side of pilot at about shoulder height.
(34) With present light facilities in pilot's cockpit, pilot is unable to see flap position indicator at night.	Redesign lighting in the pilot's cockpit to include visibility of the flap position indicator at night.
(35) Frame for fixed gunsight obstructs pilot's vision of top instrument panel.	Redesign this installation to give pilot unobstructed vision of instruments.
(36) Instruments on panel in pilot's cockpit are poorly arranged.	Rearrange instruments: as follows: *Top panel, top row, left to right*: altimeter; visual accelerometer; rate of climb. *Second row, left to right*: engine unit gauge; airspeed meter; turn and bank; engine tachometer; manifold pressure. *Bottom panel, left-hand group*: landing and take-off check-off list; fuel quantity gauge; outside air temperature; ignition switch; remote compass indicator. *Centre group*: artificial horizon and directional gyro (or automatic pilot). *Right-hand group*: cylinder head temperature; hydraulic pressure; carburettor air temperature; clock.

In summary Commander E. C. Parker stated quite bluntly that:

This accelerated Service Test has revealed entirely too many discrepancies in the production model of this airplane, and it is believed that a high percentage of these discrepancies are due to the use of materials of insufficient strength or inferior quality, poor workmanship, and careless methods of erection, installation, and inspection. Unless this condition is corrected prior to placing these airplanes in service it is felt that the maintenance problems to be overcome by operating personnel will be entirely too great and that the combat value of the airplane will be seriously reduced.

Air Vice Marshal Jones of the British Air Commission advised the director general in a memo dated 2 March 1943, in response to a query of when the Royal Navy was likely to receive their Canadian-built Helldivers, thus:

Reference your enquiry re-Curtiss aircraft being built in Canada. This is the SBW, known here as the SB2C. One or more of the early American production aircraft developed very serious tail trouble. One at least came off in the air. This has entailed the complete re-stressing in the tail unit by Curtiss Columbus. It is understood, though not confirmed, that the U. S. Navy are holding up the commencement of Canadian production so that Canada may fit the redesigned tail *ab initio*.[38]

Next day Jones was following this up with the news that '... Smeeton is busy trying to find out the information you desire. I understand that the Royal Navy would accept some of the initial production of these aircraft with the old type tail, since they will be used for training purposes.'[39] On the same date E. P. Taylor, the Canadian deputy member of the Combined Production and Resources Board in Washington DC, spelt out the situation, as the Canadians saw it, in a long memo:[40]

I gather that the major trouble is with the tail unit which, owing to a weakness in the leading edge, developed 'buffeting' and has wrecked aircraft under test. I am now informed that the engineering of the new stabilizer is completed and will be under test next week. Following a satisfactory test, the particulars are available for release to the Canadian Car and Foundry Company. How much delay will be involved, assuming the company gets the particulars next week, is largely a matter for them as they will have to ascertain in detail the scope of the changes involved.

Over and above this trouble I learn that a number of changes, sometimes as many as sixty in a week, have been issued by the Curtiss Company in Columbus, which Cancar are expected to incorporate in their production line, with corresponding effect on their plans. This point had previously been brought to the attention of the US authorities by Lieutenant Commander Smeeton of this mission, and they have taken the matter up with the firm.

On the other hand, there are complaints from the American side of insufficient liaison with Cancar, of which I do not know the rights and wrongs. Apparently Curtiss have asked that a permanent technical liaison officer should be stationed at Columbus in order to facilitate the transmission of information, etc. Possibly Cancar feel that so many modifications should not be necessary.

As a result of all these tests, in April 1943, the US Navy prevailed upon the company to make modifications in the field to the first 200 aircraft This done, they tested them and insisted that yet a second modification was needed to make them combat-worthy, Mod II, with some 800 changes, mainly minor ones such as the deletion of the DF loop aerial. A whole new modification centre had to be set up at the plant in yet another new facility, Building 3B, to carry out these alterations, involving yet more cost and delay. But this was still to prove insufficient.

Truman Committee Report

As if all this was not enough, the long delay from the placing of the original order in 1939 and the failure to produce a combat-worthy aircraft by 1943 led to Curtiss-Wright being included in the notorious Truman Committee, set up to examine a whole series of similar delays and problems being revealed in the whole National Defense Programme. Senator Harry S. Truman, a former farmer from Missouri, now a Democrat trouble-shooter, began to investigate Curtiss-Wright in January; and the replacement of J. A. Williams as general manager at Columbus followed, his place being taken by J. P. Davey who had filled the same post at both Buffalo and St Louis earlier. By 10 July 1943, the Special Committee had delivered a devastating attack on the company, which pulled no punches:

'Hopelessly behind schedule, Curtiss-Wright to date has not succeeded in producing a single SB2C which the Navy considers useable as a combat airplane ...' The report stated that there was a widespread knowledge of the failure of the company in this respect, and this in turn led to an adverse psychological effect on the workforce and on general morale in the Columbus area, and an incalculable loss of production by their sub-contractors: Truman was especially critical of Curtiss-Wright's advertising which had proclaimed the Helldiver to be, 'the world's best dive-bomber' when it had not yet flown in combat, and that '... such eulogistic self-praise was intended to give the public an ... erroneous impression'. Nor did the power gun turret escape his wrath, and he ordered the Navy to conduct 'vigorously' a detailed investigation of the matter to decide whether the current design was the best one or not in comparison with three others. It was not until 18 December 1945 that the Navy was finally to cancel all work on the Curtiss-Norge turret.

Not only did the Special Committee fail to take into account the many extra features that had been added to the original specification down the years, all of which imposed increasing weight penalties, they also overlooked the effect on the brand-new plant's efficiency of the compulsory call-up – the drafting into the services of many of the skilled workers, already limited in number, in effect ham-strung Curtiss-Wright's efforts by another government agency! Choosing to ignore such factors, the Truman Committee came down hard calling the company's performance to date, 'thoroughly unsatisfactory'.[41]

The US Navy was equally criticized for allowing matters to get so out of hand, with the resulting expenditure for no result; but they remained firmly committed to the Helldiver. Indeed, by the time that Truman was vigorously castigating the company, they and the Navy had at last begun to solve most of the outstanding problems, many of which were unforeseeable in 1939, and the SB2C was starting to come right. On 15 July 1943, Artemus Gates sent a telegram to the company stating just that, and that the programme was to be continued. The Navy, he told them, '... are confident that the design difficulties can be overcome ...'. They were equally sure that the Helldiver would eventually become the 'Fist of the Fleet' as Curtiss-Wright had been trumpeting to the world, and that it would, '... play an important part in the defeat of the Axis'.

CHAPTER SEVEN

A New Beginning

The SB2C-1 was also being tested by British pilots at about this time. On 23 March 1943, Wing Commander H. J. Wilson, RAF, flew the SB2C-1 and submitted a detailed report on the Helldiver which makes interesting reading as a comparison with the US test results.[42] The machine he flew was fitted with the Wright-Cyclone R2600-8. The figures provided here show the engine rating (production carburettor setting) (see box, top right); the fuel carried for combat-loading conditions (see box, middle right); and weights for typical loading including all combat (gunfire protection) items (see box, bottom right)

The aircraft is fitted with a dive flap, which takes the form of a split trailing edge flap, opening both up and down, 26 degrees each way, giving a total opening of 52 degrees. When used as a lift flap, only the bottom half moves, lowering to 52 degrees. A single control lever actuates either of these flaps on selection, and the operation is simple and straightforward. This aircraft is also fitted with slots which are automatically in the 'out' position whenever the undercarriage is down.

The cockpit layout is *quite satisfactory and simple*.[43] The only criticism is the position of the undercarriage and flap controls on the starboard side; but I understand that, should the Army adopt this aircraft, they intend to move these controls to the port side. The air screw revolution control lever is placed below the throttle quadrant, upside down, and actuates in the opposite direction to normal.

The pilot's view for taxiing is recorded as being, '... exceptionally good', and the taxiing characteristics, '... quite satisfactory with the brakes working well'. He noted that prior to take-off it was necessary to lock the tail-wheel, but that is was not necessary to use the flaps for take-off. He recorded that the take-off run was '... very short and in no way tricky'. All controls handled satisfactorily, and no tendency to swing directionally was noticed.

The climb was made at 150mph (241km/h) and at this speed he found the

Engine Rating Data

	BHP	RPM	Man. press.	Blower	Press. alt. (ft)	Period
Take-off	1,700	2,800	43.0	low	SL	1 min
Normal	1,500	2,400	38.5	low	SL	cont.
Normal	1,500	2,400	37.5	low	5,900	cont.
Normal	1,350	2,400	41.0	high	8,900	cont.
Normal	1,350	2,400	40.5	high	1,300	cont.
Military	1,700	2,600	43.0	low	SL	5 min
Military	1,700	2,600	42.0	low	3,000	5 min
Military	1,450	2,600	45.5–44.5	high	8,000–12,000	5 min

Combat-loading Data (Gallons)

	Wing tanks	Fuselage tanks	Auxiliary tanks
Scout (normal)	200	90	–
Bomber (1 × 500/ 2 × 500/ 1 × 1,000lb bomb loads)	200	–	–
Scout (overload)	200	90	90

Weights Breakdown Figures

Bomb load (lb (kg))	Mission	Fuel (gallons (litres))	Guns	Weight (lb (kg))
1 × 500 (227)	bomber	210 (955)	4	13,219 (5,996)
2 × 500 (227)	bomber	210 (955)	4	13,737 (6,231)
1 × 500 (227)	bomber	300 (1,364)	4	13,789 (6,255)
1 × 1,000 (454)	bomber	210 (955)	4	13,730 (6,228)
1 × 1,600 (726)	bomber	210 (955)	4	14,330 (6,500)
1 × 2,000 (907) torpedo	torpedo bomber	300 (1,364)	4	15,359 (6,967)
nil	scout	300 (1,364)	4	13,271 (6,020)
nil	scout	430 (1,955)	4	14,196 (6,439)

ailerons to be '... fairly heavy and sluggish ...', but the response was satisfactory and his verdict was that they were adequate. The rudder and elevators he also considered '... slightly heavy ...', but correctly responsive, and he found the controls, in general, to be '...satisfactorily harmonized'. During the climb he gained the impression that the Helldiver '...is very easy to fly'.

Cruising at 200mph (322km/h) the aileron forces were about 25lb (11kg). The 'feel' was positive, and the response, 'good'. He reported that the rudder was

fairly light '... and very responsive'. Again, the harmonization of the controls was satisfactory. With regard to the biggest bugbear, stability, he found that at 200mph the aircraft was neutrally stable laterally; longitudinally, '... it is just neutrally stable'; and directionally, it was stable. When he initiated a rudder disturbance causing yaw, the aircraft returned to its original path after one or two oscillations. The aileron control forces were about 30lb to apply a 45-degree bank, and were '... very responsive'. He also found the rudder light and very responsive, while the elevator forces were about 30lb which he considered, '… correct for this speed'.

At high speed, 350mph (563km/h), he reported that all the controls were satisfactory, '… and the response is good'. The forces he considered reasonable in view of the size of the aircraft. He noted a tendency for '… directional snaking to set up if the rudder is used. This snaking is bad and will obviously become worse at higher speeds'.

The wing commander then put the SB2C-1 into a test dive with the dive brakes open at 280 to 300mph (450 to 483km/h). 'On opening the dive flap, no change in trim is apparent, and the entry into the dive is very easy. A very steep angle (ie about 80 degrees) can be obtained'. He recorded that, during the dive, '… manoeuvrability is obtainable, although the ailerons feel heavy'. He noticed no blanking effect from the dive flap on either rudders or elevator, and finally observed '… it is thought that the dive-bombing characteristics of this aircraft are quite satisfactory'.

During the landing approach, on lowering the lift flaps, there was no noticeable change of trim. The approach glide '… is pleasingly steep', and all controls at 105mph (169km/h) were adequate, '… although as before, the ailerons feel heavy, but there is no dead area in their movement'.

Wilson concluded with the usual RAF indoctrinational reservation that, '…the necessity for a dive-bomber is debatable', but that, '… should it be found necessary to adopt this operational type of aircraft', then, 'I feel the A-25 (sic) is very good. The control forces are rather heavy but are not considered unduly so, in view of the size and the weight of the machine'. Apart from the snaking in the dive at high speed when the flaps were up, 'no adverse criticism can be made about the A-25 (sic)'. He concluded that:

This aircraft is very easy and straightforward to fly; and for its type, quite pleasant. Also, the view-out is good, and the landing is very simple,

and I can foresee no reason why it should not operate very satisfactorily off a carrier.

The Newport and Dahlgren Torpedo Tests

The origins of these experiments dated back to 15 December 1941, when Lieutenant Commander J. N. Murphy, the head of VSB design, gave his thinking on the complex issues regarding future US Navy bomber policy in the light of lessons so far learnt. Fighter protection was obviously essential, and if that was to be the case, then dive-bombers could be made single-seaters as they did not require flexible gun defence and gunners. This would give dive-bombers the speed of fighters. However, this in turn led to the premise that torpedo bombing and scouting duties could best be combined, in which case this type of aircraft would continue to require flexible gun defence and be two-man machines. Experimental work by VT-3, which had tested dropping their torpedoes in a diving attack, influenced his opinion, but strength and dive brakes were required for this, and the aircraft which had both was the dive-bomber!

Consideration was therefore given for the scout dive-bomber (VSB) to become a dive-bomber torpedo plane (VBT), and the torpedo bomber would have scouting added to its duties to become a VTSB, although this latter was never adopted.

His thoughts confirmed and developed by the head of the VTB design, Lieutenant A. B. Metsger. He looked at the strength and range requirements for such aircraft, and made the same point, '… that since both dive-bombing and torpedo attack used the diving approach, a strong airplane equipped with dive brakes was a necessity'.[44]

He argued that it would therefore be quite logical to build a strong aircraft for the dive/torpedo bomber role, and on 21/22 January 1943 a conference was held at the planning and engineering division of BuAer to discuss these, and other thoughts on the subject. The result was a programme for Curtiss to develop a single-seat dive-bomber with a speed of 390mph (628km/h) carrying a 1,000lb (454kg) bomb for 1,600 miles (2,574km). Admiral Towers approved, and on 4 February 1942 Curtiss was so instructed. In fact the complications with the SB2C-1 fully occupied Curtiss-Wright until the end of 1942, together with the the XSB3C-1, also the

'The Fist of the Fleet' was how the Curtiss-Wright publicity machine tried to publicize the arrival of the SB2C with the active combat fleet. While, in practice, the Helldiver ably lived up to the hype, sinking and damaging a greater tonnage of the Japanese navy than any other method in 1943–45, this was not the name that stuck with it!
Curtiss-Wright via Stanley I. Vaughn Archive

XBTC-1 in 1943 and the XBT-2 at the end of the war.

But meantime war experience was daily emphasizing the value of the aerial torpedo as a warship sinker, and BuAer decided to equip the Helldiver for torpedo dropping; tests were therefore conducted at the Naval Air Station, Quonset Point, Rhode Island and at the Naval Torpedo Station, Newport, Rhode Island, commencing on 30 November 1942. The aircraft employed was Helldiver No. 00013. All the torpedo launchings were initially made with the plane in normal horizontal flight, however no diving approaches had been made before bad weather caused a postponement of the trials.[45] The weapons used were the Mark 13-1 and Mark 13-2 torpedoes, and these were dropped at speeds of between 140 and 155 knots, with drops at normal horizontal flight.

No take-off, flying or landing difficulties were encountered with the torpedo installed, nor did the release of the torpedoes have any adverse effect upon the flight of the Helldiver, but considerable difficulty was found in their loading, which was found to be very time-consuming. A large number of recommendations were made in this preliminary report to try and improve matters, and the preliminary conclusion was that: 'Subject to the incorporation of the recommended changes, the torpedo installation is considered satisfactory for service use. The SB2C-1 is satisfactory as a torpedo plane using the Mark 2 stabilizer at speeds from 140 to 155 knots'.

Further tests at Newport followed in January 1943, with the torpedo fairing completely removed when Mark 13-1 torpedoes were launched at speeds of between

SB2C in flight. Stanley I. Vaughn Archive

140 and 230 knots.[46] The Mark 2 air stabilizer was used on all the drops and the air performance of the torpedo was satisfactory at all speeds for either condition. 'The removal of the torpedo fairing does not have any adverse effect on the air performance of the torpedo ...', and nor did its removal '... affect the flight characteristics of the airplane. However, the removal of the fairing does affect the speed of the airplane,' were the main conclusions. The following comparison figures are given (see box below).

For combat use, the obvious need was for speed in switching over, in the cramped confines of a carrier hangar, under fire, from a dive-bombing to a torpedo-bombing mode, and vice versa. Here the time difference in making this switch, with the

torpedo fairing and without it, was decisive. The removal of the fairing, under test conditions with trained staff, was 2½ hours, and obviously this would be completely unacceptable at sea with upward of sixteen aircraft requiring a rapid turnround from one mission role to another. With the fairing removed, the loading time was reduced from 12hr 51min to 6hr 8min. Moreover the Mark 5 torpedo truck and the Mark 6 bomb and torpedo skid could be used without any modifications for loading torpedoes.

The final summary on these production inspection tests was made on 6 February 1943.[47] During the period 30 November 1942 to 30 January 1943, at both Quonset Point and Newport, Mark 13-1 and 13-2 torpedoes were dropped at speeds of 140 – 235 knots with the Mark 2 air stabilizer BuOrd Dr. No. 226108, and the Mark 2 stabilizer plus a drag ring. The majority of the drops had been made with the split flaps closed, but a few later ones were made with the spit flaps fully open, and open 30 degrees, in order to determine the effects of the flaps on torpedo air performance. The air stabilization was satisfactory for all these tests, and the split flaps had little or no effect either way on this. No torpedoes were dropped in a dive, although a diving approach was sometimes used to reach the horizontal dropping position.

It was found that, as the speed was increased to about 180 knots, the torpedo was carried more nose down to the horizontal – the initial nose-down attitude

Torpedo Fairing Installation Comparison Figures				
FAIRING INSTALLED				
	With Torpedo		Without Torpedo	
Speed (knots)	RPM	Man. Press. (in)	RPM	Man. Press. (in)
150	1,800	26.0	1,700	26.0
175	2,200	28.0	1,900	28.0
200	2,400	33.0	2,400	33.0
FAIRING REMOVED				
150	1,800	26.5	1,800	25.5
175	2,200	28.0	2,200	27.0
200	2,400	35.0	2,400	35.0

All these speeds were indicated air speeds at an altitude of about 300ft (90m).

caused it to oscillate, but did not affect the air performance too much. The recommended dropping speed was 200 knots; with a drag ring in place, this could be increased to 250–275 knots.[48]

The times taken to change from torpedo to bombing condition were, however, stated to be 'excessive'. Three-man teams were used – one Curtiss-Wright man and two others familiar with the equipment – three men being the maximum that could physically work on the installation together; all the necessary tools were available in the hangar. The results show the problem.

Time to Convert to Torpedo Bomber Data	
Task	Hours
Removal of torpedo fairing	2.5
Removal of other torpedo equipment	1.5
Installation of bomb equipment	3.5
Total	7.5
Removal of bomb equipment	2.0
Installation of torpedo equipment	1.5
Installation of torpedo fairing	2.5
Total	6.0

Without the torpedo fairing these times were reduced by 2½ hours, but it was still far too long, and it was considered that the problem '... greatly reduces the tactical value of the airplane'. The removal of the bomb-bay doors was the most time-consuming operation, as they were badly designed. Removal of the fairing naturally affected the speed of the aircraft in flight, nonetheless this was what was recommended. Another was the investigation of the design of a torpedo-support structure with retracting sway braces, which would ensure that the bomb-bay doors could remain on the Helldiver for the torpedo condition; this would eliminate much time and effort in changeovers. The flaps could be utilized in any position for release, although less than 30 degrees of flap was the final recommendation. Otherwise they thought the SB2C-1 would make a satisfactory torpedo dropper.

Further Trials at Dahlgren

From 27 May 1943, further tests to adapt the Helldiver as a torpedo-dropping aircraft were undertaken at the Naval Proving Grounds, Dahlgren, Virginia. The aircraft

used was Helldiver No. 00149, and a Mark 13 Mod. 1 torpedo was employed with a Mark 2 wooden stabilizer. Experiments were also conducted with the Mark 17 Mod. 2 depth charge, a Mark 1 1,600lb bomb and two Mark 33 1,000lb bombs. The results were not initially encouraging: 'The bomb-bay of the subject airplane is extremely poorly designed for conversion between bomber airplane and torpedo airplane'.[49] This is not surprising, as torpedo carrying was not part of the original concept at all.

It took five experienced ordnance men 2½ hours simply to remove the bomb bay doors or torpedo fairing. The torpedo mounting assembly required up to three hours to mount into location, because of

Front cover of Curtiss Fly Leaf, Vol. XXV, No. 6, January–February 1943 edition. Curtiss-Wright via Stanley I. Vaughn Archive

(Below) **Resplendent in their three-tone Navy paintwork of 'non-specular' sea-blue, 'intermediate' blue and white, this quartet of SB2C-1s flies over the flat Ohio countryside.** Curtiss-Wright via Stanley I. Vaughn Archive

having to remove the bomb displacement gear forward arms '... from a very inaccessible location', and then having to fit the forward arms of the torpedo mounting assembly. Finally, all the actual loading tests were made with the torpedo fairing completely removed, and then no difficulty was found. The conclusions were that a complete redesign of the torpedo mounting assembly was essential, and methods to achieve this were detailed.

This was eventually done to complete satisfaction. However, the Helldiver was never used as a torpedo dropper in combat despite the fact that Admiral Halsey wanted to take the combining of the dive- and torpedo-bomber functions of his carrier aircraft much further some time later in the war. In November 1944 he proposed the *total removal* of the Grumman TBM Avenger, a most successful torpedo and glide bomber, from his fast carriers, proposing instead to rely entirely on Helldivers for torpedo attack. This idea had its advocates and its opponents amongst his carrier captains at this time: Captain C. D. Glover of the *Enterprise* was all for it, as was Captain W. W. Litch of the *Lexington*. However, in the short term the chief of naval operations strongly disagreed, and the idea was not taken further:

> As long as Avengers made up a part of the fast carrier complement it was only natural that they would make all torpedo attacks, if for no other reason than that the Helldiver was a more effective bomber than the Avenger. The fact that dive-bombers were never used for torpedo attack did not disprove the belief that dive-bombers could carry torpedoes, but only reflected a peculiar wartime situation.[50]

In fact Halsey was ahead of his time, because his suggestion became the US Navy's official post-war policy, as we shall see.

CHAPTER EIGHT

The Helldiver Goes to Sea

The urgent need to get the *Essex* to sea and into action led to a re-think. Her faulty SB2Cs were flown to NAS Norfolk where they were taken in hand for modification, while the two VS and VB squadrons were re-equipped with SBD Dauntlesses. Following this, the SB2C-1s originally allocated to them were switched to VB-17, commanded by Lieutenant Commander W. L. Blatchford, and VS-17 (soon renumbered as VB-7) commanded by Lieutenant Commander J. E. 'Moe' Vose. It was hoped that these units would have more time available to study the continuing faults before joining their assigned carrier.

They commenced the familiarization with the Helldiver in February 1943, and then moved on to field landing practice and tactical training with the aircraft. During this period they lost a total of six Helldivers in accidents, five in collisions. In March the pilot Art Taylor was killed trying to land his aircraft after his gunner had bailed out, the pilot of the second aircraft, George Kish, was severely injured.[51] In April, during a re-forming after a combined practice attack, the squadron's Corsair fighters inadvertently pulled up from below the Helldivers, and in attempting to avoid them, pilot Guy Frank was also forced to pull up in turn. His propeller sliced through the tail of Gus de Voe's aircraft, '... taking off the entire empennage just aft of the rear seat. The

cowling of Guy's plane was flattened against the cylinder heads to the breaking point. Ten inches were clipped from the tip of his prop, but he managed to get back to the field and land safely'.[52] Both aircrew of de Voe's aircraft managed to survive – just!

The most serious incident, from a design point of view, was one which took place on 10 April 1943. The SB2C-1 was put into a dive at Langley and failed to pull out. It was later established that, on entering the dive, the lower part of the split flap had failed to function correctly and remained shut. When the pilot noticed this he attempted to recover, but the excessive loads imposed on the wing tips and tail caused their structural failure. Another design fault was the lack of a lock on the wing-fold handle, which on one occasion started the folding procedure in mid-air. This was not rectified until early in 1944, and in the interim a temporary arrangement was made utilizing bungee cords.[53]

The two squadrons then duly conducted full carrier qualification trials aboard the escort carrier *Santee* (CVE-29) off Chesapeake Bay in April 1943. These trials were generally satisfactory, and so, to enable proper fleet trials at sea, the air groups of the next carrier being readied, the *Yorktown II* (CV-10), were also allocated SB2C-1s, these being VB-4 and VB-6. The first twelve aircraft were collected from

Columbus and flown to the Marine air base at El Centro, California on 15 February. Training commenced in March; later the unit transferred to NAS San Diego, and then all the way to Norfolk, Virginia, where they arrived on 20 April to continue their familiarization programme. Carrier trials took place aboard the escort carrier *Charger* (CVE30) in April and May.

This turned out to be a very different story. When the *Yorktown* embarked a full complement of new Helldivers in the Mod. II configuration, it was with high hopes that the worst was behind them. Alas, however, it was not to be. During two weeks of operations in May and June 1943, '... less than half of the thirty-six could be flown, and of these, two usually had to be restricted from diving. The folding wings wrinkled under the stress of carrier landing; the entire hydraulic system could be put out of operation by the failure of any of 900 different lines; the turret did not operate satisfactorily; the tail-wheel collapsed during hard landings; the brakes were not heavy enough, and so on ...'

The captain of the *Yorktown*, Captain J. J. Clark, had a special interest in the Helldiver, having been a Navy inspector at Buffalo during the early design work. He wanted the best out of his new carrier and air group, and was a perfectionist in everything, having no tolerance whatever of any failure or mistake.

Below decks of an Essex-class carrier shows the narrow headroom available to the SB2C with wings fully folded. Stringent size limitation on the Helldiver design enforced by the Navy to enable the aircraft to fit the carriers had a great deal to do with the protracted development of this aircraft. Curtiss-Wright via Stanley I. Vaughn Archive

During the period 6–18 May 1943, carrier qualifications took place which resulted in one flight-deck collision, and two crash-barrier pile-ups. Two more Helldivers were wrecked on 26 and 28 May respectively during the carrier's shake down cruise in the Caribbean, while a third lost hydraulic power and ground-looped while landing at Waller Field, Trinidad. Tail-wheel locks were fitted by this time, but some pilots failed to ensure that they were locked on before hitting the deck, resulting in some spectacular barrier prangs. The hydraulics leaked everywhere, and especially bad was the turret system, which most *Yorktown* aviators and mechanics agreed had to go.

Edward J. McCarten was assigned to VB-6 at the Marine Air Base, El Centro, California in February 1943, where VS-3 was also located, and when both squadrons received the SB2C in March, he recalled that several pilots from these squadrons went to Ohio and flew the planes back to California. 'I know one was left in a cornfield in the mid-west and I think it may have been two.'[54] His first flight in the '2C' was piloting Bu. No. 00080 on 28 March 1943. Both VB-6 and VS-3 were on the shakedown cruise aboard *Yorktown*, and their pilot wrote that, 'We had lots of mechanical problems, and "Jocko" Clark, the *Yorktown*'s skipper, refused to take the SB2C into combat. By 18 June 1943 we were back flying SBD-5s'. After this, both were decommissioned and then recommissioned as one unit, VB-5.

Indeed, Captain Clark was so disillusioned with the SB2C-1 that he flatly refused to take the Helldiver into battle. He even went so far as to recommend that an immediate enquiry should be made as to whether the whole contract should be cancelled, and wrote to BuAer expressing the opinion that he had '…grave doubts as to whether or not the airplane can ever be made into a satisfactory efficient combat plane…'.[55]

The Helldivers were promptly dumped back ashore; Clark's ship was reequipped with the faithful little SBD-5, and so went off to war. Back at Columbus, BuAer and Curtiss-Wright had already sat down to ponder yet another series of modifications, and the Mod. III programme had been set up which it was hoped would rectify many of these faults; meanwhile the SB2C-1 was restricted from carrying out clean, high-speed dives completely.

The Mod. III programme involved a lot more strengthening of the aircraft's structure to withstand the stresses of carrier operations. The collapsing tail-wheel problem was tackled by making it a fixed structure rather than retractable. The fuselage fuel tank was made self-sealing, the .50 turret was dropped in favour of the old-style flexible twin .30 machine guns first trialled in the XSB2C-2, and a bob-weight was fitted in the longitudinal control system in order to beef it up during the crucial dive pull-out phase.

The Full Modernization Programme

On 16 August 1943, the 9th Revised List of Modification #3 was issued, and all manufacturing plants, in both Columbus and Canada were supposed to keep to it.[56] There were three variants: (1) which involved the modification of the fifty 'B' programme aircraft up to serial number 00200; (2) the Modification 3 'C' programme involving the aircraft up the serial number 00200; and (3) the Modification 3 'C' programme involving aircraft *above* 00200. These are listed here to show just how much work was involved.

Modification Programme							
DESCRIPTION OF MODIFICATION (X = TO BE INCORPORATED, – = NOT TO BE INCORPORATED)							
	1	2	3		1	2	3
A: Wing folding mechanism				9. Check and test displacing gear valves to eliminate sticking.	X	X	X
1. Cancelled by item R 13.	X	X	X				
2. Cancelled by Item R 13.	X	X	X	10. Install turtle-deck on an auxiliary hand-pump system with an emergency vent to provide for collapsing.	X	X	X
3. Cancelled by Item R 13.	X	X	X				
4. Cancelled by Item R 13.	X	X	X				
5. Provide and install large wing-fold cylinder.	X	X	X	11. Cancelled by Item R 5.	X	X	X
6. Install steel track rollers.	X	X	X	12. Change check and Restrictor valve poppets to Dural.	X	X	X
7. Cancelled by item R 13.	X	X	X				
8. Cancelled by item R 13.	X	X	X	*C: Tail-wheel*			
9. Cancelled by item T 13.	X	X	X	1. Incorporated in Item C 4.	X	X	X
10. Cancelled by item R 7.	X	X	X	2. Cancelled by Item R-5.	X	X	X
11. Cancelled by item R 13.	X	X	X	3. Provide fairing for fixed tail-wheel installation.	X	X	X
				4. Redesign oleo to provide increased stroke.	X	X	X
B: Hydraulic system				5. Incorporate two provisions in axle.	X	X	X
1. Replace flap valves with reworked units having 'C' ring packing and proper porting.	X	X	–	6. Cancelled.	–	–	–
2. Cancelled by item R 8.	X	X	X	7. Inspect tail-wheel lock.	X	X	X
3. Install 'Electrol' unloader valves.	X	X	X				
4. Install 'Purolator' filters.	X	X	X	*D: Fuselage structure*			
5. Install AN 5200-8AS relief valve.	X	X	X	1. Install six bulkheads from stations #54 to #127 between present bulkheads. Add skin over old.	X	X	X
6. Replace all damaged or leaking lines with lines using two-piece fittings.	X	X	X	2. Reinforce fuselage structure in region of turret.	X	X	X
7. Provide a separate accumulator and check valve for brake system.	X	X	X	3. Tail-wheel drag link support.	X	X	X
8. Provide flexible liens on brake control valves.	X	X	X				*(continued overleaf)*

Modification Programme (cont.)

	1	2	3		1	2	3
E.: Wing structure				11. Install additional tail-wheel oleo name-plates on fuselage.	X	X	X
1. Remove skin and replace with one gauge heavier skin between rear beam and lap beam (for re-work, add skins on outside of present skin).	X	X	X	12. Cancelled by item F.	–	–	–
2. Reinforce narrow panel in lower surface between gas tank and longeron. Add skin.	X	X	X	13. Cancelled by item F.	–	–	–
3. Add gussets and stiffeners to upper surface in the region of the ammunition and gun openings.	X	X	X	14. Wingfold doors: reinspect, and if necessary replace to latest tools.	X	X	X
4. Provide a tension link between landing gear truss and nose rib.	X	X	X	15. Reinforce rudder tab control bracket and actuator mounting.			
5. Revise and secure solenoid access doors.	X	X	X	16. Inspect: (a) arresting gear oleo installed backwards;			
6. Inspect and replace nut plates around flotation door.	X	X	X	(b) arresting gear mechanism complete;			
7. Reinforce wing at inboard end of slat plus extended stiffener.	X	X	X	(c) tail wheel lock slide and fittings in fuselage;			
8. Revise flap control mechanism to stiffen in airplane 201 and up.	–	–	X	(d) complete armament check (this includes torpedo);			
9. Revise wing structure to provide for removal of cannon with wings folded.	–	–	X	(e) displacing gear installation.			
				17. Eliminate looseness of arresting gear handle.	X	X	X
F: Gun turret				18. Provide manual arming of wing racks.	X	X	X
1. Remove present hydraulic turret and replace with twin .30 cal. Douglas mount for approximately 979 airplanes. Includes a new cabin track.	X	X	X	19. Remove fuselage fuel tank boost pump and XCO2 system (Ref G-1).	X	X	
2. Disconnect heater and cap liens.	X	X	X	20. Reinforce to eliminate elevator and rudder counterweight breaking loose.	X	X	X
				21. Revise hydraulic tank vent line.	X	X	X
G: Fuel system				22. Reverse trim tab control wheel.	X	X	X
1. Provide and install fuselage and self-sealing tank. (See item R 29 for addition.)	X	X	X	23. Replace Dural piston rods in bomb displacing gear hydraulic struts with steel.	X	X	X
2. Make provision for droppable wing tank.	X	X	X	24. Replace lower oil-tank support.	X	X	X
3. Revise vent system.	X	X	X	25. Remove flare installation.	X	X	X
4. Replace fuel mixture fitting with Stewart-Warner part 472919.	X	X	X	26. Remove antenna reel.	X	X	X
				27. Remove landing light.	X	X	X
H: Arresting gear				28. Reinforce station #30 in fuselage to stiffen displacement lock.	X	X	X
1. Install snubber struts with cast iron piston rings.	X	X	X	29. Visual indicator for tail-wheel oleo.	X	X	X
2. Paint visual indicator on arresting gear control unit.	X	X	X	30. Install dowel in engineering cowl.	X	X	X
				31. Add support to exhaust tail-pipe.	X	X	X
I: Electrical system				32. Torque rigid flaps for 201 and upwards.	–	–	X
1. Rewire starter and auxiliary power unit to bureau drawing R1020.	X	X	X	33. Revise conduit for pilot's light.	X	X	X
2. Inspect and revise electrical system to conform with drawings.	X	X	X	34. Revise elevator bobweight to reinforce.	X	X	X
3. Provide indicator light to show wheels 'down' and armament switch 'on'. (See item R 33 for changes.)	X	X	X	35. Revise bomb-door relock.	X	X	X
				36. New aileron trim tab control rod.	X	X	X
4. Radio track change.	X	–	–	37. Neoprene tail-wheel boot (first twenty-five installed in Service).	X	X	X
				38. Guard on slat-actuating drum.	X	X	X
J: Miscellaneous				39. Supercharger name plate.	X	X	X
1. Revise turtle-deck guide tubes and push rods.	X	X	X				
2. Replace displacing gear bell cranks.	X	X	X	**K: Install mid-span aileron balance tab**	X	X	X
3. Tighten all trim and balance tab control.	X	X	X	**L: Replace Oleo orifice in landing gear oleo**	X	X	X
4. Re-set propeller low pitch steps. (Not to be done in B and F.)	X	–	–	**M: Rebore-sight airplanes in accordance with the revised boresight diagram**	X	X	X
5. Revise leading edge fillet attachment.	X	X	X	**N: Add ATC radio**	–	X	X
6. Replace neoprene guard on bomb door bungee.	X	X	X	**O: Ratchet type sway brace**	X	X	X
7. Check rudder tab hinge brackets.	X	X	X	**P: Paint and camouflage: new insignia**	X	X	X
8. Replace cabin crank handle springs.	X	X	X	**Q: Rudder reinforcement**	X	X	X
9. Replace landing-gear lock control rod.	X	X	X				
10. Revise low target releases.	X	X	X	**R: Modification item #3 extension**			
				1. Fuselage – station #248 – reinforce. (See item 28 for addition.)	X	X	X
				2. Fuselage – arresting gear longeron – redesign.	X	X	X
				3. Fuselage – station #261 – reinforce.	X	X	X
				4. Fuselage – station #235 – reinforce.	X	X	X

Modification Programme (cont.)									
		1	2	3			1	2	3
5. Tail-wheel locking rod – remove oleo retracting strut, and bolt lock into place permanently.		X	X	X	20. Change tail-wheel centring spring.		X	X	X
					21. Gun stowage tie-down.		X	X	X
6. Tail-wheel drag link – redesign (use same forging).		X	X	X	22. Air speed line adapted after fifty aircraft.		–	X	X
7. Wing fold handle (see item R 334 for redesign).		X	X	X	23. Reroute bomb-door vent line after fifty aircraft.		–	X	X
8. The 5in (13cm) diameter accumulators for the main hydro system; become mandatory.		X	X	X	24. Stiffen ribs in elevator after fifty aircraft.		–	X	X
					25. Carburettor Air Scoop Reinforcements.		X	X	X
9. The rear wing fold doors – redesign.		X	X	X	26. Wing tank nipples. Effective when parts available.				
10. Arresting gear control in front cockpit – push-button type to be changed to hand-grip type.		X	X	X	27. Provide electric bomb racks on wings. (Service to install aircraft #00020, 00021, 00024, 00026, 00028, 00032, 00033, 00035, 00037, 00039, 00041, 00042, 00044 and 00046.)		X	X	–
11. Fuselage – reinforce at flare doors.		X	X	X					
12. Replace two 17-amp batteries with two 34-amp batteries.		X	X	X	28. Additional fuselage reinforcement at station #248.		–	X	X
13. Hydro-wing pin-locking mechanism.		X	X	X	29. Added fuselage link supports.		–	X	X
14. Flap-stop for aircraft 00001 to 00200 (ten installed in service).		X	X	–	30. Remove radio shelf per EO895327 against 84-25-503.		–	X	X
15. Wing-fold strut restrictors.		X	X	X	31. Weighted container and water canister combined in Gunner's cockpit.		–	X	X
16. Wing-fold stop washers.		X	X	X					
17. Add washers to Lord Mounts on remote compass.		X	X	X	32. Interlocking flap device.		–	–	X
18. Redesign tail-wheel cam.		X	X	X	33. Landing gear and armament warning light relocated (Columbus 76, Buffalo 26 and Fairchild 26).		–	X	X
19. Remove trim tab instructions plate (not on first fifty).		–	X	X	34. Redesign Wingfold handle (Columbus 76, Buffalo 26 and Fairchild 26).		–	X	X

Equipping of VB-17 with Mod. II Aircraft, 1943

Meanwhile in June 1943, aircraft were allocated to VB-17 which was destined to equip the air group's new carrier, the *Bunker Hill* (CV-17); these were still Mod. II machines, and the same problems as with the *Yorktown* were encountered. Captain Robert B. Woods gave me this description of events:[57]

Alton Chinn and I were both in Bombing 17 which was commissioned early in 1943, together with Scouting 17. In April 1943 the two squadrons were consolidated into one with Lieutenant 'Moe' Vose as the commanding officer. We had thirty-six SB2Cs and were a component of Air Group 17.

The air group was assigned to the USS *Bunker Hill* (CV-17). In June 1943 we deployed to the Gulf of Paria, off Venezuela, for operational training. I was the operations officer of the squadron, responsible for scheduling and training. During our training period it became obvious that the SB2C had major structural problems. On a hard landing, the aircraft buckled just behind the rear seat compartment. In one incident, the whole tail section separated and the aircraft, with the pilot and gunner, ended up

in the barrier while the tail section was trapped by the arresting wire.

Then Captain Wood went on to give me these personal observations of the Helldiver:

Our training programme progressed smoothly, and the SB2C proved to be an excellent aircraft for dive-bombing. In a 70-degree dive, with dive brakes open, it attained a speed of 310 knots. Under ideal conditions we would push over into a 70-degree dive at about 20,000 feet. At this angle of dive, we were able to compensate on the way down for wind effect and target course and speed. We would release our bombs at about 5,000 feet, and would be out of our dive and level off at about 1,500 feet, or would continue on down to tree-top level if we were under attack. We found the SB2C to be very stable in a dive, and all of the pilots became very proficient and accurate in delivery of bombs.

The early SB2Cs, with which Bombing 17 was equipped, had screw-operated dive brakes, but this proved to be a design flaw. If you lost hydraulic pressure because of battle damage when the dive brakes were open, the gunner had to crank them closed with a hand crank. With the brakes open in level flight, one lost speed and manoeuvrability very quickly – not a good

situation to be in when under attack! This was remedied very quickly so that if you lost hydraulic pressure when the brakes were open, they closed automatically. As I recall, this fix was accomplished before we deployed to the Pacific.

The Wright engine, with which the SB2C was equipped, had a tendency to load up when idling on the flight deck waiting for take-off. As a general rule, the engines missed firing and sounded horrible on the take-off run. In order to compensate for this situation, it was the practice to give us about a hundred feet more deck-run for take-off. We did not lose any planes on *Bunker Hill* as a result of this deficiency, but one of the other carriers lost four SB2Cs, one right after another, when it followed the manufacturer's recommendations for take-off deck-run length.

The SB2C had a bomb ejection arm which prevented the bombs on the forward bomb racks from hitting the propeller when released. A sequence valve on the hydraulic system prevented the pilot from closing the bomb-bay doors on the extended bomb ejector arm, However, this sequence valve failed repeatedly, with the result that the bomb-bay doors often closed with the ejector arm extended. When this occurred, the bomb-bay doors were damaged beyond the capability of the crew to repair. We had inadequate spare doors, so that after about

(Left) In May 1943, the SB2C went to sea again for further combat testing aboard the Yorktown II, which embarked the aircraft of both VB-4 and VB-6. Curtiss-Wright via Stanley I. Vaughn Archive

Preparing for take-off from the Yorktown II, this Mod. II Helldiver proved totally unreliable for combat duties in the first trials in May 1943; the type was soon replaced by the faithful old SBD Dauntless. Curtiss-Wright via Stanley I. Vaughn Archive

a month of operations, most of the SB2Cs flew minus bomb bay doors.

While this problem affected the range and manoeuvrability of the aircraft, it had no effect on their availability for combat operations, nor did it have any effect on the accuracy of our bomb delivery since we always dived with the doors open.

As I have said, the aircraft was very manoeuvrable. It had good speed. It carried 20mm guns in the wings and the gunner had two .30 guns at his station. We were never trained in air-to-air combat. After I became a fighter pilot myself and commanded a fighter squadron, I reached the conclusion that the SB2C would have done very well in air-to-air combat. Unfortunately this capability was never exploited.

The SB2C remained the main offensive weapon of the carriers until it was replaced by the AD dive-bomber in the late forties and early fifties. The SB2C proved itself in the carrier operations of World War II. It was a rugged aircraft. It could handle much battle damage and get back aboard ship. All-in-all it proved to be a great aircraft and the backbone of the attack carriers during the last two years of World War II.

That indeed was to be the case, but in the early summer of 1943, the fate of the Helldiver was in the balance. As one historian put it: The failure on the *Yorktown* would have speeded the Helldiver's end but for the advocacy of Captain J. J. Ballentine, commanding officer of the *Bunker Hill*, and Lieutenant-Commander J. E. Vose, Jr, commanding officer of VB Squadron 17. They insisted that once the folding wing mechanism was improved, the arresting hook adjusted and the tail-wheel strength-

ened, the Helldiver would 'live up to its name.'[58]

As the first Mod. II machines became available at the end of May 1943, the old models were flown back to the factory and exchanged for some of the new ones – Bu. nos. 00102 – 0199 – which now had the 15,000lb (6,804kg) stabilizer and the heavy-duty wheel brakes fitted. The two merged squadrons, as VB-17 under the command of Vose, with a total of thirty-six Helldivers, were thus embarked aboard *Bunker Hill* when she conducted her working-up trials from 15 July onward. Also aboard was a Curtiss field service representative, William J. Clark, who worked

closely with the air group commander, Commander M. P. Bagdanovich, and VB-17 pilots in ironing out problems, and he liaised closely with Raymond C. Blaylock once the carrier had docked back in the States again. These tests were marked by one of the most spectacular of all Helldiver landings when, on 19th July, the squadron's executive officer, Jeff Norman, made a very heavy landing indeed. The Helldiver split in half, the break occurring just behind the heavy turret and the two halves spread themselves across the flight deck; but incredibly there were no fatalities. On that same day there was a fatal crash, many tail-wheel failures, and one

Specification – Model 84A SB2C-1C	
Engine type:	R-2600-8
Armament:	Two 20mm forward-firing fixed cannon; two 30mm rear-cover flexible machine guns; maximum bomb load 2,000lb (907kg); torpedo load 1-Mk 13
Dimensions:	Length 36ft 8in (11.17m); height 13ft 1½in (4.01m); wingspan 49ft 8⅝in (15.15m); wing area 422 sq ft (39.2 sq m)
Weights:	Maximum weight 16,607lb (7,533kg); empty weight 10,114lb (4,588kg)
Performance:	Maximum speed 273mph (439km/h) Take-off power 1,700hp (1,268kW) Range as dive-bomber 1,375 miles (2,213km) Climb height 10,000ft/11.4min (3,048m/11.4min) Service ceiling 21,200ft (6,462m)
Number completed:	778
Serial numbers:	00201/00370; 01008/01208; 18192/18598 (Bu. No.-Navy – Serial Army)
Contract:	Order date; 19 Nov 40 Delivery date; Aug 43–Mar 44

Helldiver went straight into the sea after losing power on launch.

The new, non-retracting tail-wheels were finally installed on the aircraft after they had flown ashore to Waller Field, Trinidad on 20 July, and they were given a thorough testing from then until 10 August. The most important alteration at this time was the strengthening of the wing-folding mechanism by the fitting of a second hydraulic strut to ensure the locking pins were firmly driven home once the wing had unfolded. This helped solve the problem of locking the wings in their extended position on the carrier flight decks prior to take-offs, a failure in function which had plagued the early trials. When the carrier returned to her home port, Curtiss delivered a full complement of Mod. IIIs to VB-17 on 20 August, and it was with these, her third full set of Helldivers, that she sailed to war on 10 September 1943, the first carrier to take the Helldiver into action.

The aircraft on which Curtiss-Wright had modified the hydraulic flap, and which they had fitted with the 20mm cannon, also began to come off the production line, and then had to be made over as Mod. IIIs before joining the fleet, where they were designated the SB2C-1C. Likewise the first of the Canadian-built Helldivers, of both the SBW-1 and the SBF-1 designations, arrived, the first flight of the former taking place on 29 July 1943, and this was followed by thirty-seven more (Bu. Nos. 21192–21200 and 21302–21231). All these incorporated the above alterations.

Next up was VB-8, under the authority of Lieutenant-Commander R. L. Shifley, working up for the carrier *Intrepid* (CV-11); this also received SB2C-1 Mod. IIIs in September 1943. Although the shake-down cruise was far more successful, VB-8 losing no aircraft in this period to accidents, they again encountered the same problems with the notorious hook bounce, even though other faults appeared to have been largely overcome. This was put down by some pilots to the basic landing technique differences between the SBD and the new SB2C-1. The former required a nose-down final drop after the engine cut from the LSO, but the latter did not, and the inability to adapt led to many heavy landings when the hook failed to engage the wires.

Four Helldivers went into the sea later through engine failures, while a fifth lost power during take-off for the same reason and also went in, although all the aircrew were saved. The difficulties of servicing the hydraulic system also still remained, and it required a high mechanical attendance to keep them flying. When the *Intrepid* reached Hawaii, VB-6 was flown ashore and replaced by VB-6 equipped with the SB2C-1C, with which model VB-8 was also later re-equipped before re-embarking (in the *Bunker Hill*) for combat service as late as March1944.

Helldiver Acceptance Procedure

By November 1943 the first Helldiver to feature all the various alterations and changes, the 601st SB2C, came out of the factory gate – by which time the dive-bomber had been 'blooded' in action. The procedure was for each Helldiver, as it left the final assembly area – by now a fully automated and finely tuned line under the supervision of R. W. Cramer, working round the clock on a three-shift system – to be taken to the ramp area north of Building 3B. Here flight testing was conducted.

Specification – Fairchild SBF-1	
Engine type:	R-2600-8
Armament:	Two 20mm forward-firing fixed cannon; two 30mm rear-cover flexible machine guns; maximum bomb load 2,000lb (907kg); one 2,000lb torpedo
Dimensions:	Length 36ft 8in (11.17m); height 13ft 1½in (4.01m); wingspan 49ft 8⅝in (15.15m); wing area 422 sq ft (39.2 sq m)
Weights;	Maximum weight 16,607lb (7,533kg); empty weight 10,114lb (4,588kg)
Performance:	Maximum speed 273mph (439km/h) Take-off power 1,700hp (1,268kW) Range as dive-bomber 1,375 miles (2,213km) Climb height 10,000ft/11.4min (3,048m/11.4min) Service ceiling 21,200ft (6,462m)
Number completed: 49	
Serial numbers:	31637/31685 (Bu. No.-Navy – Serial Army)
Contract:	Order date; 31 Dec 42 Delivery date; Dec 43–Jun 44

Specification – CC & F SBW-1	
Engine type:	R-2600-8
Armament:	Two 20mm forward-firing fixed cannon; two 30mm rear-cover flexible machine guns; maximum bomb load 2,000lb (907kg); one 2,000lb torpedo
Dimensions:	Length 36ft 8in (11.17m); height 13ft 1½in (4.01m); wingspan 49ft 8⅝in (15.15m); wing area 422 sq ft (39.2 sq m)
Weights:	Maximum weight 16,607lb (7,533kg); empty weight 10,114lb (4,588kg)
Performance:	Maximum speed 273mph (439km/h) Take-off power 1,700hp (1,268kW) Range as dive-bomber 1,375 miles (2,213km) Climb height 10,000ft/11.4min (3,048ft/11.4min) Service ceiling 21,200ft (6,462m)
Number completed: 40	
Serial numbers:	21192/21231 (Bu. No.-Navy – Serial Army)
Contract:	Order date; 23 May 42 Delivery date; Sep 43–Mar 44

Group portrait of the air and deck crews of VB-17 prior to their wartime deployment.
Commander J. Alton Chinn, USN, Rtd

Before being taken aloft by one of the test flight team of pilots, each Helldiver at the Columbus facility was given a rigorous pre-flight checking. An inspector from the company under Superintendent E. L. Bradley put each machine through a set static routines, which included cockpit tests of all instruments and controls, starting and revving of the engine, checking the propeller at varying revs, the smooth operation of both pilot and gunner cockpit canopies, and the testing of the hydraulic systems.

If a machine checked out thus far, the same inspector conducted a series of taxi tests to test the functioning and reliability of the landing gear, and the brakes and tracking were similarly tested. A two-hour minimum period of in-flight testing then followed, under the direction of the chief test pilot, (in the latter period Bill Webster) with the aircraft taken up in steps

through the whole range of the engine power, which was matched against a critical chart to see if it came up to scratch. Yaw and pitch were closely monitored, given the SB2C's history. Rate of climb tests, spin tests and engine variations were conducted over the flat Ohio countryside. The climax was, of course, the high-speed dives which were made from 22,000ft (6,705m) altitude. Opening the dive brakes gave graphic confirmation or otherwise of their ability to hold the SB2C steady at a maximum plunge speed of 250mph (400km/h), and vibration was monitored, with the pullout giving a final confirmation of the sturdiness or otherwise of the aircraft's structure. With production in full swing, three such tests a day were conducted by some of the test pilots. The record number of Helldiver test flights was eventually claimed by Charles Betzler, while Ralph Charles is reported to

have flown for eighteen months on these duties with only a single malfunction.[59]

After both Curtiss-Wright and the US Navy representatives were completely satisfied with the performance and finish, each SB2C was towed to Building 3.[60] Here they were painted in the camouflage pattern of the day, cleaned and made ready for aerial delivery to their assigned units by ferry pilots, which more often than not, were women.

Despite the settling down of production procedures, experimentation continued nonetheless, and from November 1943 a long, on-going series of detailed tests was begun in a determination to get to the bottom of the still outstanding problems once and for all. Two SB2C-1s from the production line were modified at Buffalo especially for these exhaustive trials, and equipped with a myriad of instruments to record all aspects of the aircraft's performance in its

Canadian Car & Foundry, their Fort William (now Thunder Bay) plant assembly line in full swing on the SBW-1 dive-bomber.

(Below) The Canadian Car & Foundry plant assembling SBW-1 dive-bombers, 24 September 1943.

(Below) **Further along the Canadian Car & Foundry plant assembly line. The numerous access points under the folded section of the wing can clearly be seen.**

Looking down the line of the Canadian Car & Foundry plant. The tail of No. 389 can be seen in the foreground. Note the leading edge flaps on the wings and Yagi antennae under the wings.

All pictures courtesy Ken Johnson via the National Aviation Museum of Canada, Ottawa

designed role. They were then put through their paces under what was termed 'final flight investigation' from November 1943 onward. These dive tests evaluated both air loads and surface flutter stresses and they concluded that compressibility effects were the culprit. Yet further testing, utilizing No 00140 as an air-load research flying test-bed, was conducted into the summer of 1944, with zero lift and open dive brake configuration, and these were extended to both the Canadian built types also, using the same Curtiss test pilot team.

In February 1944, the US Navy asked for the production of the Helldiver to be speeded up still further, requiring twenty-three aircraft per day to be completed, such was the demand for the now-proven-in-combat dive-bomber as the Pacific War worked up to its climax. Just how much the 880 plus alterations had changed the weight of the Helldiver from concept to combat debut is summarized below.

WEIGHT INCREASE BREAKDOWN (lb)

Item	Original guaranteed weight	Revised guaranteed weight	Edicted estimated	weight calculated	To horizontal actual	armed	Datum moment	To vertical arm	Datum moment	Over- or Under-weight	Comments
Wing Group											
Wings	1,907	1,926	52	48	2,101	45	94,160	−4	−8,404	175	Struc. G.F. Prot. and prod.
Ailerons	47	61		100	62	65	4,030	8	496	1	balance tab
Flaps	101	106	100		106	99	10,494	−4	−424		
Tail Group											
Stabilizer	89	92			127	255	32,385	33	4,191	34	Estimated increase
Elevator	52	57	100		64	276	17,664	33	2,112	7	due to larger tail
Fin	18	19	100		33	262	8,646	58	1,914	14	area
Rudder	27	27	100		28	279	7,812	59	1,652	1	Total +57
Body Group											
Fuselage	753	780	29	71	823.5	93.5	76,922	12	9,882	43.5	M. fit skin, mg to Dural
Main landing gear (Up)	621	642	8	92	688.8	15.8	10,843	−8	−5,510	47	barrier crash, tyres, wheels
Aux. landing gear	44	44		100	48.8	242	11,737	3	146	4.9	door
Engine section	261	263	5	95	292	−57	−16,615	−2	−584	29	engine mt. isolators
Power Plant											
Engine	2,000	1,994			1,994	−77.5	−154,535	0	0		
Engine accessories	126	133	35	65	154	−53.7	−8,267	−8	−1,232	21	Oil cooler and air intake
Power plant controls	17	15.8	80	20	19	−14	−266	18	342	3.2	cowl flap cont.
Propeller installation	430	424.5	100		426.5	−106	45,165	0	0	2	
Starting system	44	51		100	45.5	−35	−1,593	−2	−91	−5.5	crank in overload
Oil system	44	44		100	44	−25.2	−1,110	1	44		
Fuel system	88	125	10	90	125	46.3	5,798	6	750		
Fixed Equipment											
Flotation	53	0									
Instruments	56	53		100	53	10	530	23	1,219		
Surface controls	183	186.4	50	50	204	97.7	19,914	−1	−204	17.6	Wing flap controls
Furnishings	377	328	70	30	361	96.8	34,891	2	722	33	flex. gun chutes and mount.
Hydraulic equipment	239	225	50	50	269	44	11,805	−3	−807	44	accumulator, tank and oil
Auto. pilot installation		36		100	36	38	1,368	8	288		
Electrical equipment	240	252.2		100	256	9	2,302	−1	−256	3.8	
Arresting gear	31	36		100	37	169	6,253	8	296	1	
De-icing gear	20	13	95	5	13	34	442	−1	−13		
WEIGHT EMPTY	7,868	7,934	30	70	8,411	15.5	130,446	0.8	6,529	477	GROSS WEIGHT
USEFUL LOAD							203,139		−10,510		CENTRE OF GRAVITY (C.G.)
1–500lb BOMB		3,074			3,074						= 29.0 − 0.34 = 26.0% M.A.C.
GROSS WEIGHT		11,008				29	333,585	−0.3	−3,981		109.3
											M.A.C. 109.3in located 0.334in
											aft lew

CHAPTER NINE

Into Battle

The prolonged trials and experimentation by VB-17 of the new dive-bomber had finally been completed, and it was now time to put the Helldiver to the final test of battle. The SB2C-1 was not to be let in gently either, her first target being one of the toughest of the Japanese bases in the south-west Pacific, Rabaul. Just how did the pilots of VB-17 feel about their new aircraft on the eve of such a momentous initiation? Bill Palmer recalled his thoughts on the SB2C-1 to me thus:[61]

I was a pilot in VB-17, the first squadron to take 'The Beast' into combat. We formed in Norfolk, Virginia, and were part of Air Group 17 assigned to a new carrier designated CV-17, the USS *Bunker Hill* of the 'Essex' class.

Initially, the problems with the Helldiver were many. As I recall, one of our first losses was when the dive flaps only opened on the top while the bottom half remained closed – in our case the pilot lost control in the dive. He bailed out and was able to relate his version of the problem to the manufacture representatives working with us. As a result, the worm gear system of opening and closing the dive flaps was replaced with a hydraulic open/close system. This proved to be much better time-wise, being a quicker open/close, and it was much safer and more reliable.

The original rear seat 50-cal, hydraulically operated gun turret was also replaced due to maintenance problems – and it was slow to operate and too heavy. Its weight added to the tendency to buckle the fuselage just forward of the gunner's rear seat. On our SB2C-1s we had a large patch on both sides for added strength, and the 50-cal gun and turret was replaced with twin 30-cal guns, manually operated by the gunner.

There were so many problems to solve, but most of them were conquered by close teamwork with the Curtiss engineers assigned to our squadron, our engineering officers and men, and the pilots themselves. As you know, the landing time interval was very important, especially in the combat zone. One particular problem was getting our arresting hook retracted after the roll back and wire release. One of our pilots suggested cutting a notch in the existing gear so we would only have to raise the hook half way, enough to clear the deck, thus clearing the landing area in about 25 per cent less time. A small item, but it really worked.

The Helldiver was not an easy plane to fly, but as far as I was concerned it was worth the added effort for the additional speed (a big plus) and bomb load. I liked to fly the Helldiver. I guess you would say that it was heavy on the controls. I liked to start my flaps closing about two-thirds of the way to the target, release and pull out as low as possible. This required the use of elevator trim tab, which is not usual procedure, but the 'The Beast' held together and the added 'going away' speed surely put distance from those shooting at us very quickly!

Combat Debut: Rabaul

Rabaul, on the north-west tip of New Britain and the main Japanese base in the south Pacific, had a sinister reputation among the US aircrew, and was considered a very tough nut to crack. Early American carrier operations of the 'hit and run' category had not gone so deep into enemy waters before. Shore-based Navy and Marine Corps dive-bombers had been pounding islands up the Solomons' chain as a steady advance was made to get within range of this important target,[62] but carrier-based dive-bombers were the only ones capable of attacking the heavy concentration of enemy shipping expected to be found there, provided they could surprise the fighter defences on the five Japanese airfields that guarded magnificent Simpson Harbour. This made Rabaul the toughest land target yet taken on by the American carriers, and the risk was considerable. The first strike was made on 5 November by SBD Dauntlesses launched from the *Saratoga* and *Princeton*; the second strike was a more impromptu affair, occasioned by reports of large concentrations of shipping targets still left there.

As part of the main fleet 5th Fleet, Rear Admiral Alfred E. Montgomery's Task Group 50.3, which included the carriers

Curtiss SB2C-1Cs on their way for an attack on Rabaul, November 1943.
Emil Buehler Naval Aviation Library, Pensacola

61

Essex, Bunker Hill and *Independence*, were scheduled to make their combat debut later that month, much further north in the central Pacific, covering the invasion of the Gilbert Islands. However, 'Bull' Halsey was not a man to miss the opportunity of carrying out his sole purpose in life, which he summarized as 'Kill Japs! Kill Japs! Kill more Japs! Sink ships! Sink ships! Sink more ships!'[63] He immediately despatched Montgomery's force to 'Get the cripples', before returning to the main task without delay.

Thus it was that the SB2C-1 made its entry into the 'real' war on the morning of 11 November 1943, a significant date in our story. 'Moe' Vose led the first SB2C strike from the decks of the *Bunker Hill*, with twenty-three Helldivers, launching 160 miles from the target. The incoming American air groups were met by defending Japanese Zero fighters, and a fierce mêlée broke out. The Japanese fighters gained altitude above the Helldiver formation and dropped phosphorous bombs on them. Despite this, the Helldivers manoeuvred themselves into their dive positions over the crowded anchorage and commenced to select their targets.

Present in Simpson Harbour, other than naval auxiliaries and merchant ships, were the remnants of a Japanese cruiser squadron which had been decimated in the earlier attacks: there was the heavy cruiser *Maya*; the light cruisers *Agano*, *Noshiro* and *Yubari*; and the destroyers *Amagiri*, *Fujinami*, *Fumizuki*, *Hayanami*, *Kazagumo*, *Makinami*, *Naganami*, *Suzunami*, *Umikaze*, and *Urakaze*. As they were fully alerted, they proved to be much more difficult targets than those surprised in the earlier raid.

Composition of Air-striking Force for Attack on Rabaul, 11 November 1943				
Unit	Carrier	Commander	VB	No
VB-17	*Bunker Hill* (CV-17)	Lt Cdr. James E. Vose	SB2C-1	33
VB-9	*Essex* (CV-9)	Lt Cdr. A. T. Decker	SBD-5	36

The Helldivers made their dives down from 12,000ft (3,660m) to 1,500ft (460m) through flak at 09:00, and within twenty minutes it was all over; then they had to fight their way back out again through the waiting Zeros.[64] Despite heavy assault, the rear-seat men claimed to have shot down several of the Japanese fighters with no losses in return. Although many of the SB2C-1s were badly shot up, none were lost over the target. Two Helldivers were forced to ditch close by the screening destroyers on return to the task force, but all the others landed safely aboard, despite their condition. They claimed to have inflicted very heavy damage on enemy cruisers and destroyers, and the general summary of their debut was that it had been a resounding success. In fact the final enemy loss tally was that the destroyer *Suzunami* had been dive-bombed and sunk at 09:20, and the destroyer *Naganami* took a direct hit amidships and a near miss alongside, the combined effects of which crippled her so that she had to be towed into harbour for repairs. The light cruisers *Noshiro* and *Yubari* and the destroyers *Umikaze* and *Urakaze* received slight damage from near misses, but that was all.

Nonetheless, even if the actuality of the damage they inflicted on the enemy was far less than they thought, the SB2C-1 had performed very well, bombing accurately and with very low losses despite the enemy being fully alerted. 'The Beast' had also shown herself capable of absorbing enormous punishment and yet still being capable of getting her aircrew back safely. The toughness of the big new dive-bomber did much to endear her to her crews, and made up for the heartache she had caused them in the preceding months. No doubt the Navy made the most of this attack; after all, it was totally committed to introducing the SB2C whatever happened, and Rabaul had proven that it was an effective weapon at last.

The Japanese reaction was fierce, and a very heavy counter-attack was made, which caused the second Helldiver strike to be cancelled. Some Helldivers took part in the aerial defending of the fleet against the incoming bombers, claiming to have destroyed three Zero fighters and a Val dive-bomber, with the loss of one Helldiver; no ship was hit, although there were some close calls. The task group was then recalled north.

Operation 'Galvanic'

The landing of the Third Marine Division on the island of Betio in the Tarawa atoll, in the Gilbert Islands on 20 November 1943, was a bloody business. Despite

A Helldiver from the Operational Training Unit, Fort Lauderdale Naval Air Station, Florida in 1944. National Archives via National Museum of Naval Aviation, Pensacola

A SB2C-1C Helldiver of VB-1 operating from Yorktown off Guam in May 1944. US Navy via National Museum of Naval Aviation, Pensacola

intense pre-landing air and naval bombardment which reduced it to a moonscape of craters, enough Japanese defenders survived among the rubble and coral to inflict enormous casualties on the marines wading ashore. The SB2C-1s of VB-17 were fully stretched in these preliminary bombings and in the follow-up operations which followed before the atolls were finally taken.

Initially Betio was VB-17's target, and a model of the island had been closely studied beforehand. Take-off was at 05:45 on D – 2, 18 November, in the darkness, and attacks were made through cloud down to 800ft (245m) against the coastal guns and the anti-aircraft gun batteries and pillboxes which were their main targets. Lieutenant Commander J. E. Vose led the first Helldiver wave, Lieutenant Commander Geoffrey P. Norman the second. Return fire from the defenders damaged eight SB2C-1s, though none seriously. The same pattern was repeated the next day, 19 November. Again, two Helldivers were heavily hit by AA fire but managed to land back aboard, one, piloted by Lieutenant (j.g.) William 'Puffy' Kornegay, receiving a shell in the engine itself, as Robert Olds described:

> Mechs found two five-inch pieces of shrapnel in Puffy's engine. Everyone crowded around later to look in through the big hole in the cowling and see the exposed reduction gears. Oil leaking out had covered the plane and created smoke when it struck the hot manifold. If the shell had

struck further forward, the prop would have been knocked off.[65]

At 08:20 on D-Day itself the SB2C-1s made an initial attack to cover the actual touchdown of the marines ten minutes later. They targeted trenches, pillboxes and buildings in the immediate rear of the landing beaches, as well as machine-gun nests and known

artillery positions, even though the island appeared completely devastated. AA fire again scored damaging hits, however, showing, that there was still plenty of fight in the enemy; one victim was Lieutenant Philip J. Rusk's aircraft which received a major hit in the engine cowling as well as having the big tail riddled with machine-gun fire, and he had to make an emergency landing.

Curtiss Helldiver I (SBW-1B), May 1944. Fleet Air Arm Museum, Yeovilton

Curtiss Helldiver I (SBW-1B) under test in the USA. Fleet Air Arm Museum, Yeovilton

From then on, with the marines precariously ashore but pinned down, a standing air support force of six Helldivers and four Hellcats was maintained over the island. Strikes, both dive-bombing and low-level strafing runs, were called in by radio by a controller code-named 'Dynamo' on the beach itself. Photographic missions were also undertaken, with large cameras hand-held by the rear-seat men. On landing back aboard the carrier, the films were quickly printed off so that further target evaluation could take place. In addition the Navy had equipped a special photographic squadron (VD-2) which utilized up to six SB2C-1 and -1Cs from January 1944 onwards.

The same type of sorties followed the next day, though this time an additional target was the hulk of a sunken ship on the reef which was being used as a machine-gun nest by the enemy; this was taken under attack on two occasions. In addition to these main tasks, routine anti-submarine patrols were maintained, as well as long-range scouting missions in case the Japanese main fleet ventured out to aid their comrades ashore. Another innovation was to utilize depth charges as blasting weapons by the Helldivers against dug-in enemy soldiers, to good effect.

Nauru Island

VB-17's next target was Nauru Island, Japan's easternmost garrison of the Gilbert Islands. On 8 December 1943 the Helldivers made two strikes against the runways under construction on its southern coast, also the radio station, barracks, ammunition and supply dumps, the anti-aircraft batteries guarding the vital phosphate plants, and other buildings. They made their dives from 10,000ft (300m) in conjunction with a battleship bombardment. This time one aircraft, that of Ensign E. D. Williams, was heavily hit by flak and crashed in the sea offshore. Lieutenant Commander Geoffrey P. Norman gave this eyewitness description of the attack:[66]

Twenty-three SB2Cs, led by Moe Vose, Bags Bagdanovitch in his TBF, eight fighters and nine TBFs from our little carrier comprised this strike. The composition of the SB2Cs was as follows: 1st section, Vose, Dave Martin, William Gerner and Willie Palmer; 2nd section, Rip Kline, Chief Balenti, Bucky Harris and Red Dog Shearon; 3rd section, Larry Madden, George Glass, Bob Temme and D. L. Thompson; 4th section, Bob Wood, Cliff Van Stone, Leo Martin and Dilbert Digman; 5th section, Bob Friesz, Nels McGuire

and Johnnie Walker; 6th section, Gus De Voe, Les Fry, T. C. Thomas and W. P. Harris.

Moe led the boys in from the north, down the east side of the island during the last of the shelling. He turned right and paralleled the south coast heading west, then gave the signal 'Open bomb-bays' and, 'Let's go, Helldivers!' and the boys started peeling off to the right from 10,000 feet, each making his own run on the targets previously assigned. Moe, Dave and Flip hit buildings at the west end to the north of the west runway. Will Palmer got a direct hit on the radio station in the same area.

Rip's section dived on the AA guns along the ridge, back of the runway, that is north of the runways – this was where the big guns of the AA battery were located; and two hits were observed. AA fire was in fact rather light, as though still shocked from the terrific pasting of the big guns.

Larry Madden's section dropped on storage dumps in the trees north of the runway. George Glass went on to take pictures after the attack. All the bombs landed in the area but results couldn't be observed through the trees.

Bob Wood's section attacked revetments and buildings north of the runways, and scored at least two good hits in the revetment area. Bob Frisz's section hit at buildings, revetments and suchlike, between the runways and the beach and Bob strafed parked planes in the path of his pullout. Gus de Voe's section went last and hit at guns and buildings west of the west runway – Gus hit near the AA guns he was aiming at. Thomas's bomb went wide and ended up in the phosphate diggings, doing no damage, and Les Fry dropped close to buildings but the damage was slight.

All hands strafed either during or after their dives, then proceeded north to rendezvous five miles north of Nauru, and return to the ship. All joined up safely and returned to base with only a few holes to indicate that the enemy was still manning the guns.

Kavieng, New Ireland

On Christmas Day 1943 the SB2C-1s were back in action as part of an air-striking force which hit Kavieng Harbour, on the north-western tip of New Ireland, north of Rabaul. Few targets were found, but they managed to sink a 4,861-ton freighter and damage another, as well as the minesweepers *W-21* and *W-22* (which they mistook for destroyers). The Helldivers had no losses.

On New Year's Day 1944, there was another celebration when reconnaissance sighted two Japanese cruiser forces, comprising the heavy cruisers *Kumano* and *Suzuya*, the light cruisers *Oyodo* and *Noshiro*, each with their escorting destroyers. These had just completed a run to Rabaul taking troop reinforcements to Kavieng and Rabaul and were returning to the main fleet anchorage, Truk in the Carolines. In attacks made that day the SB2C-1s again got through, despite strong fighter protection and the *Oyodo* and *Noshiro*, as well as the destroyer *Fumitsuki*, were all damaged to varying extents.

Kwajalein Eniwetok

In January, Lieutenant Commander Vose left the squadron and Lieutenant Commander Norman assumed command. Aboard the *Bunker Hill*, VB-17 sailed as part of Rear Admiral Marc H. Mitscher's Task Force 58; this time their initial targets were the Marshall Islands. VB-17 was still the only unit equipped with the Helldiver. On the morning of 29 January, Norman led the first strike in from 13,000ft (3,960m), dividing their attacks between enemy AA batteries, ammunition dumps and the island's runways. When the ammunition dump was hit it blew with such a force that it stripped

Curtiss Helldiver I (SBW-1B) in early US markings. Fleet Air Arm Museum, Yeovilton

most of the fabric from the tail of Lieutenant (j.g.) Robert L. Temme's aircraft – but he got her back safely. Again, the

ability of the 'Big-Tailed Beast' to take tremendous punishment had been most graphically demonstrated.

Similar strikes were made on Engebi and Parry islands, part of the Eniwetok atoll, some 355 miles (570km) further north, to prevent it being used as a staging post for Japanese aircraft disputing landings on the Marshalls. Attacks continued until 3 February against airfields, AA positions and barracks, and the only losses suffered by VB-17 were three aircraft in operational accidents, but none to enemy fire.

Truk, Carolines

Even more feared than Rabaul was the Japanese main fleet anchorage of Truk in the Caroline Islands. Known as the 'Gibraltar of the Pacific', stories of its defensive strength and the great fleet lurking there abounded. It is a measure of the burgeoning strength of the US Navy that it now prepared to tackle this bogey with its carrier forces head on. Bill Palmer told me how they approached it:[67]

Moe Vose – the C.O. of VB-17 - was the first to use the four-plane 'fighter' section for Navy dive-bombers. It gave us a faster dive order, more fire power concentrated from our rear-seat gunners,

The aircrew of VB-17 await their briefing aboard the USS Bunker Hill **(CV-17).** Captain Robert L. Temme, US Navy Rtd, via J. Alton Chinn, USN, Rtd

old

new

Four plane attack section, as introduced by VB-17.

and made it easier for our own fighters to cover us going into, or leaving, the target area.

After preliminary fighter strikes on the 17 February to whittle down the powerful Japanese fighter defences, the Helldivers went into action the next day. Hoping to find the juicy targets of the main Japanese combined fleet at anchor there, the '2Cs' of VB-17 were disappointed to find that almost all of them had left for Singapore. Only a few cruisers and destroyers, plus the usual auxiliaries and freighters were present, and no battleships and no aircraft carriers whatsoever, despite repeated claims to the contrary both then and fifty years later![68] When the Helldivers tipped over there were the light cruisers *Katori* and *Naka*; two armed merchant cruisers, the *Aikoku Maru* and *Kiyosumi Maru*; five destroyers, the *Maikaze*, *Fumizuki*, *Oite*, *Tachikaze* and *Nowake*; two submarine tenders, the *Rio de Janeiro Maru* and *Heian Maru*, plus a mass of merchant shipping.

Both the submarine tenders, six oil tankers and seventeen freighters – some 137,019 tons of shipping – were sunk in the mass air strikes, the VB-17 Helldivers claiming ten of them (plus the non-existent carrier!). The light cruiser *Naka* was caught by the SB2C-1s as she was trying to escape this carnage and was also sunk by them, as was one of the destroyers. The

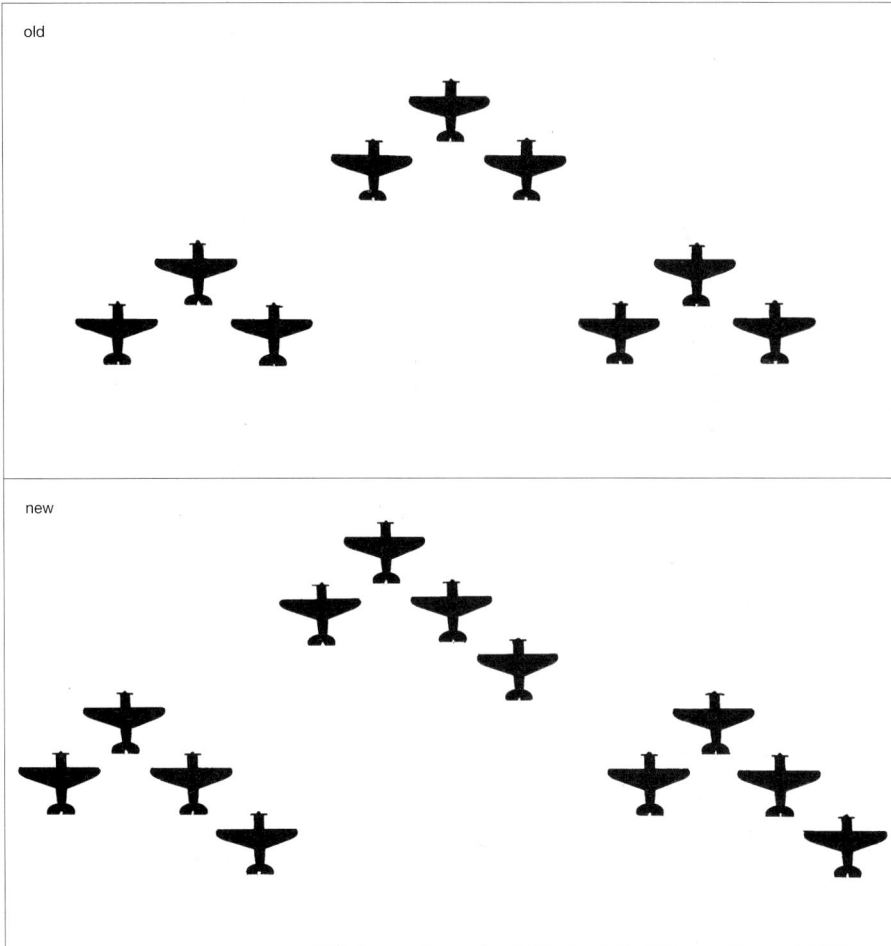

Three SB2Cs from VB-17 at dusk.
Commander J. Alton Chinn, US Navy, Rtd

(Above) **A group of VB-17 pilots with '108' aboard the USS** Bunker Hill **(CV-17).** Commander J. Alton Chinn, US Navy, Rtd

A formation of SB2C Helldivers heads out in the early morning, low over the Pacific Ocean on another sortie, 28 February 1943. Robert Olds Collection courtesy Peggy Olds

total loss to VB-17 was one Helldiver, yet again one piloted by Lieutenant (j.g.) Robert L. Temme, and which was hit and forced to ditch; but both aircrew were rescued. Many other SB2C-1s took flak damage, but got back to base safely.

Tinian and Rota

On 22 February, VB-17 had one last tilt at the Japanese when they hit targets on Tinian and Rota in the Mariana Islands. The first wave made their approach through clouds at 10,000ft (3,000m) and selected a freighter as their target, claiming four direct hits and four near misses on her; they also claimed to have destroyed one of the small escorts with two direct hits. The second wave attacked through rainclouds and bombed the radio station. When the cloud lifted later in the day a third Helldiver strike was sent in and bombed the sugar factory and the airfield, using delayed-action bombs to keep it out of action while the fleet withdrew. Some sixty-five Japanese aircraft were claimed destroyed on the ground. It was a fitting finale to a really successful combat initiation for the Helldiver – Curtiss and the Navy had finally been vindicated, and there were now few doubts about 'The Beast's' ability to take on the enemy. This was to be the last combat mission of the SB2C-1. Furthermore VB-17 was to be relieved by VB-8 aboard the *Bunker Hill*, as their tour of duty had come to an end.

The Curtiss Helldiver in classic pose with bomb doors open. The crutch which swung the bomb out and down is retracted or obliterated by an over-zealous censor, but the unit markings F-B-T 5 are clear enough. US National Archives, College Park

Arrival of the 'Dash 1C'

Joining the active fleet now was an airplane which was the result of all the many modifications of the Mod. III programme, but with added enhancements: the SB2C-1C. The prototype (Bu. No. 00016) had rolled out of Columbus sporting its two single 20 mm wing cannon, hydraulic flaps and a baffle fitted to the rear sliding canopy. Identified only by a straight pitot tube on the left wing, she was followed by a production run commencing with Bu. No. 00201, which featured the standard right-angled bent pitot tube. In all, some 778 of the 'Dash-1C' variant were to be produced in Ohio: (Bu. Nos 00201 to 00370, 01008 to 01208 and 18192 to 18598). These were joined in the front line by twenty-eight Can-Car SBW-1Bs (Bu. Nos 21201 and 21202, and 60010 to 60035), most of which were allocated to the Royal Navy, and fifty Fairchild SBF-1s (Bu. Nos 31636 to 31685), built to almost the same specification. The speed with which these now began to join the fleet is shown in Appendix 5, and although Truk saw only the original SB2C-1s afloat, by June, SB2C-1Cs had re-equipped the SBD-5 squadrons of VB-1, VB-2, VB-8, VB-14 and VB-15.

In February 1944, the US Navy asked for the production of the Helldiver to be speeded up still further, requiring twenty-three aircraft per day to be completed, such was the demand for the now proven-in-combat dive-bomber as the Pacific War worked up to its climax.

Specification – CC & F SBW1-B	
Engine type:	R-2600-8
Armament:	Two 20mm forward-firing fixed cannon; two .303 rear-cover flexible machine guns; maximum bomb load 2,000lb (907kg); one 2,000lb torpedo
Dimensions:	Length 36ft 8in (11.17m); height 13ft 1½in (4.01m); wingspan 49ft 8⅜in (15.15m); wing area 422 sq ft (39.2 sq m)
Weights;	Maximum weight 16,607lb (7,533kg); empty weight 10,114lb (4,588kg)
Performance:	Maximum speed 273mph (439km/h)
	Take-off power 1,700hp (1,268kW)
	Range as dive-bomber 1,375 miles (2,213km)
	Climb height 10,000ft/11.4min (3,048m/11.4min)
	Service ceiling 21,200ft (6,462m)
Number completed:	26
Serial numbers:	60010/60035; (JW100/125) (Bu. No.-Navy – Serial Army)
Contract:	Order date; 23 May 42
	Delivery date; Sep 43–Dec 43

CHAPTER TEN

Stillborn: The Australian Experience

Among the many myths that have been perpetuated down the years are two in particular which need laying to rest. These are that both the RAAF and the Royal Navy rejected the Helldiver outright, after testing them and finding them inadequate.[69] In *neither case* is this story true, and we will examine both in detail to show the real reasons, both of which were a result of decisions at the highest level and *not* because of adverse flight test reports.

In April 1943 the Australian government was pressing for an increased supply of American aircraft in order to build up the strength of the RAAF from its existing thirty-one squadrons to forty-five and then to seventy-five; but it was concerned at the slow rate at which combat aircraft orders were being fulfilled, including the 150 Shrike dive-bombers it had asked for. Vultee Vengeance dive-bombers were arriving and soon a whole wing was built up ready for combat in the Nadzab area, where they performed outstandingly for a period.[70] But as early as 30 March, only 160 aircraft from a total of 397 American-built aircraft of all types had been shipped. Dr H.V. Evatt was sent to Washington DC from Canberra ACT, to try and speed things up. Although they received the backing of General Douglas MacArthur, the American C-in-C Southwest Pacific, his subordinate, General George C. Kenney, USAAF, took the opposite view completely. He did all he could to discourage any increase of RAAF strength, preferring instead to build the USAAF strength instead, and he showed a marked and unreasonable bias against dive-bombing and dive-bombers, in both nation's services.[71]

While the Australians were pleading for more aircraft, Kenney was telling Washington that the RAAF did not have the men to fly them, a statement which was strongly refuted by the Australians themselves when they got to hear of it.[72] Nonetheless Kenney was to get his way, abruptly ordering the Vengeance Wing out from New Guinea, despite their high accuracy, success rate and almost minimal casualties;

moreover he did this without any prior consultation with the Australians themselves. Furthermore he issued a statement that, '… the operational requirement for the dive-bomber squadrons had ceased to exist in the Southwest Pacific Area.'

As a result, the RAAF representative in Washington DC was told, in January 1944, to negotiate for the cancellation of the balance of 140 A-25 aircraft and spares diverted behind the aircraft, and to arrange the return or transfer of the ten already sent. The Combined Assignment Board MAC (Air) approved these cancellations, but decided that as the ten Shrikes already shipped would not be required elsewhere, their return to the USA was not justified.

These first deliveries of the A-25A Shrike had already been made to Australia from the original order. Ten Shrike dive-bombers, which the RAAF classified as the A-69, arrived crated in Australia at the end of November 1943, along with a number of cases of spares which formed part of the 1st and 2nd echelon back-ups for the first twenty aircraft. The aircraft assigned, manufacturer's block 15-CS, were as shown below.

These were all unshipped and went straight to No. 2 Aircraft Depot at Richmond, New South Wales. In December 1943, one of these aircraft – the A69-4 – was then issued to No. 1 Air Performance Unit, based at Laverton, Victoria, for trials. No flying had been conducted up to January 1944 because – according to the unit history, '… the lack of spare tail wheel tyres has prevented any flying with this aircraft'. So it is quite clear that the Shrike was cancelled because of a decision by US Air Force Commander Kenney and *not* by the RAAF, and *before* any trials by the RAAF had taken place and *not*, as is often stated, as a result of trials!

On 23 August 1944, John Curtin, Minister for Defence, wrote to General Douglas MacArthur, C-in-C, Southwest Pacific Area about the Shrike. After tracing the history of the order he went on to state that: 'Subsequently, advice was received from the Commanding General, Far East Air Force, to the effect that the operational requirement for the dive-bomber squadrons had ceased to exist in the Southwest Pacific Area.'[73]

Now of course, thanks to Kenney's decision, the Australians had no use for the ten

Allocation of A-25A Shrike to RAAF A-69s					
Curtiss No.	USAAF Br.	RAAF No.	Date Rec'd	Movements	Date
4583	42-79683	A69-1	23 Nov 1943	Stored 2AD, returned FEASC	18 Jan 1945
4584	42-79684	A69-2	23 Nov 1943	Stored 2AD, returned FEASC	26 Nov 1944
4585	42-79685	A69-3	25 Nov 1943	Stored 2AD, returned FEASC	18 Jan 1945
4586	42-79686	A69-4	25 Nov 1943	2AD, 26 Dec 1943 1 APU 16 April 1944 2AD 6 August 1944, returned FEASC	7 Dec 1944
4587	42-79687	A69-5	24 Nov 1943	Stored 2AD, returned FEASC	18 Jan 1945
4588	42-79688	A69-6	25 Nov 1943	Stored 2AD, returned FEASC	18 Jan 1945
4589	42-79689	A69-7	25 Nov 1943	Stored 2AD, returned FEASC	26 Nov 1944
4590	42-79690	A69-8	24 Nov 1943	Stored 2AD, returned FEASC	26 Nov 1944
4591	42-79691	A69-9	24 Nov 1943	Stored 2AD, returned FEASC	18 Jan 1945
4592	42-79692	A69-10	24 Nov 1943	Stored 2AD, returned FEASC	26 Nov 1944

An RAAF A-25A under test in Australia. Note the turtleback is fully lowered although for test purposes no rear seat man is carried. Emil Buehler Naval Aviation Library, Pensacola

aircraft, and the Minister for Air suggested that in this, '... and any other similar cases that may occur,' the aircraft should be transferred to the Far East Air Forces and held so that aircraft, engines, components and spares would be available for allocation by the commamder-in-chief.

Three days later Brigadier General Fitch, USAAF, adjutant general, sent a memo[74] to the commander, Allied Air Forces enclosing copies of the Prime Minister's letter. Colonel Perry G. Ragan, Air Adjutant General, the HQ, Allied Air Forces, responded to MacArthur on 8 September thus: 'No operational requirement exists for these aircraft within this command', and recommended that the Australians be advised that the airplanes could, '... be disposed of in any manner that they may desire'.[75]

MacArthur wrote to the Australian prime minister on 16 September, saying that he agreed with his disposal suggestions and had notified Kenney accordingly. 'The War Department recently reiterated to me the official United States attitude, which is that the title to lend-lease goods remains in the United States Government.'[76]

Brigadier General B. M. Fitch recommended to MacArthur on 16 September

that '... it is desired that the commander, Allied Naval Forces be queried as to whether there is an operational requirement for subject aircraft and spares for the Seventh Fleet ...'[77] This was done on 29 September,[78] and they in turn passed it on to the Seventh Fleet,[79] but received no joy there. D. S. Fahrney replied on 12 October that, 'This command cannot employ the Army Shrike (Navy SB2C-1A) dive-bomber in this area, as they are an obsolescent type and are not designed for shipboard use'.[80] And the acting chief of staff, Charles J. Maguire, duly passed this on to Kenney, concurring.

With the buck duly back in their court, HQ Far East Air Forces informed the commanding general, Far East Air Service Command, that he was to notify them '... when these aircraft and parts have been delivered. Report will indicate location, model, type, series, serial number, condition, and number of man-hours to place in AFO or AFNO status in order for the headquarters to determine whether the airplanes are reportable or non-reportable to Headquarters, AAF, under provisions of AAF Regulation No. 65–85, September 1944'.[81]

Also in September, the directorate of technical services, special duties and perfor-

mance flight, RAAF, issued a report on the abbreviated tests conducted with the solitary Shrike (A69-4) that was uncrated and assembled.[82] This gave a good insight into the A-25A variant. When configured by the RAAF as 'a reconnaissance bomber with external bomb load ...', the maximum 'all out' level speed was 271mph (436km/h) at 400ft (120m), and the maximum initial 'all-out' rate of climb was 1,940ft (590m) per minute. The time to 10,000ft (3,000m) on rated power was 9.1 minutes. With external (sic) [this should have read internal] stowage of bombs (dive-bomber) maximum 'all-out' level speed was 291mph (468km/h) at 400ft. The estimated 'useful range' at 5,000ft (1,500m) was 1,030 air miles (1,660km) or 690 air miles (1,110km) as a dive-bomber.

The handling of the Shrike as a reconnaissance bomber was '... *quite satisfactory*, the heaviness of the ailerons in this role being considered *acceptable*'.[83] As a dive-bomber the ailerons and rudder appeared to be, '... *heavy*'.

The actual abbreviated trials conducted on the Shrike were divided into three parts: performance, handling and serviceability. The scope of the performance trials was considerably shortened, mainly because of lack of flying hours. Although the Shrike remained for three months in the unit, only thirteen-and-a-half hours flying time were done.

Performance

The two nominal categories under which the figures were given for the Shrike were as a reconnaisance bomber and a dive-bomber:

(1) Reconnaissance bomber, with a take-off weight of 15,145lb (6,870kg) a mean weight for level speed and range performance of 14,500lb (6,577kg) and with bomb-bay fuel tanks fitted with a 500lb bomb load carried externally.

(2) dive-bomber, with a take-off weight of 14,756lb (6,693kg); a mean weight for level speed and range performance of 14,100lb (6,396kg), with 1,000lb (450kg) of bombs carried internally with full dive-bombing equipment.

These were broken down further as follows:

(a) Position error determination: determined by the normal tower-aneroid method.

(b) Level speeds at heights from sea level to 15,000ft (4,570m) using normal rated and war emergency power as a reconnaissance bomber. The engine cooling with the cowl gills closed appeared adequate at all times for atmospheric conditions 10°C above ICAN standard atmosphere air temperatures.

(c) Climbs from sea level to 15,000ft using normal rated and war emergency power as a reconnaissance bomber. No allowance was made for weight change during the climb. With partially opened gills there was ample engine cooling, as above.

Those conclusions relating to the Shrike's major use, as a dive-bomber, in the same three categories, were only 'estimated performance' figures.

The main external fittings which affected the performance of the Shrike were listed as follows:

a. A 4ft (1.2m) radio mast mounted on the port side of the forward fuselage with an aerial to the fin.
b. Flush-type, by-passable air filter on top of the engine cowl.
c. Large, bulging fairing around the exhaust manifold exit.
d. Muzzle of rear .50 calibre flexible gun projecting through the cabin roof.
e. One 250lb (113kg) GP bomb with a modified universal carrier under each main-plane. A large air-screw spinner was fitted.
f. Non-retractable tail-wheel.
g. Paint finish of a fine matt surface of fairly uniform texture.

An A-25A Shrike with the markings of the Royal Australian Air Force. Initial heavy orders for this machine all eventually fell through when the Australians were forced to fall in line with US Army Air Force thinking, and abandoned dive-bombing in 1944. Existing squadrons of Vultee Vengeances, already most successful in combat in the south-west Pacific, were abruptly sent home and the wholesale cancelling of Shrike orders also took place. *Curtiss-Wright via Stanley I. Vaughn Archive*

The engine details of the Wright R2600-8, two-row, fourteen-cylinder air-cooled engine with two-speed, single-stage super-charger were as follows:

Serial No:	94Y41
Propeller gear ratio:	0.5625:1
Impeller gear ratio:	7.06 and 10.06:1
Compression ratio:	6.90:1
Fuel:	100 octane
Carburettor:	'Holley' model 1685HA, with manual mixture control.

The performance figures for tests under static conditions are given below.

The air screw details were:

Type:	Curtiss electric, three-bladed, model no. 05325D-A28
Serial no:	47642
Blade drawing no:	89324-12
Blade serial nos.	101039, 101038, 101043.
Diameter:	12ft (3.6m)
Pitch Range:	21.7 degrees to 51.7 degrees (at 42in station)

For the blade section details, see the figures given below.

A-69 Performance Figures					
Rating	RPM	Absolute manifold pressure (in Hg)	Blower gear (L/H)	Bhp	Alt. ft
Take-off	2,800	44.3	Low	1,700	sea level
War emergency	2,600	49.5	Low	1,850	*sea level
	2,600	48.0	High	1,500	*11,200
Maximum continuous	2,400	37.5	Low	1,400	sea level
	2,400	37.5	Low	1,500	*6,700
	2,400	41.0	High	1,350	*13,000
Maximum cruise	2,180	30.5	Low	980	sea level
	2,180	30.5	Low	1,125	*10,300
	2,180	32.3	High	1,012	*16,400
* = full throttle					

A-69 Blade Section Details		
Item	At 42in section (maker's reference)	At 50.4in station (0.7 radius)
Chord (in)	11.03	10.53
Thickness (in)	1.06	0.88
Thickness ratio	0.096	0.083
Blade section	Approx. a Clark Y	

The still-air range estimates were obtained using 800bhp at 5,000ft (1,525m) as listed above; available fuel was estimated by allowing 20 gallons (91 litres) for warm-up and climb to height, and also a reserve of thirty minutes maximum rich-mixture cruising.

A-69 Range Figures		
Item	Reconnaisance bomber (actual)	dive-bomber (est.)
Available fuel (imp. gal)	277	169
Fuel flow (imp. gal/hr)	51	51
True airspeed (mph)	190	207
Air miles per gallon (imp.)	3.7	4.1
Still air range (miles)	1,030	690

A Curtiss Shrike of the Royal Australian Air Force undergoing trials and testing.
Royal Australian Air Force

Handling notes make for some interesting comparisons with those of the Helldiver also contained within these pages. The cockpit layout, for example, was described by the RAAF as good, having comfortable seating, with sufficient adjustment to both it and also the rudder pedals '... for both large and small pilots'. The view from the cockpit was described as '... good for an aircraft of this type, especially when in flight'.

The hydraulically operated undercarriage, flaps and diving brakes were situated '... well within reach ...' on the starboard side of the cockpit, and the elevator, rudder and aileron trim controls on the port side of the cockpit were also '... handy to the pilot'. Engine controls situated in one quadrant on the port side, with throttle, two-speed blower control, mixture control and pitch control. The hydraulic brakes were '... very effective ...'. The only adverse comment was that there was no provision made for locking the brakes when the Shrike was parked.

Taxiing was found to be very rough due to the limited travel of the oleos. Take-off was described as quite straightforward, '... there being only a slight tendency to swing to the left. But this was, '... quite controllable. The run was short and the aircraft climbed away well'.

Above 240mph (386km/h) the ailerons became very heavy and they, '... tighten considerably as speed increases. At low speeds the controls are good down to stall.'

In dives with the dive brakes extended '... there is considerable buffeting over the tail unit which can be felt in both the elevator and rudder'. The allowable maximum speed with diving flap was 280mph (450km/h) which was quickly reached in a 70 degree angle dive, '... and the aircraft has to be brought into a shallower dive so as not to exceed this maximum allowable speed. With the diving brakes out, the ailerons and rudder become excessively heavy and cannot be used effectively for manoeuvring or sighting'.

The stability was '... quite stable ...' in normal flight and dives without the flap, but with flaps out in a dive '...the aircraft hunts directionally if bumpy air is encountered. This is not a desirable feature for dive-bombing'.

The stall was generally quite gentle, with the left wing and nose dropping, although with landing flaps and undercarriage down, and the power on, the aircraft '... stalls suddenly with buffeting over the tail plane and snatching on the ailerons; this is followed by the left wing and nose dropping rapidly ...'. The behaviour at the stall was still described as '... quite satisfactory', however.

Serviceability

No particular maintenance or serviceability trials were made during that time; nonetheless, even limited experience brought forth the following comments on this aspect of the A69:

(1) For only sixteen take-offs and landings, at an average weight of approximately 14,500lb (6,577kg), three new tail-wheel tyres were required; this was mainly due to wall failure after landing and taxiing.
(2) The ball races in the tail wheel hub became loose, due to the hub expanding under load.
(3) The very small oleo travel of the main undercarriage – approximately ¾in (20mm) – made taxiing over rough ground very 'harsh' and it was considered that it would have caused deterioration of the airframe in only a short period.
(4) In general, maintenance was easy, and servicing, apart from those items above, proved to be, '... up to normal standards'.

They concluded that the Shrike's undercarriage was not suitable for normal aerodrome operation, even at normal loads. 'Apart from this the ease of maintenance and serviceability of this aircraft is straightforward ...'.

Meanwhile the issue of what to do with these ten Shrikes had resurfaced back in Washington DC. On 2 November, Wing Commander A. N. Hocking for air marshal RAAF representative wrote a long memo on the subject to Lieutenant Colonel G. B. Brophy, the acting chief,

Two views of the A-25 Curtiss Shrike, the only one of a batch shipped to the RAAF and which underwent trials in Australia but was ultimately rejected due to a higher policy switch. Royal Australian Air Force

international branch, material division, AC/AS. M & S.[84] He pointed out that the ten Shrikes were '... still new airplanes and in the paralcatoned state, ie in the condition in which they were processed for deck-loaded overseas shipment'. He went on: 'The situation now arises as to the method of effecting adjustment in the lend-lease accountability.' The Shrikes had already been debited to the Australian lend-lease account by means of the shipping tickets which were raised when they were shipped. After discussion with Major Ramsey he understood that '... when this material is physically handed over to the appropriate USAAF authority

in Australia, the receipt by that authority will be recorded by means of a "receiver's report", and that the transaction will ultimately be reported to the AAF international officer'. On receipt of this report Australia's lend-lease account would be credited for them and the account adjusted accordingly. He asked Brophy to confirm this was the case so that he could inform Canberra.

On 11 November Major E. C. Ackerman, adjutant general, HQ Far East Air Service Command, wrote to the commanding officer, forward echelon, RAAF HQ, Brisbane on the final details of the hand-back. He requested that the Aus-

tralians arrange for the transfer of the aircraft to Depot #2 at Townsville, prior to 1 December, '... transportation cost to be for your account'.[85] A final minute of 15 November from the officer in charge, aircraft status to the overseas indent officer, confirmed this would be done. 'Nine aircraft are held at two aircraft depots unerected and packed in special protective wrapping. They are not in cases. One aircraft is erected and held serviceable at 5 Aircraft Depot Storage, Cootamundra. In respect of the unerected aircraft at the two aircraft depots, the extension mainplanes and airscrews are not fitted but are held in cases. The number of cases is eight.'[86]

CHAPTER ELEVEN

Strangled: The British Experience

Just as the Australian Shrike rejection was a myth, so too was the Royal Navy's Fleet Air Arm disappointment with the Helldiver, always put forward as being *solely* the *direct* result of unfavourable trials. This supposition has been reinforced by the strongly anti-Helldiver comments made by that most respected of British naval aviators, Captain Eric Brown, CBE, DSC, AFC, RN, the test pilot who flew every type of dive-bomber during World War II and whose views must command great respect. In his post-war account of his flight testing of the Helldiver he was scathing in his distaste for the aircraft, ending his summary with the words, '… the Helldiver would never have been acceptable for deck landings by British

standards, and indeed, was *never*[87] flown onto a British carrier by an FAA pilot'.[88]

Even though much the first part of that statement might be true, in the latter comment he is patently mistaken, for the Helldiver *was* flown by several Fleet Air Arm pilots onto the decks of a British carrier, and a small one at that. In addition, US Navy dive-bomber pilots made deck landings aboard the escort carrier HMS *Avenger*, and Helldivers were flown from a *British* light carrier post-war (the former HMS *Colossus*), albeit by French Navy pilots.[89] Furthermore, no matter how respected Captain Brown's views must be, the official documentation from the British archives shows that, far from the

Admiralty being eager to rid themselves of what few Helldivers they had obtained, they mounted a strong and determined campaign up to the very highest level of governments, asking premier Winston Churchill himself to intervene with the US authorities to obtain the many more Helldivers that they wanted for the British Pacific Fleet. That this plea was in vain is due less to the opinion of test pilots, however distinguished, than to the fact that the American Navy chief, Ernest King, was a strong Anglophobe and, like Kenney with the Australian Shrikes and Vengeance dive-bombers, did everything he could to obstruct the allocations already agreed in order to further the needs

A good view of Curtiss Helldiver JW117, one of a batch of Canadian-built aircraft supplied to No. 1820 Squadron RN in 1944. Ray C. Sturtivant

of his own service. Let us examine the history of the Royal Navy's involvement with the Helldiver in more detail.

Despite the Blackburn Skua's success as a dive-bomber, and its other notable 'firsts' in British naval aviation, it was gradually phased out from 1941 onwards. Its most logical home-built replacement, the Fairy Barracuda, followed the usual pre-war British practice of having to combine torpedo dropping and reconnaissance as well as dive-bombing. In effect it was not a very good dive-bomber, and it was used more as a glide bomber. Ironically it never carried out a combat torpedo attack either – but more than just these failings, it suffered from even worse delays than did the Helldiver, and when finally it did arrive, it proved a sad disappointment, being underpowered and aerodynamically absurd! The Baracudas were soon replaced by the Grumman Avenger as the Royal Navy's general-purpose workhorse.

The hiatus that followed the relegating of the surviving Skuas (plus a few purchased Vought Vindicators, which the British named the Chesapeake) to second-line duties from the summer of 1941, was not at all to the Admiralty's liking. They, like their opposite numbers in the US Navy, firmly believed in dive-bombing, but just as their hands had been tied pre-war by the RAF's control over production, so, as the air war in Europe became increasingly a war of heavy bomber versus civilian target, the Ministry of Aircraft Production (MAP) gave ever greater priority to the mushrooming needs of the RAF's fighter and heavy bomber forces. The Royal Navy therefore had no choice but to look to America, at that time still not involved in the conflict, as a ready source of supply. They took the Chesapeake, tried to make it an anti-submarine aircraft, and abandoned it; they rejected the fixed-wing Douglas SBD Dauntless; but they pinned their hopes on obtaining quantity deliveries of the new-generation dive-bombers, the Brewster Buccaneer (known to the British as the Bermuda) and the Curtiss SB2C Helldiver. In both cases they were to be bitterly disappointed.

What the Royal Navy was looking for to re-equip its dive-bomber squadrons was spelt out to the British Air Commission (BAC) representative in Washington DC by the MAP as early as January 1941:[90]

Following are the Admiralty requirements in connection with dive-bombing and navigational

facilities. We request that information may be sent as to how nearly the Grumman XTBF1 can meet these requirements.

The observer must have a good all-round view, particularly in the lower hemisphere. When seated, he must have a clear view of the horizon from 45 degrees to 135 degrees each side. He must be able to take compass bearings from 45 degrees to 180 degrees each side. He must be able to transmit and receive and have access to the radio set for tuning from his normal position. He must be able to pass messages to the air gunner who normally operates the radio set. Aircraft must be capable of dive-bombing up to 55 degrees to the horizontal. Dive speed is to be limited by flaps or dive brakes to about 250mph. Flaps must be capable of quick operation, and a quick pullout must be possible from a bombing dive with flaps down. The operation of the flaps should cause the least possible change of trim.

As the Grumman aircraft in question was designed as a torpedo bomber it was not surprising that it failed to fit the bill, but by June the two dive-bomber types under construction were beginning to attract British attention. The basic specifications of both the SB2C1 and the SB2A1 were despatched from the US Navy department on 23 June to London by air at the request of the U.S. naval attaché. The figures the British received on 28 June, were listed thus:[91]

The details are as follows: where details differ, the first figure is for the SB2C1, the second for the SB2A1.

This was followed up by an examination visit by the Royal Air Force Group Captain H. W. Heslop of the BAC, to the Curtiss-Wright plant at Buffalo, New York on 10

Manufacturer's Figures Given to the BAC Comparing SB2C and SBA 1940		
	SB2C	SBA
Engine:	Wright R-2600-8	Wright R-2600-8
All-up weight:	11,622lb (5,272kg)	11,455lb (5,196kg)
Span:	49ft 8½in (15m 22cm)	47ft (14m 3cm)
Length:	35ft 3in (10m 11cm)	38ft 10½in (11m 32cm)
Static height:	15ft 2in (4m 10cm)	15ft 2in
Folded width:	22ft 7in (6m 15cm)	22ft 7in
Top speed:	313mph (504km/h) at 15,000ft (450m)	315mph (507km/h) at 15,000ft
Stalling speed:	70mph (113km)	70mph
Climb to 10,000ft:	4½min	4min
Max. endurance – bomber:	6½hrs at 13,000ft (3,960m) at 135mph (217km/h)	7.7hrs at 7,000ft (2,130m) at 115mph (185km/h)
Max. endurance – scout:	9.8hrs at 13,000ft at 135 mph	11.6hrs at 7,000ft at 115mph
Max. range as bomber:	1,000 miles (1,609km) at 13,000ft at 175mph (290km/h)	1,000 miles at 7,000ft at 160mph (257km/h)
Max. range as a scout:	1,500 miles (2,400km) at 13,000ft at 175mph	1,500 miles at 7,000ft at 160mph
Take off against 25kt wind:	315ft (96m)	270ft (82m)
Service ceiling:	27,000ft (8,230m)	26,700ft (8,140m)
Dive check:	Required in SB2C1	Dive brakes to be fitted in SB2A1
Fuel tanks:	Two self-sealing integral wing tanks; one auxiliary fuselage tank fitted with CO_2 purging in both; one droppable tank.	
Armament :	Two 0.5 wing guns 400rpg; one rear flexible gun, 400 rounds.	
Bomb load:	One or two 500lb, or one 1,000lb bomb.	
Navigating equipment:	Chart board, drift sight and smoke grenades; W/T transmitter covers 350–1,500kcs and 3,000–9,050. Special coils available for 2,000–3,000kcs. Master oscillator A1, A2, A3.	
Armour protection:	Pilot's and rear seat, windscreen, oil tank and forward panel; pilot's headrest.	

November 1941, when the XSB2C-1 was shown to him by Raymon (sic) C. Blaylock himself. Heslop later reported back favourably on what he had seen,[92] stating quite plainly that: 'Examination of the aircraft indicated that the specification held by the BAC did not represent the aircraft, and that the main differences were:

a. It carried two cannons or 4 × .50 guns firing forward.
b. Its bomb load can be 1,600lb, i.e. 1 × 1,000lb or 2 × 500lb in the fuselage, and 1 × 300lb on each wing.'

He went on to say that there were, in fact, '... three models of the XSB2C-1:

a. The naval model with folding wings and accelerator and arrestor gear.
b. An Army type, duplicate of Navy type, with the folding mechanism etc. removed, with an approx. saving of 175lb (79kg).
c. A marine type, with alternate float w/c armoured floor. This type is not gear cleared.'

Heslop concluded that in his opinion, the SB2C1 '... in its present state would be equal to the Brewster or Vengeance, and with modifications to the fuel tank arrangements by deletion of the fuselage tank, the aircraft would be much improved'. He added that by the time the aircraft became available to the BAC direct from Curtiss, '... it is considered that the twin-gun rear turret would be introduced'.

He considered that the turret layout was good '... as it would allow the use of the radio while sitting in the turret, due to the particular type of canopy fitted which covers the turret when not in use'. He considered the angles of fire of the existing turret, '... most satisfactory'. He also noted that: 'The Curtiss Company are turning over their Columbus plant to the production of this aircraft. However, the plant is not yet jigged up for its production, and the first aircraft is not expected until the middle of 1942 ...'.

This was followed by a memo from Colonel H. Burchall, BAC, to the MAP on 28 November.[93] He told them that the US Navy were expanding production of the SB2C for Navy purposes to fullest possible extent so that supplying British needs would involve the following options:

a. Creation of a new facility.
b. Re-allocation of capacity at present reserved for other types; or
c. Use of Vultee facilities after completion of Vengeances. [Then on order for the RAF and to be subjected to yet the same delays and problems as the other American-built dive-bombers.][94]

The latter option was not recommended, as '... continuity of production of the Vengeance with such improvements as may be practicable would on balance be preferable to changing over to build SB2Cs'. This would seem to indicate that the RAF themselves were interested in the SB2C1, rather than the Royal Navy. He noted that Brewster's capacity would be allocated entirely for naval type aircraft on completion of the Bermuda. Finally, on an up-beat note, Burchall reported to London that: 'The first experimental SB2C is now flying, and the first production model is due out next spring.'

But as we have seen, that did not happen. However, throughout 1942 the Admiralty kept pressing for allocations, despite the setbacks. In May 1942, a report by Lieutenant Commander Smeeton RN, the senior British naval representative on the BAC, convinced the naval staff that: 'Although criticisms remain, in that they would be difficult to employ afloat, they should prove of great value in meeting our increasing commitment for shore-based operational squadrons.'[95] On 13th July 1942, the naval secretary was reporting to the First Sea Lord and Chief of Naval Air Service that:

SB2C – Deliveries have been taken of 90% of estimated production to cover loss in transit, and have been lagged three months to cover time in transit and possible production delays. Estimated deliveries:
1-4-43: 7
1-7-43: 52
1-10-43:81
1-1-44: 81
1-4-44: 90
1-7-44: 94
270 required by 1-7-44.[96]

At the end of that same month it was recorded that: '450 SB2C on order, of which 65 expected by the end of June 1943 and 30/35 per month thereafter.'[97] By January 1943 this had become: 'SBW deliveries are expected to commence in March 1943.'[98]

British anxiety only increased as problems gave way to production, even with Royal Navy allocations being assigned to the Canadian factory. In April 1943, Lieutenant Commander Smeeton was advising the MAP on contract changes:[99]

Recent information obtained from the Bureau of Aeronautics indicate that the sixty-sixth and

Seen during protracted trials at Worthy Down during 1945, one of the Canadian-built Helldivers supplied to the Royal Navy. Ray C. Sturtivant

Front (top) and rear cockpits of Canadian-built Helldiver I JW117, during testing at the A&AEE during flight testing conducted between May and July 1944.

Public Record Office, Kew, London

subsequent aircraft produced by 'Canadian Car' will have the R2600-20 engine installed in lieu of the R2600-10. The bureau estimate this change will increase the weight of the aircraft by approximately 50lb and will counteract part of the reinforced tail. These aircraft having the -20 engine will be designated as the SBW-2.

He added that in addition to the above increase in weight, '… a self-sealing fuel bag will be installed in the auxiliary fuel tank located between the pilot and observer. This tank at present is equipped with the CO_2 purging system. This change will increase the all up weight by 308lb ; the self-sealing bag accounting for 104lb; additional gas capacity of 28 US gallons and 36lb for tank supports.'

He also reported that US Navy pilots had found it necessary to apply a 1,000lb thrust when experiencing difficulties when pulling out of dives. The final gloomy prediction was that: 'It is estimated that the first Helldiver may not be expected to arrive in the UK before January 1944.'

Another sidelight was shone by a report on rocket assisted tests being conducted in the States. It was revealed that, 'Tests on the Helldiver are due to take place in June. In order to reduce the take-off run in a 15 knot wind to 420 feet, it will be necessary to use three rockets …'[100]

In addition to the numerous modifications being made at Columbus to get the Helldivers fit for combat, yet further modifications were required over and above those to fit in with the particular British operational requirements. This involved another modification line being set up at Roosevelt Field to carry these out as they became available. The BAC informed MAP of what was required:[101]

Essential:
a.   R.1147 and watch wedge plates.
b.   British Telephone sockets.
c.   O.2 compasses.

Desirable:
a.   Stowage for observer's chart board.
b.   Stowage for observer's parachute pack and dinghy.
c.   British oxygen system.
d.   Stowage for 6 flame floats or smoke floats.
e.   Stowage for signal pistol and cartridges.
f.   Speaking tubes.
g.   ASI in rear cockpit.

On the last day of August 1943, it was reported that the latest assignment of SBW-1 and SBW-2 aircraft approved by

Curtiss Helldiver I (SBW-1B), coded JW117, under test in England. Fleet Air Arm Museum, Yeovilton

the US Munitions Assignment Board was obtained from the Canadian Car and Foundry Corps during a visit to their works at Fort William, Ontario, by H. Luby of the BAC. Allocations were agreed at:

First Three for RN
Next Three for USN
Next Three for RN
Next Ten for USN
Next Two for RCAF
Next Twelve for RN
Next Sixteen for USN
Next Three for RCAF
Next Ten for RN
Next Fifteen for USN
Next Ten for RN[102]

This meant that thirty SBW-1s would be delivered to the Royal Navy, and that all the remainder would be SBW-2s.

Lieutenant (A) S. J. Miller, RNVR, senior naval representative of the BAC at BuAer, delivered another cold *douche* in September: he reported to the MAP that, due to the third list of modifications being carried out, '… the acceptance of these aircraft by us before January will be a miracle'.

And that was only to Roosevelt Field. 'Adding the time necessary for [British] modifications which will be incorporated here, it is doubtful whether the first SBW will reach the UK before February 1944, and even then, this aircraft cannot be considered an operational aircraft, as interchangeability is non-existent.'[103]

Yet further changes were indicated a few weeks later, when the US Navy informed the British that they intended to remove the following equipment from Helldiver aircraft:

a. Parachute flare container – the US Navy intend utilizing wing bomb racks.
b. Landing light – the US Navy contend that operations in the Southwest Pacific have shown this to be non-essential.
c. Cabin heating – US Navy tend towards electrically heated flying suits.
d. Windscreen de-icing – the US Navy state that their operations in the Southwest Pacific show this to be a redundant feature.
e. Bomb-bay smoke tank – the US Navy intend carrying smoke tanks in future in wing bomb racks.

f. Two-man life raft – US Navy state that the present type is too difficult to remove, and they intend using two 'K'-type dinghies.[104]

BAC Washington informed the Admiralty in November that: 'The first fully modified Helldiver squadron is at present undergoing operational trials. Results of trials are being awaited, to determine the extent of the revision of spares contract. Revision is expected to take place about December 1st …'.[105]

Estimates were now that delivery of the first eighteen Helldivers would be: March 1944, seven; April, two; May, two; June, four; and July, three;[106] but even this modest hope was destined not to be attainable, another cable reporting that: 'Two of the Helldivers allocated to us, JW. 113 and JW. 118, have been written off charge, having crashed before delivery to Roosevelt Field. Aircraft have been reduced to spares, and salvaged parts are being sent to Brunswick.'

This was bad news, for it was said that although the British could bid for replacements, '… it is very unlikely that we will

78

get them as the US Navy have stated unofficially that they will oppose the bid on the grounds that all Helldiver production is required to feed into the Pacific as combat replacements'. Nor was this the end of the bad news. '… it appears that delivery of our last ten aircraft will be delayed due to the modification programme taking longer than expected, and so it will not be possible to form 1820 Squadron on 1 April with a full complement …'

The advice was that, '… as the retention of aircraft over here waiting for the Squadron might lead to repossession, it appears to us that it would be better to ship the six to UK as planned and to form the squadron on 1st April as planned, but with a reduced complement, rather than postpone the formation date, bringing the squadron up to full strength as aircraft become available'[107] This would leave the Fleet Air Arm and the British Pacific Fleet, both now gearing up for the assaults on Okinawa and then Japan itself, with just six front-line dive-bombers on its strength, a ridiculous situation.[108]

This dire *impasse* led to a drastic last-ditch intervention by no less a person than the First Lord of Admiralty himself, A. V. Alexander, being moved to write direct to the Prime Minister stating the facts and asking for his personal help in resolving the dilemma.[109] Alexander pointed out that, since the Minister of Aircraft Production had informed the Admiralty in January 1943, that it would not be possible to employ further British capacity for naval types, the Admiralty had had to depend on America for almost half their intake. They had hoped to get an increased share of UK production in 1944, but this had proved impossible due to shortage of manpower.

The Admiralty delegations that had visited Washington in November 1942 and again in June 1943 had compared the UK and US needs and deficiencies, and at those times the US Navy had shown itself ready to treat US production as a common pool, in accordance with the joint directive issued by Churchill and Roosevelt in January 1942. He then stated that the agreement which the Americans had signed in June had covered the first six months of 1944, and was subject to revision at the end of 1943, in relation to US production shortfalls and changed strategic needs only.

When, however, another mission went out last December, it met with an entirely different American attitude. The Americans declined to honour their June agreement, although it contained no clause providing for unilateral denunciation. They made no attempt to compare the needs and deficiencies of the two navies, and made an offer of a reduced number of aircraft, sufficient only to cover wastage on aircraft already delivered…

The Americans, Alexander explained, argued that the carrier offensive in the Pacific having begun, it must be increased in tempo and power, and that, to ensure success, US carriers should be fully manned with the best types. This was fair enough, but they then went much further and demanded that their requirements also extended to the provision of manned replacement groups '(100 per cent in the case of fleet and light fleet carriers and 25 per cent for escort carriers) together with large shore-based establishments of carrier-type aircraft both for the Marine Corps and the "Fleet Air Wing"'. They were due to meet all those requirements to the extent of 100 per cent, in the next few months, reported the First Lord, '… but they would not assign aircraft to us, without which our carrier strength will be much below complement and without

manned reserves'. He went on: 'The implication was that the US Pacific offensive has absolute priority over RN operations in the West and over anything that the Royal Navy may be planning in the East, whether this year or in early 1945.'

Alexander realized that, '…we cannot reasonably expect any help from the RAF…', although with the end of the European War in sight, he hoped British production might switch back towards the Navy's needs a bit more. However, what he did want was '… a revival of the joint directive of prime minister and president that resources are to be treated as a "common pool". Otherwise, if the US Navy follows the principle that its own requirements are to be a prior claim, to the extent of 100 per cent, our assignments will vanish.

Helldiver I JW117. Fleet Air Arm Museum, Yeovilton

He added the opinion that the Americans, '… had lost less than they expected in their Pacific operations, and we believe that they could go some way to meet us without hardship. Instead they have already reduced the production of one type of aircraft and thereby are unable to meet our needs even to the extent agreed last June…' He summed by telling Churchill that: 'I think it is clear that unless we can get early accommodation from the Americans in

Royal Navy Helldiver I JW107 of No. 1820 Squadron, 1944. Fleet Air Arm Museum, Yeovilton

The SBW-1 Curtiss Helldiver JW117 in Royal Navy markings, one of twenty-six finally assigned to the Fleet Air Arm. Built by the Canadian Car & Foundry Company from July 1943 onward, they were the equivalent of the US Navy's Columbus-built SB2C-1Cs, and were powered by the Wright R-2600-8 Cyclone 14 engine. Another of the batch, JW115, was flight tested against the Vultee Vengeance IV and the Douglas Dauntless I, at Farnborough in October 1944, but was given a very poor comparison report. Imperial War Museum, London

The alternative choice, the Brewster Buccaneer (Bermuda), which also failed to reach Squadron Service due to factory problems. They were finally used as Target Tugs. Imperial War Museum

this matter we shall be unable to deploy our full carrier strength in 1945.'

This was all in vain. As the First Sea Lord, Admiral Andrew Cunningham was already finding out in his meetings with his opposite number, Ernie King, that here was not a man to be thwarted. Even with Roosevelt's backing for the British Pacific Fleet he remained determined: 'He was resigned to the use of our fleet in the Pacific, but made it quite clear that it must expect no assistance from the Americans. From this rather unhelpful attitude he never budged.'[110]

The Royal Navy therefore had to resign itself to the fact that it was not going to get a large allocation of Helldivers. In September 1944 it was revealed in a secret memo that, 'A total of twenty-six Helldiver I (SBW-1) aircraft have been allotted to the Royal Navy. It is understood that further allotments and replacements are likely to

be SBW-3 or SBW-4 aircraft'. The question was then asked '... whether it is intended to use the Helldiver I operationally, or only for training, and what modifications are likely to be required.'[111]

The following month, after extensive trials, the MAP was reported to be, 'reasonably satisfied that, from the point of view of handling and *deck landing*, this aircraft has *no snags*.'[112] The armament was considered unreliable, but by this time the first Royal Navy Helldiver squadron had been training for some time and, '... were fully aware of the difficulties'. The memo continued that: '... I think that, provided a sufficiently important operation is envisaged, the use of the aircraft is fully justified.'

But that was to be that. That same month the director of Naval Air Organization replied that he had raised the question of what was to be done with 1820 Squadron, and had been directed to loan it

to the C-in-C Home Fleet. 'This has been done, and the squadron will be available and may be used operationally despite all its disadvantages until the end of the German war.' He went on, 'As 1820 is our only Helldiver squadron and *we will not form any more*[113] it is not considered justifiable to undertake extensive modifications or improvements to the aircraft.'[114]

The final word was written in February 1945: 'In view of the fact that the allocation of Helldivers to the RN is not being used operationally it is proposed to cancel the Helldiver mod. system together with the supply of retrospective mod. sets.'[115] So it would appear that the Fleet Air Arm's failure to use the Curtiss SB2C Helldiver in combat in the last drive on Japan owes as much to the American Admiral King's attitude as to any failings of the aircraft itself, despite continued claims to the contrary.

CHAPTER TWELVE

The Helldiver in the Fleet Air Arm

During all this period the Helldiver was flown extensively by British naval and Air Force pilots who viewed it from a different perspective to their American cousins.

BAC Tests at Patuxent

In April 1944, Helldiver No. JW 112 was flown from the Patuxent NAS by Group Captain C. Clarkson, RAF, to test its general handling and to obtain data for pilots' notes. He made no attempt whatsoever to assess the operational value or efficiency. He pulled no punches on some aspects of what he found, although he did concede that '… many of these may be considered as academic and ones which must be put up with in the interests of wartime production. They are emphasized, however, in the hope that in future designs, or in modified designs of the SBW, some improvement may be undertaken in the interests of the pilots who have to operate these aircraft.'[116]

He found the cockpit was, '… extremely badly laid out, and with the exception of the undercarriage control, practically every single unit that has to be moved or is under the control of the pilot is either placed where it is difficult to get at, or it is so far away that shoulder straps have to be released in order to reach it.'

He found that the 'enormous' arrestor gear handle interfered with both throttle operation and the trimmer tabs; that the rudder and aileron trimmer tabs were both badly placed; and that the engine switch was too far away from the pilot. 'To obtain any view at all …' he had to raise the pilot's seat to full height, but then the top of the cockpit obscured some of the important instruments. Although he found the landing flap and dive flap controls 'easy to operate …', he felt it was, '… not very clear in its mode of operation, and that mistakes might be made by tired or flustered pilots.' He found that the electrical switches were on, '… a dark and not easily visible panel …'; that the cowling flap was operated by means of, '… an awkwardly placed handle which is difficult to reach and difficult to turn'; while the cockpit he dismissed as, '… terribly draughty and very cold'.

He was equally unhappy with many other aspects of the SBW-1. The view over the nose on the ground was bad, and when in the air with the hood shut, '… there are too many hood frames in the way'. He found the take-off, 'sluggish', and the lateral control, '… too heavy; particularly to the right, which needs two hands above 230 knots.' There was a tendency for the ailerons to, '… take charge if applied rapidly to the left …', and there was no willingness of the stick to return to neutral after application on either bank, possibly due to the high mechanical friction in the system. The rate of roll was, '… terribly slow'.

On the trim changes he found some change of longitudinal trim with alteration of power, and a '… very considerable …' directional change with varied throttle setting. 'This is not helped by bad rudder tab position.' The left wing dropped into the stall at 83 knots, and the right wing at 74 knots, but on the whole he considered the stall characteristics as '… quite good'.

Carrier trials of the Royal Navy Helldivers during 1944, with British pilots landing the British Helldiver on a British aircraft carrier deck! Here JW104 (18205) of No. 1820 Squadron, Fleet Air Arm, takes an arrestor wire on the training carrier.
Ray C. Sturtivant

Other factors to the aircraft's credit were that he found very little change in trim when the flaps of the undercarriage were opened or lowered.

When he put the aircraft into the dive he noted, '… some buffeting on the tail as flaps are opened, which is most noticeable at slower speeds.' In dives of 5,000ft (1,500m) at an angle of between 60 and 70 degrees, he found that, despite the dive flaps, '… high speeds (240 knots and 250 knots) were obtained …' He found that lateral and directional control in the dive were very heavy, but that the Helldiver was, '… steady with no signs of hunting or buffeting.' The landing was, '… easy, with no apparent tendency to swing.' His final conclusions were:

> Despite the fact that this aircraft has spent many months in development, it does not appear that any attempt has been made to try and give the pilot a good instrument layout or a comfortable cockpit, and for a machine which has done as much flight testing as this type, the lateral control is a disgrace …

Strangely, the overall verdict, despite all the above, was that the Helldiver was, '… easy to fly, and inexperienced pilots could probably take considerable liberties with it. The stall characteristics are satisfactory and the landing and take-off very easy.'

Also in April the Helldiver was flown at Patuxent by Lieutenant Commander G. R. Callingham, RN, who submitted a much more detailed report on his findings.[117] He began with listing the various combat loads the Helldiver was by then capable of carrying from the deck of a carrier (figures provided below).

Combat Loads – Royal Navy Test Figures

Internal bomb bay
2 × 1,000lb AP
1 × 1,000lb GP
1 × 1,600lb AP
2 × 500lb
1 × 650lb depth charge
3 × 120lb cluster bombs

Wing racks
2 × 250lb
2 × 325lb depth charges
2 × 100lb

Externally slung

The defensive armament was two fixed 20mm cannon in the wings and a twin .30 calibre rear mounting. He then went on to describe the flight characteristics thus:

'The airplane behaves normally on the ground. It is easy to handle, but suffers from some undesirable features.' These he specified as a, '… very bad …' cockpit layout, with the *exception*[118] of flap and engine controls. He experienced, '… considerable difficulty in operating the various knobs, levers and switches', and found that the view from the front cockpit was hindered by the hood frames. With regard to stability, when the Helldiver was equipped as a dive-bomber with maximum bomb loads he thought it, '… markedly unstable longitudinally'; he also recommended that inexperienced pilots should not be allowed to fly the Helldiver in this condition. He found the directional stability, '… positive at all loads and speeds …', while lateral stability he considered, '… weakly positive'.

Like Group Captain Clarkson, he made much of the frictional forces in the control circuits as being '… much too high, they are approximately 6½lb in the aileron circuit'. This applied to the Can Car produced aircraft, and he went on to emphasize that, '… in the SBF (Fairchild Helldiver), friction is at a minimum and it is possible to move the rudder trimmer with one finger. Similarly the control column friction is at a low value.' In the actual dive-bombing tests fifty 4½lb (2kg) practice bombs were dropped and he summarized his conclusions as follows:

> The approach view is moderately good over the leading edge of the wing root, however I consider that little use can be made of the downward target approach vision panel as it is too small to cover the necessary area. We found that after losing sight of the target over the wing or cowling it was extremely hard to pick it up in the panel owing to drift or small alterations of course. In addition the panel is liable to become obscured by oil leaks and scratches. We used differing methods of approaching the dive – nose down, pushover, half roll and gentle diving turn – and found the latter was the best because, speed builds up least quickly, and corrections to aim can be made before control becomes too heavy. After trying out varying angles of dive from between 30 and 90 degrees, we considered a *release*[119] angle of 75–80 degrees to be the most satisfactory.

With regard to trim in the dive it was found that with dive flaps split to the maximum 25

degrees ten to twelve divisions of left rudder trim were required at 280 knots. Callingham recommended that this trim be applied just prior to entry into the dive. The longitudinal and lateral trim did not appear to alter appreciably while in the dive, and experienced just a '… moderate amount' of tail buffeting with the bomb-bay doors open. 'The dive is steady and at the recommended angle of 75–80 degrees, at 2,100rpm and 15in Hg. manifold pressure, the speed stabilizes at approximately 285 knots IAS with a 1,000lb bomb load.' The control forces met during the dive were not serious enough to prevent him making corrections to the aim, and sighting was done through the gunsight. There was a normal pullout, and he reported '… no violent changes of trim …' when the dive flaps were shut, while rudder trim could be adjusted as required.

He also conducted low-level strafing and found the aircraft satisfactory in this role, '… having regard to its probable vulnerability to ground defences'. In a simulated torpedo attack, speed was reduced from 275 knots to 155 knots in 15–18 seconds from dives at 65 degrees.

When it came to the defensive armament, the comments were far more critical; for instance, 'Considerable difficulty has been experienced in making the 20mm installation operate satisfactorily.' And although the rear twin .30 calibre mounting '… functioned satisfactorily after minor adjustment,' he found it difficult to handle, '… especially under skidding conditions'. He thought the gunners position cramped, which did not help.

Also included in the report was a summary of discussions with US Navy pilots '… recently returned from duty with an SB2C squadron in the Pacific'. Interestingly, the overall consensus from the men who had actually flown the Helldiver in combat was somewhat different from those who had only test-flown it. The Helldiver, the Americans told him, '… is well liked operationally and is considered to be a useful weapon.' They made the following points as guides to the Fleet Air Arm pilots who might join them out there:

1. It is essential that, before a squadron operates, the maintenance crews should have a thorough knowledge of their aircraft.
2. The aircraft require a most thorough inspection on receipt, with particular reference to the hydraulic system. (This is only too well borne out by experience with the Patuxent sample of the SBW.)

3. The most required spares are those for the hydraulic system, particularly jacks and pipelines.
4. When operating with 1,000lb or 1,600lb bombs, one cylinder of the displacing gear should be disconnected. If this is not done, it is possible for the two cylinders to work against each other, thus preventing the gear from returning to the stowed position. (This occurred at Patuxent when the writer dropped a 1,000lb bomb.)

The final verdict on the SBW-1 in this report was as follows:

'... the Helldiver is a useful military weapon with good striking power but not outstanding performance. Easy to handle, it should present few problems in the formation and training of squadrons. The Fairchild-produced aircraft appear to be better made than the Curtiss or Can Car models in general workmanship and finish. This point is made in view of possible future allocations to Britain.

Aircraft and Armament Experimental Establishment, Boscombe Down

The Helldiver-I No. JW.117 was put through full performance and handling trials at Boscombe Down. Here we shall confine ourselves to their purpose, and to a summary of their conclusions.

Cockpit layout: This report was on the general suitability of the controls and instruments and the comfort factor both on the ground and in flight.[120] **Verdict**: 'It is considered that the layout of the cockpit and the layout and operation of the controls are good'.

Preliminary handling trials: This dealt with handling trails made at a take-off weight of 13,700lb (6,214kg), centre of gravity 50.4 in (128cm), aft of the datum point (with undercarriage down).[121] **Verdict**: '... the Helldiver has been found to be satisfactory for Service use at speeds up to 350 mph (563km/h)ASI.'

Further handling trials: From September to November 1944, further tests were made to determine the most aft acceptable CG position, to assess the handling characteristics at a forward CG position, and to obtain static stability measurements at each of these CG positions.[122] **Verdict**: The aft-most acceptable CG position was found to be 50.4 ins (128cm) aft; the handling on the climb was unacceptable with the CG further aft. At the forward CG loading, CG 45.9in (116.6cm) aft, the aircraft was considered to be acceptable for Service use, but there was just sufficient elevator range for landing.

Weights and loading: The results of weighing and centre of gravity determination were made to verify the 'weight empty' and CG, and calculation of CG range due to dissipation of load across a wide range.[123]

So much for test pilots' conclusions; but what of the Royal Navy's limited operational experience with the Helldiver?

No. 1820 Squadron, Fleet Air Arm

Although destined never to see combat, No. 1820 Squadron was the Fleet Air Arm's only operational squadron to utilized the Curtiss Helldiver, and their experiences, along with various test pilots' reports above, give a good insight into how the British viewed this dive-bomber. The aircrew for what was to become No.1820 Squadron actually began training at Jacksonville, Florida, where they underwent the standard US Navy VS dive-bombing course at Cecil Field from 1 February 1944, through to early April.[124] Don Sidnell recalls an early experience of this time:

We were joined by Lieutenant Commander (A) Grant-Sturgis, RNVR, soon after beginning the course, as he was destined to be our C.O. when the Helldiver squadron was eventually formed.

However, on one of his first flights leading us on a dive-bombing exercise, he did not return and his Dauntless was found in pieces only a short distance from the target – he never came out of the dive. That was on 16 February 1944. He was then replaced by Ian Swayne. It was, of course, easy to become so engrossed with getting the correct aim in the dive that not enough attention was given to altitude, and you travel quite a long way down in a few seconds at dive speeds.

The squadron was allocated their aircraft under the lease-lend programme, and were accompanied by a technical representative of the Canadian Car and Foundry Company all the time they were in North America. In the words of the late F. Denys Walter,[125] a TAG (telegraphist/air gunner) who had retrained as an observer and who was on his first appointment in that category: 'The aircraft arrived from Canada, with

Landing at Squantum in 1944 is Helldiver JW121, one of a batch evaluated by No. 1820 Squadron FAA.
Ray C. Sturdivant

Royal Navy Curtiss Helldiver I (SBW-1B) touching down at US NAS Squantum, 1944. Donald F. Sidnall

USA markings painted over British roundels. All kinds of chemicals failed to remove the USA markings – and then soap and water was tried ...'

The squadron formed at NAS Brunswick, Maine, on 13 April. Here the aircrew first got to know their machines, whose reputation had gone before. What did *they* make of this formidable new dive-bomber? Don Sidnell described its handling thus:

> Heavy and slow at slow speeds, but good in the dive using dive flaps – and after all, that was its main purpose. Forward visibility for deck landings, in the 'nose-up' attitude, wasn't bad. Pushover into an apparently vertical dive gave about 80 degrees angle which was about right and reasonable accuracy could be achieved. It was important to avoid 'cork-screwing' in the dive and to get the allowance for wind right, so that the 'pipper' in the bomb-sight (projected on the cockpit windscreen) drifted during the dive to the correct position in relation to the target, when the bomb is released, again allowing for wind and speed.

While at Brunswick the British joined the station's first anniversary birthday party with Richard Hallet the radio commentator as MC; among the many guests were Rear-Admiral J. Cary Jones Jr, Sumner Sewall, governor of Maine and Margaret Chase Smith, a congresswoman.

Familiarization with the SBW was followed by various dive-bombing exercises throughout the remainder of April. Also the rear cockpit armament was replaced by twin .303 British Browning machine-guns, a conversion which was not undertaken without some difficulty; moreover the US authorities did not like the idea. Another local modification to RN standards was the fitting of navigational compasses in the rear cockpits but these '... never operated in a satisfactory manner due to a strong magnetic field around the Scarff ring and seat mounting.'[126]

On 1 May the squadron personnel moved to USNAS Squantum, Massachusetts, south of Boston and Thompson Island. Here the practice dive-bombing was intensified, with exercises being carried out using a floating target off Cape Cod as well as against an armoured target boat, to give them the opportunity to try at a high-speed moving target.

'Here we lost one pilot, Jamie Dawson, and his aircraft. He was beating up the home of a girl friend near Boston, pulled up into a wing-over turn, stalled and crashed.'[127]

Their training at this point also encompassed anti-submarine bombing, air-to-air and air-to ground firing, ADDLES and navigation exercises. This programme continued throughout May 1944 and most of June. Towards the end of that month – on

All that was left of Sub-Lieutenant Jamie Dawson's Helldiver when he was killed, crashing near Boston.
Donald F. Sidnell

the 22nd, to be precise – the squadron shifted base yet again, this time to Norfolk, Virginia, where they practised deck landings afloat. Inevitably there were mishaps: on 24 June, Sub-Lieutenants John Fenwick (a New Zealander) and Pearson both ditched their aircraft alongside the carrier; and on the 27th Sub-Lieutenant Everett put a wheel over the side of the deck, and Sub-Lieutenant Cornabe flew into the barrier. The SBW was not all bad news, however; for instance, Don Sidnell remarked on the Helldiver's sturdy undercarriage, '… tougher than most British naval aircraft, probably because of the 'clumsy' US deck-landing methods – namely, drop onto deck from several feet up, as compared with the British technique which we were using, of cutting engine power when the aircraft was virtually stalled just above deck.'

On 27th June No. 1820 Squadron moved back to Squantum again for further training, before once more returning to Norfolk on 5 July, to go aboard the escort carrier HMS *Arbiter* for passage to the UK. The voyage across the North Atlantic through the U-boat packs proved uneventful, and as the aircraft were stowed away for transit, no actual flying time was logged by the squadron for the journey.

After disembarking at Speke, near Liverpool, they transferred to Royal Naval Air Station (RNAS) Burscough in Lancashire, on 25 July. Once their aircraft had been readied and checked the squadron was soon involved in intensive exercises again; these involved low-flying and night-flying, formation tactics, anti-submarine bombing, and air firing on towed drogue targets over the Irish Sea. A high-altitude exercise using oxygen apparatus up to 26,500ft (8,077m) was followed by a dive from 21,000ft (6,400m) down to 8,000ft (2,440m). There was an endurance test which took the squadron on a 3 hour 25 minute flight. But the chief occupation was dive-bombing practice against a target moored in Morecambe Bay, followed by carrier deck-landings aboard the escort carrier HMS *Speaker*. According to Don Sidnell's logbooks, he made four deck

The CO of No. 1820 Squadron FAA, 1944. Fleet Air Arm Museum, Yeovilton

(Below) **The petty officers and ratings of No. 1820 Squadron, Fleet Air Arm, at Squantum.** Donald F. Sidnell

landings on 29 October, with Sub-Lieutenant Pearson as his passenger each time, and the following day he acted as passenger for Sub-Lieutenant Everett who made one deck landing and carried out two dummy dive-bombing attacks, one against the *Speaker* and the other against her escorting destroyer. Finally the Helldivers conducted practice dive-bombing attacks on units of the Home Fleet.

Whilst at RNAS Burscough, we lost another pilot and aircraft (Alan Neville) – practising dive-bombing a sea target in Morecambe Bay and he just did not come out of the dive. Another pilot did not open his dive flaps before diving and only just managed to pull out. However, the aircraft wings were severely stressed – another example of accidents causing a shortage of aircraft.

On 30 October, No. 1820 Squadron transferred north to the RNAS Hatson, (via a stay at Donibristle again), in the Orkneys, the very same airfield from where, 3½ years earlier the Fleet Air Arm Skua dive-bombers had taken off to make history by sinking the German cruiser *Königsberg* in Bergen harbour. The squadron would have been really pleased to have been given the opportunity to match that achievement in the new Helldivers, but, alas, no targets presented themselves during their short stay. Instead they conducted a Navex over the sea, culminating in a simulated dive-bombing attack on the battleship HMS *Rodney*, making shallow dives through eight-tenths cloud cover. Next day they carried out dive-bombing practice in Woodwick Bay.

Observer F. Denys Walter later stated that '... we flew to Hatson to embark on a carrier, but the carrier was u/s so we went on leave. While on leave I heard the RAF had sunk the *Tirpitz*, and when I returned to Hatson I found we were to disband. Possibly these two events were not co-incidental.'[128] This carrier would have been the old *Furious*, which had taken part in earlier attacks on the German battleship holed up in her Norwegian lair, using Barracudas. However, No. 1820 Squadron's Helldivers were not to get their chance to have a crack at this tempting target, for the ancient *Furious* was reported by the DNC as being completely worn out at the end of August, and on 15 September the Controller of the Navy had issued an order that she was to be paid off and laid up in reserve.

Therefore on 4 December the squadron returned to RNAS Burscough, prior to final disbandment. Don Sidnell recalled

The US Navy's BATS officer for No. 1820 Squadron at Squantum with some of the Royal Navy Helldivers. Donald F. Sidnell

that: 'When we were flying in formation from the north one aircraft had engine failure, resulting in a wheels-up landing in the Lake District. The pilot skidded to a standstill before hitting the hedge around the field, but the aircraft was, of course, damaged.' He also noted that his own last flight in the Helldiver was on 9 December. On disbandment he told me that:

My understanding was that the reason was lack of aircraft. Through accidents we lost quite a few of the twenty-six allocated to us, and the aim of having fifteen aircraft available at all times so that twelve could be put in the air was becoming difficult. Spares were also a problem, I believe. In retrospect I think we did have a bit of bad luck.[129]

Back in the UK the personnel of No. 1820 Squadron at RNAS Burscough, Lancashire. The CO (Swayne) is seated centre, front row. Sub-Lieutenant (A) D. F. Sidnell is second on the far right. Donald F. Sidnell

HELLDIVER MEN: Lieutenant Commander (A) H. I. A. Swayne, DSC, RN

One of the great young Fleet Air Arm pilots of World War II and a fitting leader of the Royal Navy's first (and what proved to be only) Helldiver squadron, Henry Ian Ashton Swayne was born on 2 April 1914, just before the outbreak of World War I. His early service was under the RAF but when they were forced to relinquish their cold and hostile grasp from the Fleet Air Arm he was one of a number of maritime aviators that transferred to the Royal Navy. This was accomplished on 13 July 1937 when Ian Swayne became Sub-Lieutenant (A) Swayne, R.N.

He joined 822 Squadron as a pilot on 17 October 1938 and served with them aboard the carrier HMS *Courageous*. On 8 March 1939 he was transferred to No. 825 Squadron aboard HMS *Glorious*, and on 13 March 1939, with Hitler swallowing what remained of Czechoslovakia and with the second great conflict clearly imminent, he was promoted to full Lieutenant.

In August 1940 he joined No. 815 Squadron embarked on the brand new armoured carrier HMS *Illustrious*, and went out to the Mediterranean theatre of war with her; here she soon earned high renown with the sinking of three Italian battleships at their mooring at Taranto in October of that year. As pilot of the Fairey Swordfish torpedo bomber L4M, Ian attacked the brand-new Italian battleship *Littorio* that night.

Hitler sent swift retribution upon *Illustrious* in the form of Junkers Ju 87 Stuka units, and in this way Ian Swayne was given a close-up view of how dive-bombing should be conducted, and also how deadly it could be, for these caught the carrier in the Sicilian Channel in January, scoring a record number of direct hits and near misses and damaging her very badly. When eventually she escaped from Malta dockyard to Alexandria in Egypt, prior to sailing to the USA for massive repairs, 815 Squadron was sent ashore to HMS *Grebe*, the RNAS at Dekheila, Egypt, and operated in the Western Desert. And not only there, but courageous and almost totally unreported missions against the Axis force in Albania took place which earned Ian Swayne the Distinguished Service Cross, awarded on 2 December 1941.

On 25 July 1941 Ian Swayne once more had the opportunity to tread a carrier deck, joining 829 Squadron aboard HMS *Formidable*, embarking on that ship for passage back to the UK after she too had been badly damaged by Stuka dive-bombers off Crete the previous May. On 22 October Ian was appointed to 767 Squadron at HMS *Condor*, the RNAS at Arbroath on the east coast of Scotland.

His tenure ashore was again a brief one, for when the *Illustrious* came back from America fully repaired, Ian was re-assigned back to 829 Squadron; he joined her on 25 February 1942. Here he remained for a year before being appointed to another shore position, this time at HMS *Saker*, the Royal Navy HQ in Washington DC. This was an eventful move, for he met, courted and wed his lovely American wife here.

One year later, on 1 April 1943 Ian Swayne was appointed 'In Command' of the Navy's first Helldiver squadron, No 1820 after the nominated CO had been killed in a flying accident. His time with No. 1820 is related in this book, but when the squadron was prematurely disbanded, Ian was transferred to HMS *President*, in Central London, for duty at the Ministry of Supply. He served two years in this job before again receiving a sea-going appointment.

The Commanding Officer of No. 1820 Squadron, RN, Lieutenant Commander (A) H. I. A. Swayne, RN. Donald F. Sidnell

On 13 March 1947, Ian was promoted to Lieutenant Commander (A), and in July joined the light carrier HMS *Ocean* as her Lieutenant Commander (Flying). Here he served two eventful years, mainly in the Mediterranean and off Palestine, *Ocean* becoming Flagship (Air) Mediterranean Fleet; after this time he was appointed back to the Admiralty in London in October 1949. On 1 March 1950, he was appointed to HMS *President* for duty inside Admiralty with the Air Equipment and Naval Photography Department.

Tragically, after surviving all that the war could throw at him in one of the most dangerous of all occupations, he was to die in a railway accident, of all things, when the train in which he was travelling whilst on active service crashed at Weedon, Hampshire, on 21 September 1951. It was a wasted end to an already brilliant career. Ian Swayne, DSC, was just 37 years old.

CHAPTER THIRTEEN

The Philippine Sea

The repeated blows of the burgeoning American fleet across the Central Pacific had already made great inroads into the Japanese defence perimeter and exposed its weakness. Yamamoto's 'decisive battle' had never come about in his day, save at Midway where the result had not gone according to the Japanese script. Now Yamamoto himself had gone and his policy of continually holding back his fleet and not committing it to combat had only resulted in the Americans being given the time to rebuild their strength, firstly to equal the Japanese and then increasingly to exceed it. There was

clearly no point in holding back forever as this gap would merely get wider, so it was decided to throw every available ship and aircraft against the next American thrust. This lead to the largest of the many carrier-to-carrier battles of the Pacific war, with nine Japanese carriers pitted against twelve American. The Japanese had hoped to even the odds up beforehand by extensive use of their land-based aircraft which it was expected would cause such damage and loss to the Task Forces that they could then be crushed.

Captain Mitsuo Fuchida, IJN, at that time senior staff officer of the First Air

Fleet in the Marianas, explained the scenario thus:

The plan was, that if the US Task Force struck the Marianas, then the strength from Palau would be moved up to the Marianas to reinforce and attack with all the forces except Air Flotilla 23. It was made very mobile to meet the expected attack. When the attack was expected, Admiral Ozawa's Task Force which was in Linga and the Homeland, were to rendezvous at Tawi Tawi and then move up to position northeast of Palau and stand by so as to meet the attack.[130] [This was the A-Go Plan.]

SB2C-3 of VB-19 aboard the Lexington (CV-16) on 20 August 1944. National Archives via National Museum of Naval Aviation, Pensacola

The Saipan landing provided the spur, and both carrier- and land-based aircraft were duly committed to battle. Back in mainland Japan, therefore, frantic preparations were put in train to make an early strike with the maximum number of aircraft. As Captain Akira Sasaki, IJN, at that time a staff officer with the Yokosuka Air Corps, was to recall:

At Yokosuka they were able to scrape up about 120–130 miscellaneous naval aircraft including Bettys, carrier bombers, torpedo planes and fighters. Of the pilots employed, only about one-third were experienced, the others being students. This conglomerate attack was launched on 18 June 1944 with the specific objective of attacking the landing beaches or landing ships. En route, however, they became involved with the United States carrier aircraft with result that few, if any, reached the assigned target.[131]

Practically all the attacking aircraft were lost either through being shot down or as a result of forced landing and landing crashes instant to battle damage. However, approximately fifty pilots survived the action. The Japanese were unable to employ regularly organized combat air groups from the mainland for the simple reason they didn't have such groups available. The campaign against the Marianas was initiated before the training programme had accomplished useful results.

The same fate was met by the shore-based aircraft of the First Air Fleet, Fuchida stating quite simply that: 'On 11th, 12th, 13th practically all of them were wiped out. It was ordered that aircraft carriers were to be the target, all the time, but we received no reports that they were effective.'

The Japanese carrier-based aircraft suffered exactly the same fate in what became known as the 'Great Marianas Turkey Shoot' – and the fleet's fighters had a field day. All this left the Helldiver crews frustrated, for Admiral Marc Mitscher had kept his carrier task groups of Task Force 58 well back in the field. Thus when the decision was finally taken on 20 June 1944, to launch a strike at the retreating enemy carriers, it was both late in the day (15:53) and at extreme range (300-plus miles (480km)). Moreover there was one other factor which was to make the battle famous, and this was known from the very outset, as Admiral Mitscher signalled to Admiral Spruance: 'Expect launch everything we have. *We will probably have to recover at night*.'[132] The composition of the Helldiver units at this time is given in the box right.

Deck party manhandling a 'Beast' of VB-17 aboard the Bunker Hill **(CV- 17), May 1944.**
Flying Magazine via National Museum of Naval Aviation, Pensacola

SB2C Units at the Battle of the Philippine Sea, 20 June 1944				
Unit	Carrier	Commander	Type	No
VB-2	Hornet (CV-12)	Lt Cdr G. B. Campbell	SB2C-1C	33
VB-1	Yorktown (CV-10)	Lt Cdr J. W. Runyan	SB2C-1C	40
VB-8	Bunker Hill (CV-17)	Lt Cdr J. D. Arbes	SB2C-1C	33
VB-14	Wasp (CV-18)	Lt Cdr J. D. Blitch	SB2C-1C	32
VB-15	Essex (CV-9)	Lt Cdr J. H. Mini	SB2C-1C	36

Two strikes were originally planned, but when further information came in and it was revealed that, if anything, the enemy was yet more distant, it was decided to hold back the second group's launch until first thing the next morning.

The Outward Leg

The launching began around 16:24 in position 13.58 N, 138.57E, and a total of fifty-one SB2C-1Cs got away, each armed with drop tanks as well as a 1,500lb

twenty miles astern of them. Course was therefore changed to 284 degrees on this new heading as the aircraft continued to gain altitude slowly in order to conserve precious fuel. It took them half-an-hour to reach 10,000ft (3,000m), and then they flew in the most part in silence, on the leanest fuel mix the Wright engine would stand.

The Attacks

The first enemy ships to be sighted were, of course, the tanker refuelling force, trailing

Force – the tankers *Genyo Maru* and *Azusa Maru*, escorted by the destroyers *Uzuki* and *Yukikaze*.

While the majority of the American aircraft tended to ignore this group, pressing on, '... to get at their Fighting Navy', the commanding officer of *Wasp's* VB-14, Lieutenant Commander J. D. Blitch, considered that a hard strike at these ships would so cripple the enemy's fuel supplies that they would be forced to withdraw at slow speed overnight and thus be well in range for the decisive second blows he thought would follow at first light. He

An SB2C-1C of VB1 coming in to land aboard the Yorktown **(CV-10).** US Navy via National Museum of Naval Aviation, Pensacola

(680kg) bomb. A corrected sighting position of the enemy ships was received by Lieutenant Commander Ralph Weymouth of VB-17 once the strike had taken its departure, and it confirmed that, as well as the carrier/battleship groups, there was also a tanker refuelling force tailing the Japanese fleet, some fifteen to

even further behind the main fleet as the latter were aware they were going to be attacked and had increased speed accordingly. This Japanese force consisted of the fleet oiler *Hayusui*, and three other tankers, *Kokuyo Maru*, *Nichiei Maru*, and *Seiyo Maru*, escorted by the destroyers *Hatsushimo*, *Hibiki*, *Tsuga* and *Yunagi*; and the Second Supply

therefore elected to strike the tanker group and led in accordingly.

The oilers were divided up between his Helldivers, and they hit the *Hayasui* with one bomb and damaged her with two very near misses. Both the *Genyo Maru* and *Seiyo Maru* were also near-missed and disabled in this attack. Two of these sank

almost immediately, but the *Genyo Maru*, which had three near miss bombs disable her machinery and open up her hull below water, lingered a while longer. Eventually she became completely unmanoeuvrable and finally had to be sunk by the *Uzuki* after her survivors had been taken off. Blitch himself commented favourably on the '2C', stating in his subsequent report that the, '... overall operation of the SB2C-1 and SB2C-3 airplanes in combat has been excellent.'

The remaining SB2C-1s split their attacks between two of the three Japanese carrier groups, which were first located at around 18:30. The American force was in three sections, which roughly correspond- ed to the three Japanese carrier groups' positions. The northern force contained the *Hornet*'s Air Group 2 led by CAG Commander Jackson D. Arnold, with the fourteen aircraft of Lieutenant Comman- der G. B. Campbell's VB-2; next came the *Yorktown*'s Air Group 1, with Lieutenant Commander J. W. Runyan's VB-1 thirteen Helldivers.

The various Japanese warship groups had become somewhat mixed up during the battle, but were in three loose groups all steering a course of 300 degrees to the north-west at 24 knots with several miles between each group. Passing by the near- er, more southerly Japanese groups, which he saw were already under attack, Arnold concentrated his attacks against Force 'A', which at this time only contained one air- craft carrier, the *Zuikaku*, her two com- panions having been sunk by submarines earlier in the battle. She was the biggest prize, however, and was escorted by the heavy cruisers *Haguro* and *Myoko*, the light cruiser *Yahagi* and destroyers *Akizuki*, *Asagumo*, *Hatsuyuki*, *Isokaze*, *Shimotsuki*, *Urakaze* and *Wakatsuki*.

The Helldivers were met by an incred- ible wall of 'Technicolor' flak bursts as they came in for their attacks, with both heavy cruisers firing their main 8in arma- ments as well as all ships' lighter weapons. Different coloured bursting charges were used as aids to the gunnery directors of each ship, with the result that the sky was filled with explosions that varied from red, white, yellow and black, to the more exotic pink, burgundy and lavender bursts and incendiary detonations. Undeterred, the SB2Cs plunged, making a dead-set at the solitary carrier.

First in, and out of sequence, was the six-plane division led by Lieutenant

Commander Harold L. Buell, of VB-2. He led down in an 80-degree dive, and later gave a graphic description of this attack, and his feelings at the time.

As I pushed over into my final dive at about 12,000ft, the AA fire became so intense in my immediate vicinity that I saw no way of getting through it. I had never encountered such flak. My dive brakes were already open and I was well into a good dive, but because of the potent defensive fire, I felt like I was moving in slow motion in quicksand. I would never make it. At this point I did something I had never done before in a dive – I closed my dive brakes. My plane responded by dropping like a stone toward the target below, leaving the heavy AA fire behind.

So much for that problem, but now I had another more serious one. My speed was building up and, in the clean condition without flaps, I could never expect to pull the plane out of the dive after firing my bomb. Shouting a prayer to my guardian angel, at 6,000ft I placed the dive brake selector back into the open position. The wing brakes did what no manufacturing specs said they would – they opened! It was if a giant hand grabbed my plane by the tail, my headlong plunge slowed and there was the enemy carrier dead in my sight below me, turning into my flight path along its lengthwise axis. At a point-blank range of two thousand feet, I fired my bombs.[133]

At the moment of pullout, as he bent for- ward to close his bomb-bay doors, Buell's aircraft took a direct hit under the starboard

The trainee pilot scrambles clear as an SB2C-3 of VB-82 vanishes beneath the waves alongside the training carrier Charger (CVE-30) in June 1944. Emil Buehler Naval Aviation Library, Pensacola

wing, some eight feet out from the fuselage. This shell passed through the cockpit and detonated above it, shrapnel hitting Buell in the back. Despite a horrendous wound he managed to retain control of his aircraft, and got out between the destroyers on the screen with the large hole in his wing burn- ing steadily away.

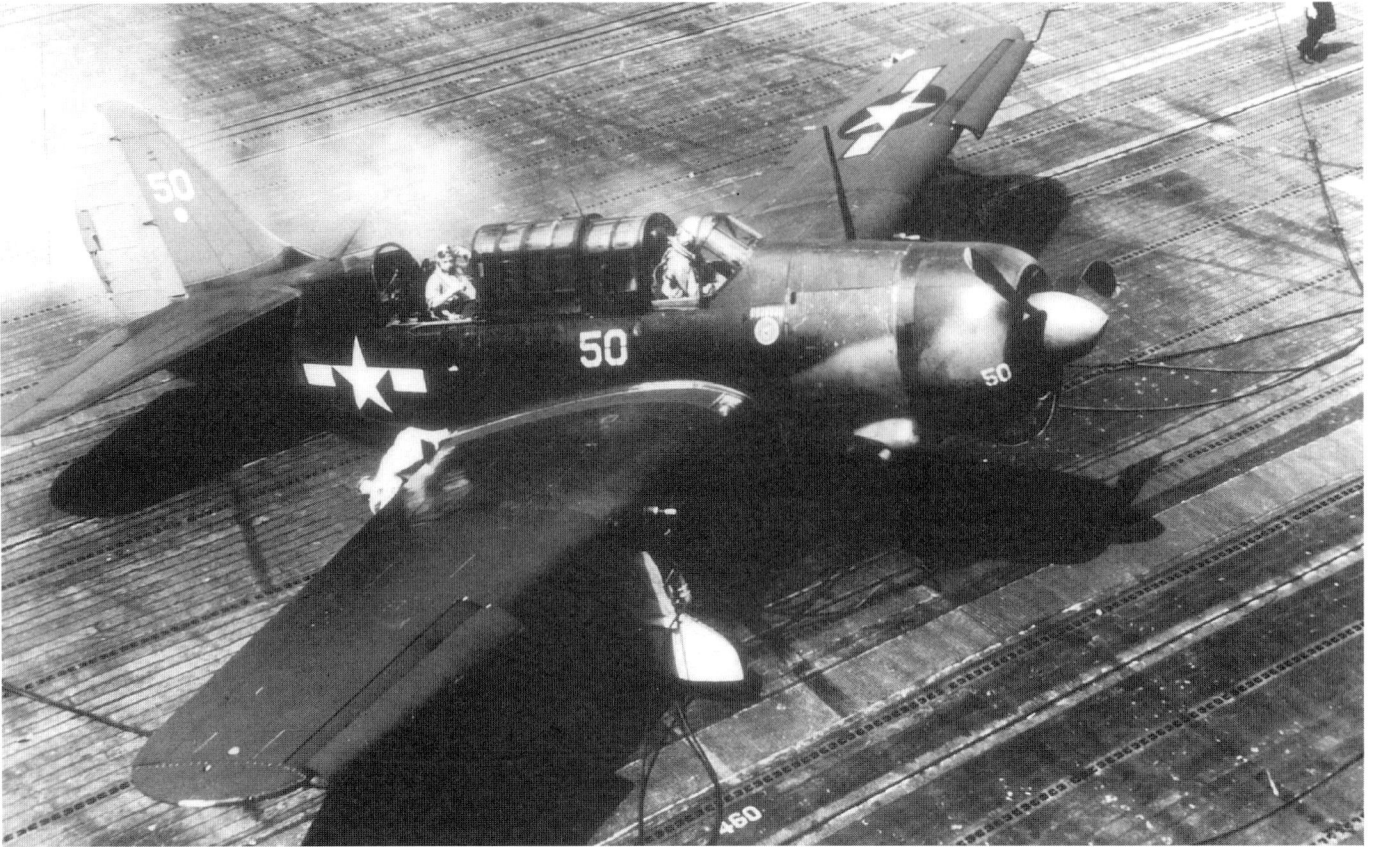

(Above) **A SB2C-1C of VB-2 'brews up' on the deck of the** Hornet **(CV-12) in 1944.** US Navy via National Museum of Naval Aviation, Pensacola

A SB2C-1C of VB-8 with folded wings on the after deck of the Bunker Hill **(CV-17) off Saipan in June 1944.** Emil Buehler Naval Aviation Library, Pensacola

These six dive-bombers claimed a total of, '… three solid hits …' on the *Zuikaku*, and observed both explosions and fires aboard her as a result. Next in were the eight aircraft of Lieutenant Commander G. B. Campbell's division which had taken advantage of a large area of cloud to get themselves ahead of the enemy force. They then commenced their dives from an altitude of 12,000ft (3,600m), releasing their 1,000 GP and SAP bombs at around 2,500ft (760m) against the same target. They too claimed several direct hits on the *Zuikaku*, and Campbell observed '… one big hole with a fire down inside near the island'.

The thirteen Helldivers of VB-1, under Lieutenant-Commander Runyan, were then directed to finish her off, and made their attacks on the smoking carrier which was turning in a full circle and ablaze. The first divisions made their dives and they also claimed to have scored at least three more direct hits on her flight deck and several more near misses. She appeared doomed, so the following divisions divided their attacks among the enemy heavy cruisers, though without success.

The Japanese admitted that there were several (they were not specific as to how many) direct hits, and at least five very close near misses, the combination of which was serious damage and dangerous fires in the carrier's hangar decks in the aviation gasoline stores. So serious did the situation become with the fires raging unchecked and the sprinkler system unable to cope with it, that at one point the order was given to 'Abandon Ship'. However, the damage control parties were veterans with cool heads, they gradually got the fires under control and the order was rescinded. Badly gutted and with heavy casualties, the *Zuikaku* lived up to her name ('Lucky Crane') and survived this time, to limp back to Kure harbour, where she arrived on 23 June, to go into dry-dock for heavy repair work.

The centre force was made up of Air Group 10 with the SBD Dauntless dive-bombers of VB-10 from the *Enterprise*, and Air Group 16 with the SBDs of VB-16 from *Lexington*.[134] They hit Force 'B', the carriers *Hiyo*, *Junyo* and *Ryuho*, the battleship *Nagato*, the heavy cruiser *Mogami* and the escorting destroyers *Akishimo*,

Hamakaze, *Hayashimo*, *Michishio*, *Nowaki*, *Samidare*, *Shigure* and *Yamagumo*. Combined dive- and torpedo-bombing attacks sank the *Hiyo* (the only Japanese carrier lost to air attack in this battle) and damaged the *Junyo* (heavily), the *Ryuho* (slightly) as well as the destroyer *Shigure*.

The southern force consisted of what remained of Air Groups 8, 28 and 31, which had only the twelve Helldivers under Lieutenant Commander J. D. Arbes VB-8 from the *Bunker Hill* left for dive-bombing strength. They tackled the heaviest and toughest concentration of the enemy, Force 'C', which consisted of three aircraft carriers: *Chiyoda*, *Chitose* and *Zuiho*; four battleships: *Haruna*, *Kongo*, *Musashi* and *Yamato*; seven heavy cruisers: *Atago*, *Chikuma*,

Flight of SB2C-2s with top cover of Grumman F6F Hellcats, the classic combination that ultimately won the naval war in the Pacific.
US Navy Official via Stanley Vaughn archives

Kumano, *Maya*, *Suzuya*, *Takao* and *Tone*; screened by eight destroyers: *Asashimo*, *Fujinami*, *Hamanami*, *Kishinami*, *Naganami*, *Okinami*, *Shimakaze* and *Tamanami*.

Some of the attackers concentrated on the carrier *Chiyoda*, and they scored at least two confirmed direct hits on this ship with 500lb bombs, both of which penetrated the flight deck aft, wiping out two torpedo bombers parked there and damaging another and a fighter. There were several near misses, and splinters and fragments of bombs coming inboard from these added to the damage and started fires. Incredibly only twenty crew members were actually killed and thirty wounded by these strikes, and her fires were eventually brought under control.

Other accurate attacks were made on the heavy ships of the force; for instance the battleship *Haruna* – so often claimed as 'sunk'

by the Army Air Force in 1942 – was hit for the first time on her fantail by a 500lb bomb. This caused damage to her propeller shaft brackets and distorted her hull. Her aftermagazine was flooded as a precaution and her speed restricted to 27 knots. There were several near misses also, but the sturdy old *Haruna* lived to fight yet again!

The heavy cruiser *Maya* was well able to take care of herself, and her only damage was caused by a near-miss bomb which started a fire in her port side torpedo deck; this was quickly subdued and she was otherwise uninconvenienced. It was only while the Helldivers were withdrawing from these attacks that Japanese fighters made interceptions, and two VB-8 2Cs were shot down before the Hellcats intervened.

Total tally for the Americans was the sinking of one carrier, heavy damage to another and light damage to three more; damage to one battleship, one heavy cruiser and one destroyer; two fleet oilers sunk and one so badly damaged she sank later. This was achieved at long range and at a cost, in battle action, of just twenty aircraft, only four of which were Helldivers.

Homeward Leg

A much higher toll was now to be paid as the American air groups turned back towards their carriers, many hundreds of miles to the south. Few had been trained in night flying and deck landings, many had damaged aircraft, some had badly wounded pilots, as Buell, most were exhausted by the long flight out, and the adrenalin of the dive was replaced by a grim realization that they had a very long way to go to safety. Few had much gasoline left, and there were reports of tropical thunderstorms between them and safety. The estimated flight time was about two-and-a-half hours. In compact air groups, in squadrons, in divisions, as lone stragglers, the US aircraft struggled back to base through a very lonely sky.

If the flight out had been relatively silent, the journey back was punctuated, for the most part, with messages that varied from desperate, through lonely to resigned as the fuel tanks slipped into.the red and soon aircraft began to drop out into the black sea below. Those that made it back by 20:45 found the fleet waiting to

1.

2.

3.

4.

5.

6.

7.

8.

9.

10.

11.

12.

Helldiver landing sequence. Curtiss-Wright via Stanley I. Vaughn Archive

Something New ... HAS BEEN ADDED

Publicity for the arrival, on Christmas Day 1942, of the A-25 Shrike. Unfortunately, all the great press release and talk of expanded production was to prove a chimera and the A-25 ended up a very unwanted airplane. Curtiss-Wright via Stanley Vaughn Archive

(Below) Covering the Philippines invasion, a SB2C Helldiver is launched from the USS Hancock (CV- 19) on 25 November 1944, for a strike on Manila Bay. US National Archives

give them all the help it could. The *Enterprise* flew off one of its night fighters to guide two groups in over the final leg, Admiral Clarke's ships turned on all their masthead and running lights, and searchlights were shone from the flagships, while cruisers and destroyers fired star shells over the polytechnics, some even tried landing on destroyers of the screen!

Soon the night sea around the task force was full of ditching aircraft: some groups elected to go into the water *en masse*, others tried desperately to find a deck, only to end up in the sea alongside when engines hook failed to connect ('a bolter') and the Helldiver bounced over two barriers, went through the third and crashed into another 2C, piloted by Lieutenant (j.g.) David Stear that had only just landed, killing the rear-seat man, AR2c W. E. Redman, and one of the deck crew. Both Buell and his

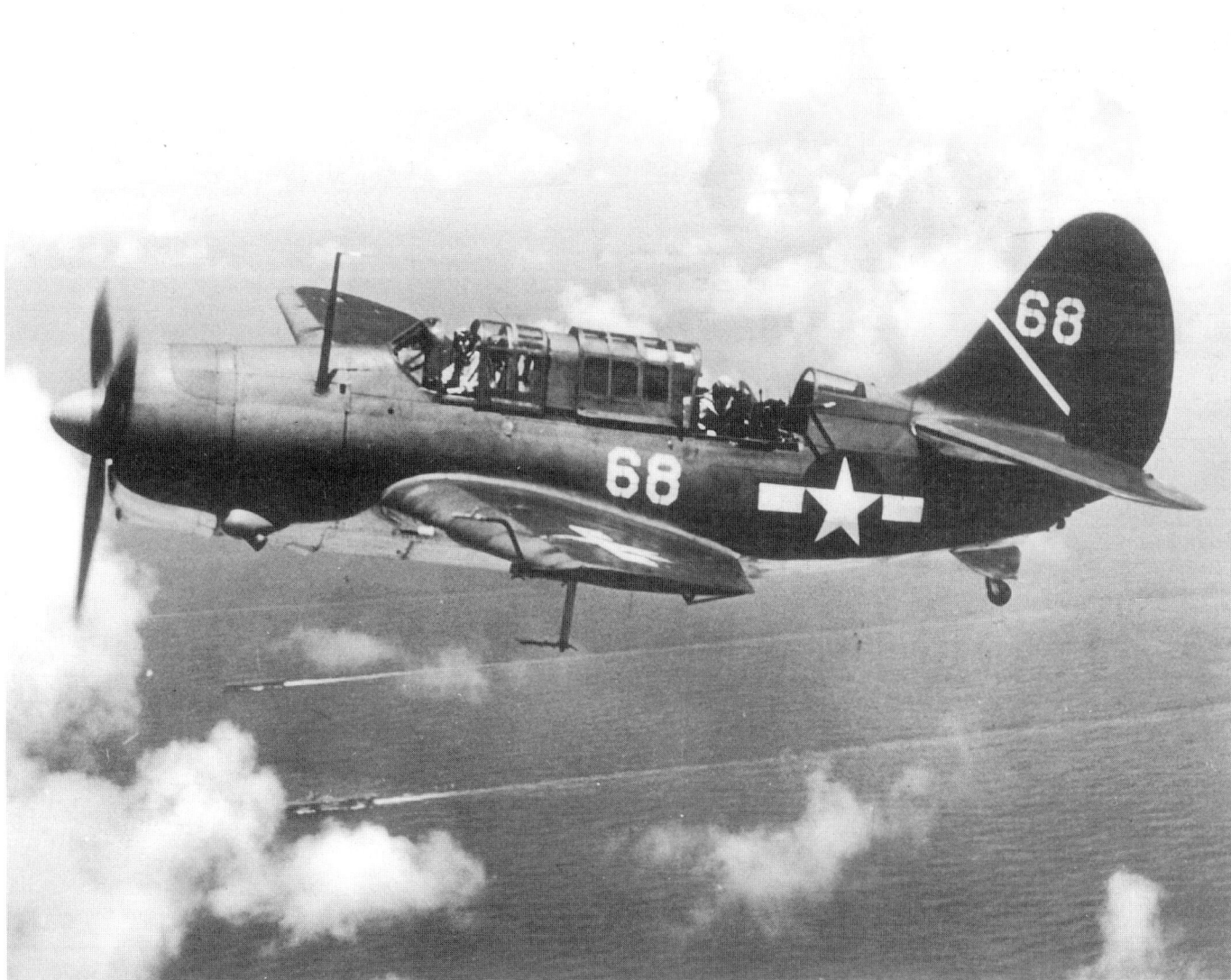

Sporting the camouflage scheme of June 1944, this SB2C-3 (Coded 68) is shown over her parent ship, the escort carrier Guadalcanal (CVE-60) and other ships of the American fleet in the summer of 1944. Jerry Crisman was the 'backseat driver' (Radio-Radar-Gunner) in this photo, and he told the author that he had ASB radar installed. **Jerry Crisman**

the carriers to illuminate them to the tired flyers. After a while there was little attempt made to find one's own carrier; with just a thimbleful of gasoline, pilots were grateful to get down on any deck at all, and in their tiredness and confused by gave up on them. Others got down in very hard landings indeed. Buell himself, on his last ounce of gas, only one flap operational and severely weakened by his back wound, got a 'wave-off' from the *Lexington*, but had no choice but to go down anyway. His rear-seat man, AR1c Red Lakey, survived the crash and Buell his wounds.

Others were not so fortunate, and in total some thirty-five 2Cs either ditched or crashed on landing and were write-offs. Some article writers have made much of

Battle of the Philippine Sea, 20 June 1944. This marked the swan-song for the Japanese carrier fleet but the coming-of-age for the Helldiver. Here the Japanese carrier Zuikaku **receives a direct hit from a Helldiver despite frantic manoeuvring.**
National Archives, College Park, MD

these losses, using this figure to bolster their argument that, '… The Navy's "Last dive-bomber" was, "Its Worst"[135] (two totally incorrect statements in one sentence, since the SB2C was neither the worst *nor* the last dive-bomber built for the US Navy). They usually fail to point out that the much longer-serving TBM Avenger also had twenty-three aircraft ditch, along with fourteen F6F Hellcats and four SBDs.

Certainly '2C' units like VB-15, flying from the *Essex*, had combat records that could not be challenged by *any* aircraft, suffering absolutely zero combat losses dur-

ing a tour of duty that included the Battle of the Philippine Sea, attacks on the Marianas, Palau, Mindanao, Manila Bay and Formosa. As for statements that the war in the Pacific was now fought, '… by big fighters and heavy four-engine bombers' so the contribution of the Helldiver was 'outmoded', not a *single* Japanese heavy ship, from carrier and battleship down to heavy cruiser, was so much as hit, let alone sunk, by any 'heavy four-engine bomber' during the entire Pacific War! The '2C', by contrast, was to go on from the battles of June 1944 to even greater achievements.

CHAPTER FOURTEEN

Leyte Gulf

This was the greatest naval battle of World War II, and it resulted in another very one-sided victory for the American Navy over the remnants of the Imperial Japanese Navy. The battle was fought as a desperate gamble on the Japanese side, after the US invasion of the Philippines with the landing at Leyte Gulf on 20 October 1944. The fall of the Philippines would cut Japan off from her oil supplies, for which she had gone to war, and the reasoning was that, without those supplies the Navy could not operate: therefore almost the whole of the Navy was thrown into the battle, as they had nothing to lose.

The C-in-C of the combined fleet, Admiral Toyoda, therefore set in train a pre-arranged plan by which four separate striking forces would converge on Leyte and inflict a crushing defeat on the protecting warships, destroy the transport ships on which the army depended, and wipe out the invading force. This was to be done in combination with all-out air attacks from the numerous air bases there. Part of the operation was an elaborate decoy, by which the surviving carriers of the Japanese fleet, themselves almost impotent as they lacked all but a skeleton air complement after the Philippine Sea slaughter, would be used to lure the massive American task force northward, thus leaving the way open for the battleships and cruisers of the other forces to penetrate through to the beachhead and decimate all American shipping concentrated there.

The plan *almost* worked in some respects, but was really doomed from the start considering the enormous American resources available to meet it. Exactly as earlier, pre-emptive strikes by the Helldivers and Avengers of the task groups had decimated the land-based Japanese air forces so that their contribution was far less than expected. They soon began to revert to the suicide tactics of the kamikaze, as conventional air attacks just could not make any impression on the US fleet's defences. The SB2Cs were heavily

SB2C Units at the Battle of Leyte Gulf, October 1944				
Unit	Carrier	Commander	SB2C type	No.
VB-7	Hancock (CV-19)	Lt Cdr J. L. Erickson	SB2C-3	30
			SB2C-3E	12
VB-8	Bunker Hill (CV-17)	Lt Cdr J. D. Arbes	SB2C-1C	17
			SBF-1	3
			SBW-1	4
VB-11	Hornet (CV-12)	Lt Cdr L. A. Smith	SB2C-3	25
VB-13	Franklin (CV-13)	Lt Cdr A. Skinner	SB2C-3	31
VB-14	Wasp (CV18)	Lt Cdr J. D. Blitch	SB2C-1C	25
			SBW-31	
VB-15	Essex (CV-9)	Lt Cdr J. H. Mini	SB2C-3	25
VB-18	Intrepid (CV-11)	Lt Cdr M. Elsick	SB2C-3	28
VB-19	Lexington (CV-16)	Lt Cdr R. McGowan	SB2C-3	30
VB-20	Enterprise (CV-6)	Cdr R. E. Riera	SB2C-3	34

involved in attacks on the various surface fleets as they moved east, and it was thought they had turned them back for good, with heavy losses, including the super-battleship *Musashi*.

The lure of the Japanese carriers, albeit with almost empty flight decks, was eagerly taken by Halsey who sped off in pursuit, and the Helldiver helped take final revenge for Pearl Harbor by sinking and heavily damaging a number of the few remaining Imperial flat-tops. The southern Japanese battleship forces were intercepted by their American opposite numbers, and bloody close-range fighting took place in which they were almost totally wiped out. But the most powerful battleship force, under Admiral Kurita, emerged from the Sulu Sea to find their way open and clear to the beaches. They were timidly handled and frustrated by a small force of escort carriers and destroyers who kept them at bay long enough for the big carriers to come back and redress the situation.

The final part of this battle was the hunting down of the Japanese survivors by the Helldivers and the various sorties this

involved them in. We cannot here cover every aspect of the SB2C's involvement in this huge and wide-ranging air battle – instead we will concentrate on an example unit's participation in each of the four facets of it; the preliminary air strikes on the Philippines; the early strikes against the Japanese fleet; the pursuit of the Japanese carriers; and the final mopping-up of the surviving squadrons as they struggled back to safety.

Preliminary Air Strikes on the Philippines

During a three-day period between 12 and 14 September, the carrier forces of the US fleet struck hard at the hitherto inviolate Japanese airfields in the Philippines to test their mettle. It was expected that strong and efficient aerial opposition would be encountered, and heavy losses were expected. Instead, the carrier air groups swept all before them, and from their results, and the reports of American airmen who were shot down and found refuge with the Filipino guerrillas in the mountains, the weakness of the Japanese defences was revealed, and as a result, the date of the actual invasion was brought forward by many weeks.

Typical of the Helldivers' work in this period was that carried out by VB-18, commanded by Lieutenant Commander M. Elsick, embarked aboard the *Intrepid* (CV-11) as part of her Air Group18.[136] On 13 September 1944, at 06:10, VB-18 launched twelve SB2C-3s each armed with one 1,000lb GP and two 250lb GP bombs. They were joined by six TBM-1C

due to lack of suitable targets there, and while returning to base, they sighted another airfield, believed to be Cadiz. On its runway and in nearby revetments they counted between twenty and twenty-three Japanese aircraft. Those Helldivers which still had their bombs selected these grounded planes as their targets, and destroyed two by dive-bombing, seriously damaging two more. They subsequently made numerous strafing runs which resulted in the further destruction of another pair, the probable destruction of another three, and serious damage to six Japanese

hole in the port wing; but none was seriously hit, and the defending Hellcats claimed to have destroyed at least eight of the enemy, plus a probable. The only casualty in the SB2C-3s was one which took a light flak hit; this punched a hole in her tail and injured rear seat man ARM2C F. P. Crevoisier, with shrapnel hits at the base of the first and second fingers of his right hand, grounding him for one day!

Similar strikes followed this first attack on Negros, with an anti-shipping sortie against small vessels which were being used as troop transports to reinforce their garri-

Curtiss SB2C-3 on 16 October, 1944. National Museum of Naval Aviation, Pensacola

Avengers also armed with bombs, and escorted by eight F6F-5 Hellcat fighters. Their instructions were to 'Destroy aircraft and installations, northern Negros, Philippine Islands'. They followed Air Group 8 to the target area, climbing to 14,000ft (4,260m) on the way. Once over the target area the SB2C-3s dived on an airstrip '… believed to be Silay, obtaining enough hits to make the runway temporarily inoperative'. Five of the Helldivers did not release

aircraft there. The enemy were really caught flat-footed in this attack. One SB2C-3 which had become separated from the rest of the strike, joined VT-18's Avengers in an attack on Alicante airfield, cratering the runway.

Enemy fighters – Oscars, Zekes, Hamps, Tojos and Petes – swarmed up in reply and two of the Helldivers were hit by 7.7mm machine-gun fire, one ending up with a hole in her right wing root, the other a

son on Leyte; these were hit in the Janobates Channel. This mission also included attacks on Bacolod, Cadiz and Alicante airfields. The SB2C-3s carried out both dive- and glide-bombing attacks, and claimed to have sunk two of the small ships and damaged four others, and also to have destroyed four two-engined aircraft at Cadiz. In yet further attacks, VB-18 bombed an oil refinery, a warehouse on Mactan Island, Cebu Harbour, and also attacked a 1,000-ton coaster – and it met no air opposition and only meagre AA fire. The lack of enemy response was incredible. It was this weak response that led to the bringing forward of the actual invasion of the Philippines, a development for which the Helldivers can take full credit.

Striking at the Japanese Fleet: First Stage

The USS *Franklin* (CV-13) was one of the new carriers involved in the Leyte

a pair of 500lb SAP bombs with .025 tail fuse settings, and had an escort of sixteen F6F-5 Hellcats. The early morning light was clear with a visibility of 15 miles (24km).[137] The dive-bombers were laden with 436 gallons (1,928 litres) of fuel, for the distance to the target was a long haul, 325 miles (523km) away. The search was divided up into two sectors, 250° to 260° and 260° to 270°. The northern search was led by Lieutenant K. R. Miller, the southern by Lieutenant (j.g.) Eisenhuth.

Lieutenant Miller's search sighted nothing on their outward leg until they reached

do, and led his unit back to them to regain contact and then attack. They relocated the Japanese ships three miles (5km) east of Manigun Island at 08:15.

Once they had closed with the small enemy squadron, they could see that Japanese ships were a light cruiser of the *Tenryu* class and two old destroyers. Commander Kibbe, leading the second hop, also reported the target and assigned two Helldivers to each of the three ships; but one of the pilots failed to pick up the orders, so the actual attacks were made as three SB2C-3s against one destroyer. Two

An SB2C-3 of VB-7 operating from the Hancock **(CV-19) off Formosa in 1944.** US Navy via National Museum of Naval Aviation, Pensacola

Gulf battle between 24 and 28 October 1944, and her Air Group 13, which included VB-13 equipped with SB2C-3s, gave a good account of themselves throughout. Their experiences can be taken as typical. Thus at 06:12, in latitude 11° 30' North, 126° 30' East, a first strike was despatched to search for and attack units of the Japanese fleet located in the Visayan and Sibuyan seas. The twelve Helldivers were each armed with

the east coast of Panay, where they spotted three enemy warships, heading south into the Sulu Sea. They were taken to be three old-type Japanese destroyers, and a contact report was immediately sent back to the carrier. Ignoring these warships in the hope of flushing out much larger game, the Helldivers continued to the end of the outward leg, but nothing else of importance was sighted. Miller therefore decided that the first sightings would have to

pilots, Lieutenant Harding and Lieutenant (j.g.) Pickens, claimed direct hits, and the third pilot, Lieutenant (j.g.) Pingrey, a near miss, and they also claimed the destroyer as sunk – certainly after the attack it was practically dead in the water, smoking extensively from the bridge aft. Although the ship was not sinking as the Helldivers departed the target, by the time the next strike arrived there remained only two ships of the group at the same position and a large oil slick; so the destroyer had apparently sunk in the interim. In fact their victim had been the *Wakaba*. Two other Helldivers made their dives against the cruiser, claiming one direct hit and one near miss, and causing major damage; and the final attack was made by a sin-

Meanwhile the southern section found no targets, but they did run into a Japanese flight of twelve Aichi D3A1 'Val' dive-bombers, escorted by five Nakajima Ki-43 Army type 1 'Oscar' fighters, apparently on their way to dive-bomb the American beach-head at Leyte. One division of Hell-

either to protect their own dive-bombers or to attack the American ones, however. After the first division had rejoined, the whole unit continued their flight and the search, but no worthwhile shipping was seen. On their return leg the Helldivers carried out attacks on La Carlotta Field on

Hunting the Musashi, **Curtiss SB2C-1C Helldivers 62 and 70 from the USS** Hancock (CV-19) **on patrol in the Central Pacific theatre in 1944. The Yagi radar antenna below the wing can clearly be seen on the nearest aircraft.** US National Archives, College Park

gle dive-bomber against the second destroyer, but without success. No damage was taken by any American aircraft, although return AA fire was described as 'accurate' and they all returned without further incident.

cats broke formation to deal with these, shooting down three 'Vals' and forcing the rest to jettison their bombs and scatter, while the Helldivers circled at 9,000ft (2,700m) covered by the other fighter division. The 'Oscars' made no attempt

Negros Island, but no important installations were observed.

The next strike to take off from the *Franklin* was of eleven SB2C-3s, again armed with a pair of 500lb SAP bombs apiece, escorted by twelve F6F-5 fighters at

US Navy Curtiss Helldivers attacking the Japanese battleship Musashi *(left)* with a destroyer of her screen *(right)* during the Battle of Leyte Gulf, 24 October 1944. US National Archives, College Park

(Below) Another Helldiver's eye-view of the giant Japanese battleship Musashi and an escorting destroyer under heavy dive-bombing attack during the Battle of Leyte Gulf. Hit many times by both bombs and torpedoes, the great vessel finally sank later that same day. Note the Helldiver on its way down to attack in the bottom left-hand corner of the photo. US National Archives, College Park

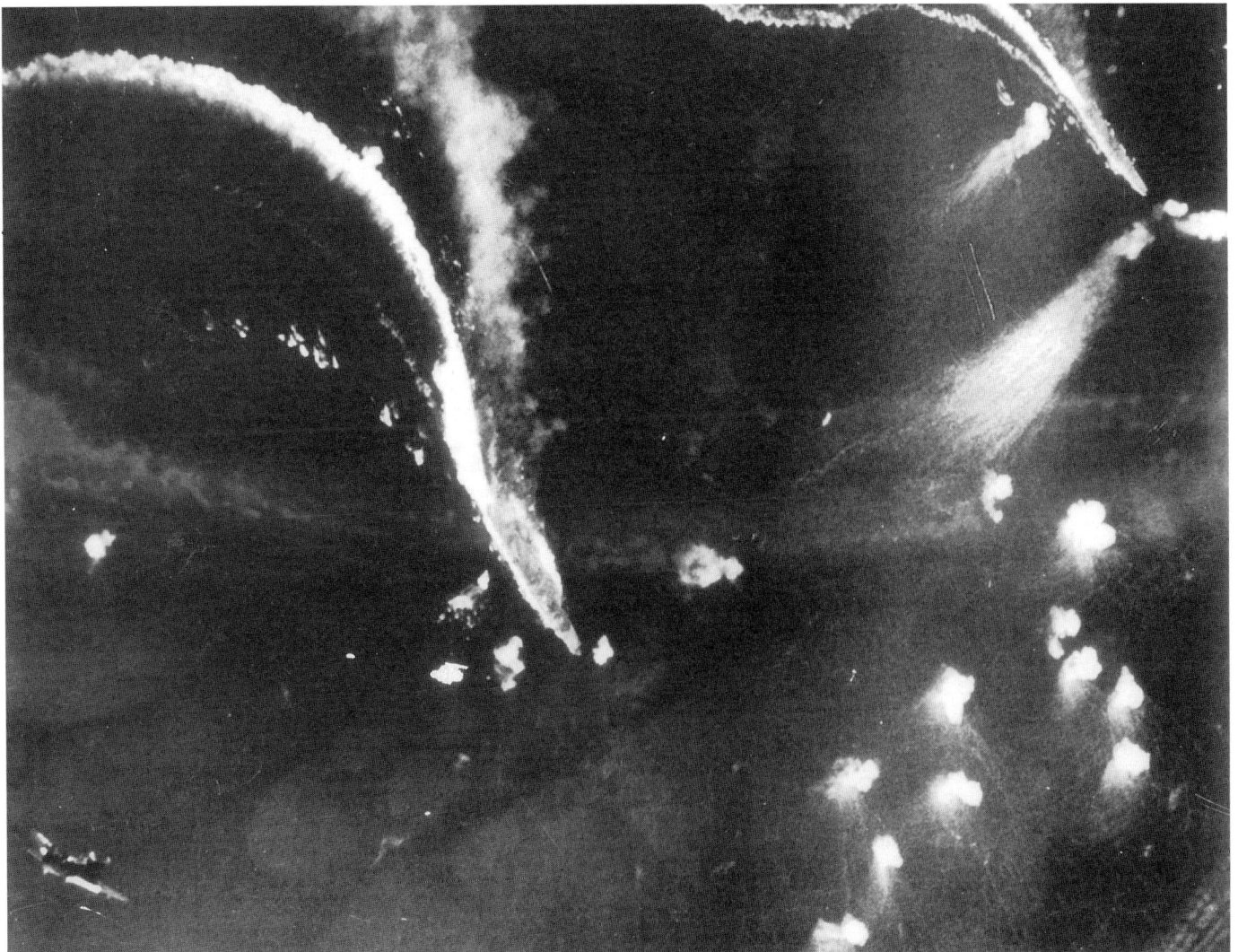

09:51. This was in response to the first group's attack reports. Their targets were, '... enemy war vessels east of Panay Island', at a range of about 240 miles (390km), but they found the light cruiser and the surviving destroyer in the same place, three miles east of Manigun Island. They were circling closely, apparently picking up survivors of the sunken vessel.

The Helldivers immediately went into the attack from 11,000ft (3,350m) altitude, roughly dividing the ships between the divisions. The Japanese warships, '... went into very violent manoeuvres, turning to the left in tight circles, and were very difficult to hit, besides throwing up accurate AA.'

Lieutenant (j.g.) H. D. Barnett, flying as No. 2 on the flight leader, Lieutenant C. A. Skinner, came down in his dive closely following the leader against the light cruiser: he was closely followed in turn by the no. 3 man, Lieutenant (j.g.) V. L. Miller. Miller saw Barnett's SBC-3 start to pull out at 1,000ft, (300m) after releasing his bombs, but when it was almost level it went into a slow roll, which it repeated – and then it dived straight into the sea on its back, at probably little better than 300 knots. There were no survivors, neither Barnett nor his ARM3e, Leonard Pickens, standing a chance. 'The plane was obviously hit by AA, although no fire or exterior damage was apparent.'

In compensation, one of Barnett's 500lb bombs smacked into the target, the light cruiser *Abukuma*, which was already burning slightly from the previous attack. Lieutenant (j.g.) Bogan scored a near miss off the same ship, which was confirmed by the following pilots: but although the fires increased in the cruiser as a result, she was not in a sinking condition when they left her. The destroyer was near-missed by both Lieutenant Wood, whose bomb detonated within five feet of her fantail, and Lieutenant (j.g.) Kehoe, off her port beam; but she was obviously not seriously damaged by these as she was still under way when the group left the scene.

Hunting down Ozawa's Carriers

On 25 October the *Franklin*'s air group was again in heavy action, against the carriers of Ozawa's decoy force. The first wave consisted of fifteen SB2C-3s which were sent off from VB-13, along with six TBM-1C

Avengers of VT-13 bombers, and twelve F6F-5 Hellcats of VF-13, all of which departed at 11:55. The Helldivers were carrying mixed bomb loads against the variety of Japanese warships likely to be encountered: nine had a single 1,000lb SAP with a .025 tail fuse; the other six carried twenty 1,000lb SAP each, with .08 tail fuses. They shared the strike with the Helldivers of Air Group 20 from the *Enterprise*, and fighters and torpedo bombers from the light carriers *Belleau Wood* and *San Jacinto*. Their orders read: 'Destroy units of Jap Fleet in Philippine Sea, East of Luzon. Good Luck.' The strike leader was Commander R. L. Kibbs, commander Air Group 15.

On reaching the Japanese fleet, Kibbs also acted as air co-ordinator: escorted by three Hellcats of VF-13, he took station above the fleet, and for five hours directed incoming flights of bombers and torpedo planes and assigned them their targets, as well as making diversionary strafing runs ahead of the dive-bomber attacks. All of VB-13's Helldivers attacked the northernmost carrier, (see opposite) which they thought was of the *Chitose* class, and they claimed a total of eight hits and six near misses on this unfortunate vessel, leaving her seriously damaged. The hits itemized were as follows:

1: by Lieutenant Bomberger – amidships, aft.
2: Lieutenant (j.g.) Horn – starboard, stern.
3: Lieutenant Borts – starboard, stern.
4: Lieutenant (j.g.) Young – port, amidships.
5: Lieutenant (j.g.) Luther – forward, starboard.
6: Lieutenant (j.g.) Bisonhuth – amidships.
7: Lieutenant (j.g.) Carlen – starboard, amidships.
8: Lieutenant (j.g.) McPhie – starboard, stern.

All of these hits were confirmed by at least three other pilots and aircrewmen, and by photographs taken at the time.

In one group of very accurate dive-bombing, five bombs were seen to fall within a 100ft (30m) diameter circle, which included about half the width of the flight deck. Two to three bombs hit the port side of the flight deck, and the others were very near misses. The hits and explosions heeled the enemy carrier sharply to starboard and swung her bow to the east.

The near misses, all within fifty feet, were made by the following pilots:

1: Lieutenant (j.g.) Einling – off starboard bow.
2: Lieutenant (j.g.) Berry – off starboard bow.
3: Lieutenant (j.g.) Bevan – off port beam.
4: Lieutenant (j.g.) Fellner – off port beam.
5: Lieutenant (j.g.) Moyers – off stern, to port.
6: Lieutenant (j.g.) Allen – off stern, to starboard.

The Helldiver crew reported that this carrier appeared completely undamaged at the time they attacked her, and, '... was manoeuvring violently at high speed'. Their victim began to slow down during the course of their assault as bomb after bomb crashed into her, and when the surviving SB2C-3s left the area she was, '... burning furiously, and starting to settle by the stern; her speed was very much reduced'.

Lieutenant John H. Finrow, the Helldiver's flight leader, was reported 'missing in action' during this attack. His wingman reported that, at the commencement of his dive, at about 14,000ft (4,260m), Finrow had not opened his bomb-bay doors. He appeared to be having difficulties and he pulled out very high, at about 5,000ft (1,500m), and was lost to the sight of both wingmen as they continued on down in their dives to the normal release altitude. He was seen to pull out in the general direction of the rally point, but was never seen again. A report of a bomber making a water landing led to intensive searches by air and surface forces, but neither Finrow nor his gunner, ARM1c H. E. Borja, were picked up. Another casualty was Lieutenant (j.g.) D. A. McPhie, the last SB2C-3 to dive. While well into the dive attack – he had just released his bombs at 2,000ft (600m) – his aircraft took a hit in the port wing; this blew away his flaps and ailerons and he never regained control or pulled out. The CAC in the aircraft ahead of him, as well as Avenger pilots astern of him, saw the Helldiver go straight in, with no survivors: his rear-seat man was ARM2c R. D. Chandler.

The second attack was mounted at 08:02 and VB-13's contribution was six SB2C-3s, three of which carried a single 1,000lb SAP, the other three two 1,000lb AP bombs; and they were accompanied by three TBM-1C

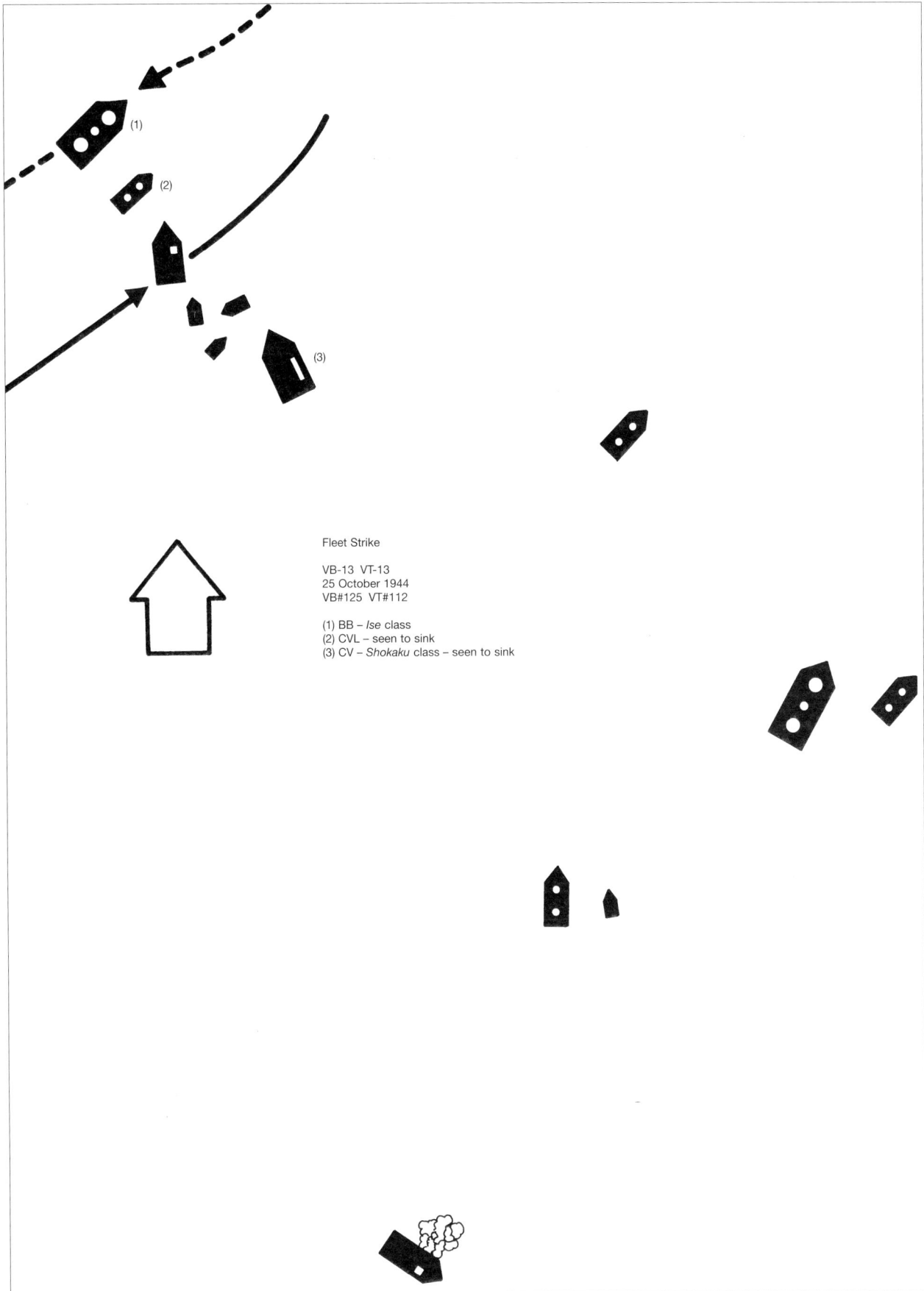

Fleet Strike

VB-13 VT-13
25 October 1944
VB#125 VT#112

(1) BB – *Ise* class
(2) CVL – seen to sink
(3) CV – *Shokaku* class – seen to sink

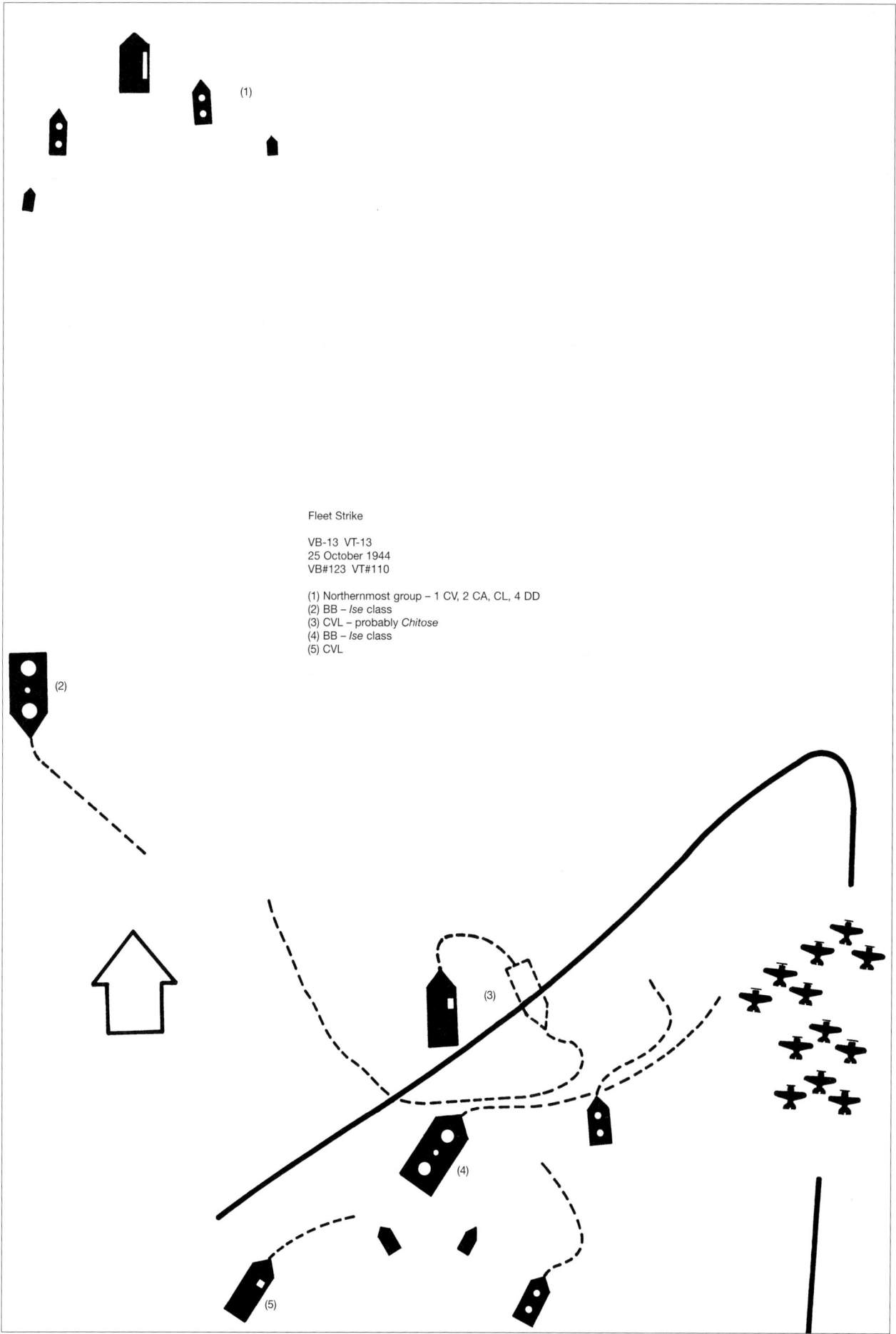

Fleet Strike

VB-13 VT-13
25 October 1944
VB#123 VT#110

(1) Northernmost group – 1 CV, 2 CA, CL, 4 DD
(2) BB – *Ise* class
(3) CVL – probably *Chitose*
(4) BB – *Ise* class
(5) CVL

(1)

(2)

(3)

(4)

(5)

Avenger's also armed with 1,000 lb SAP bombs to act as glide bombers. They launched at 08:02 and took departure at 09:05, with Lieutenant Commander L. C. French, the CO of VT-13, as strike leader. This was just one of many groups on the second strike against the Japanese fleet east of Luzon, VB-13 being joined by aircraft from the *Enterprise, San Jacinto* and *Belleau Wood* for the attack. The target range was put at 135 miles (217km) for the outer leg.

They reached the Japanese fleet at 10:05. The original target given by the target co-ordinator was one of the battleships of the force, but he switched them to a carrier target which was apparently undam-

Lieutenant K. R. Miller, dived to the south-west, pulled out straight ahead and rallied 15 miles (24km) south-west of the target. The SB2C-3s then approached the target from south to north. The pushover was made from 10,000ft (3,000m) at close intervals between aircraft – two to five second gaps only – and because of the cloud cover which, '… hampered accuracy …', a glide approach angle of 45 degrees was adopted at 300 knots speed, with bomb release made at 3,000ft (900m) and pullout at 2,500ft (760m). All the Helldivers made their dives practically across the beam of the ship.

The pilots of VB-13 claimed three direct hits and three near misses on the carrier in

The two hits scored on the stern of the ship were followed by, '… heavy explosions and thick black smoke'. The carrier was also heavily damaged by the near misses, all of which were within fifty feet of the ship; these were delivered by Lieutenant (j.g.) Harding, Lieutenant (j.g.) Heizer and Lieutenant (j.g.) Barksdale.

The rally afterwards was effected very quickly by a running rendezvous. As the SB2C-3s pulled out of their dives they were half-heartedly attacked by three enemy 'Oscar' fighters – these made only one ineffectual pass and scored no hits on any dive-bomber. As the group took departure at 10:15, the carrier was '… burning

The Japanese fleet caught by VB-14 off the east coast of Samar Island, with the battleship Yamato *(centre)*, **a** Kongo**-class battleship on either flank and a surrounding screen of destroyers at the Battle of Leyte Gulf, 26 October 1944.** National Archives, Washington DC

aged and speeding along at 25 to 30 knots. The target carrier, thought to be the *Chitose*, was heading north-west; she was about 50 per cent covered by clouds to the leader, and completely covered to the second division at the angle from which they were diving. She made a turn to the left at the start of the attack, then swung to the right (*see* opposite).

As she turned hard to starboard to commence a tight evasive circle, she and her accompanying ships threw up a wall of very accurate flak. The Helldiver flight leader,

this strike, and these were confirmed by at least three pilots and gunners in each case and by photographs taken by the lead aircraft on the pullout. These hits assisted in the subsequent sinking of the carrier. Each pilot hit with one or two 1,000lb bombs, as follows:

1: Lieutenant K. R. Miller – about 150 feet from the stern, starboard.
2: Lieutenant (j.g.) Riley – just aft of Miller's hit.
3: Lieutenant (j.g.) Cole – forward of amidships.

badly, mainly aft, and only a small wake was visible'. At 11:30 a report was received from another group that their target was dead in the water and its sinking was confirmed later.

A third and fourth strike followed. The fourth and last of the day saw six SB2C-3s of VB-13, with four Avengers and eight Hellcats, led by Commander W. N. Coleman, the C.O. of VF-13; they were launched at 13:28 and they joined with the *Enterprise* group before taking departure. They met no enemy fighters *en route*, and outside of AA range, the Helldivers circled while awaiting their targeting instructions from the co-ordinator. They were eventually directed against a light carrier damaged in earlier attacks and were split into two sections to conduct their approach. The carrier was making very little way at the time of the attack, and was

but '... intense fire was received by the nearby BB [battleship] and CA [heavy cruiser]' (*see* opposite).

The second section of VB-13, led by Lieutenant (j.g.) Kehoe, made the first dive on the sinking carrier '... in compliance with instructions from air co-ordinator', while the first section, led by Lieutenant

2: Lieutenant (j.g.) Howard – off the port beam
3: Lieutenant (j.g.) Garrett – off the starboard quarter.

One of the AP bombs was a very near miss, '... but due to the type, target damage was probably inconsequential'.

VB-14 closes in on the Japanese fleet off Samar Island. Note the heavy cloud which made attack conditions for dive-bombing difficult. National Archives, Washington DC

dead in the water by the time the second section made their dives, already settling badly by the stern with smoke issuing from her after hull. Most of the fires started by the previous strike had been extinguished above decks, '... but the ship was obviously in a sinking condition and was being abandoned'.

The Japanese ships put up what was described as, '... intense and accurate ...' defensive fire. 'It was reported that a Jap battleship fired what appeared to be a main battery salvo at retiring planes, which burst with great accuracy at 5,000 to 6,000 feet.' None of the defending fire came from the stricken carrier, and she was incapable of evasive action – a mere sitting duck in fact,

Wood, made their dives fifteen minutes later, after awaiting orders. Between dives the second section circled at the rally point, and upon joining up with the first section, proceeded back to the *Franklin* without further incident.

The Helldivers claimed one direct hit and three near misses in this attack, the direct hit being made by Lieutenant (j.g.) V. L. Miller, whose bomb hit the starboard side, amidships, below the carrier's flight deck, on the waterline; it was confirmed by two other pilots and photographs. The near misses were scored as follows:

1: Lieutenant (j.g.) Kehoe – off the port bow.

Taking on Kurita's Heavyweights

Among the many other Helldiver units in action on 25 October in this crushing American victory, was VB-14, led by Commander J. D. Blitch, and operating from the *Wasp* (CV-18).[138] The enemy carriers were known to be somewhere northeast of Luzon and were found by two squadron pilots at 09:30, about 350 miles (560km) north-west of the *Wasp*; they were already under attack by other carrier air groups at that time. However, a deck-load strike was standing by to hit the carrier force nevertheless, and the American

Fleet Strike

VB-13 VT-13
25 October 1944
VB#124 VT#111

(1) CVL
(2) CV – *Shokaku* class
(3) BB – *Ise* class
(4) BB – *Ise* class and CA
(5) CVL – burning

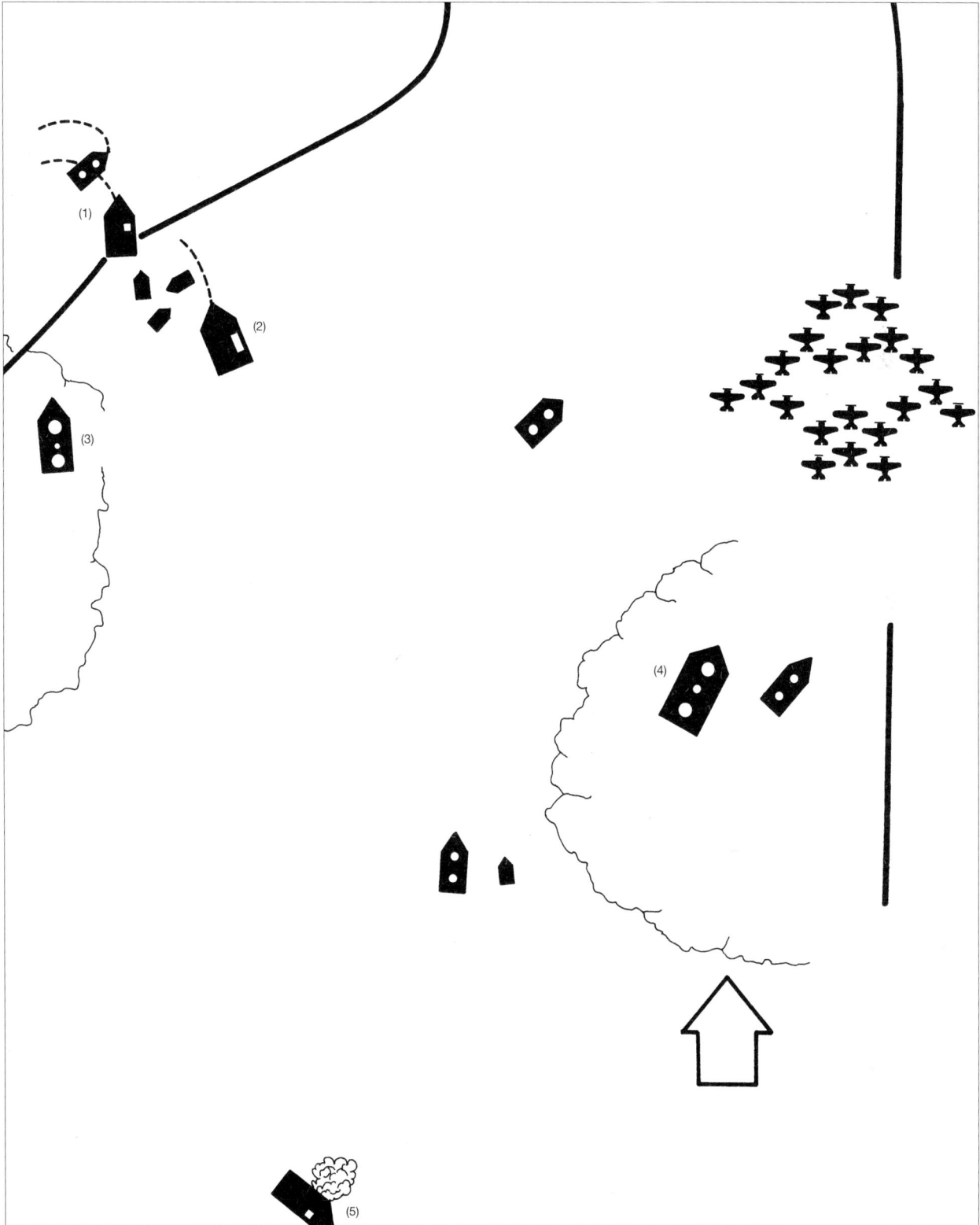

carrier force was steaming flat out on a north-westerly course in full chase. The alarming news that a powerful Japanese battleship and cruiser force had burst out of the unguarded San Bernardino Strait behind them and was attacking the few escort carrier groups off Samar immediately saw the abrupt abandonment of any further plans to hit the fleeing enemy carriers. Halsey had fallen for Kurita's sucker punch and urgent steps had to be taken to rectify a desperate situation.

Orders were therefore suddenly received for Task Group 38.1 to reverse course to the south-west and the deck-load strike was immediately sent off against this very different type of target; they did not have time to change their bomb loads, which had been set up for attacks on the wooden decks of the enemy carriers and were not suitable for penetrating the thickly armoured decks of the enemy battleships.

Two four-plane sections of SB2C-3s were launched between 10:35 and 10:41 that morning, led by Commander Blitch, four armed with one 1,000lb GP and one 250lb GP bombs, the other four with one 1,000lb SAP and one 250lb GP bombs. The reason only one wing bomb was carried was so that they could have an extra wing fuel tank fitted, for the target was over 300 miles (480km) distant at the time of the launch. Even so, for many, as at the Philippine Sea battle, this was to prove insufficient for the round trip. They were joined by six TBM Avengers and escorted by fifteen F6F Hellcats, and their orders were quite simple, 'Strike Japanese Fleet.'

The weather to the target zone was poor, and the group was forced to make several detours to bypass storms on the way. They came upon the Japanese fleet at 13:20, in approximate position 11°20 N, 125°50 E, off the east coast of Samar Island, heading north at about fifteen knots; for Kurita had already, incredibly, broken off his attack with victory in his grasp, and was retiring. At 13:30, when still some twelve miles north-west of the target, the group was intercepted by five or six Japanese fighters, Zekes or Oscars; but the Hellcats scared them away and the dive-bombers were not troubled by them.

The Japanese ships were steering north at fifteen knots and consisted of four battleships ('Yamato, two Kongo, and Nagato'), six heavy and one light cruiser and thirteen destroyers, the battleships being in staggered column with the giant Yamato second in line, with the cruisers forming an inner screen and the destroyers an outer screen around them. The Helldivers orbited this awesome mass of fighting vessels at a respectful ten-mile distance, looking them over while their targets were assigned to them. One five-plane division was told to hit the Yamato, while the other three were assigned the Kongo class ship on the port side of the column, to divide

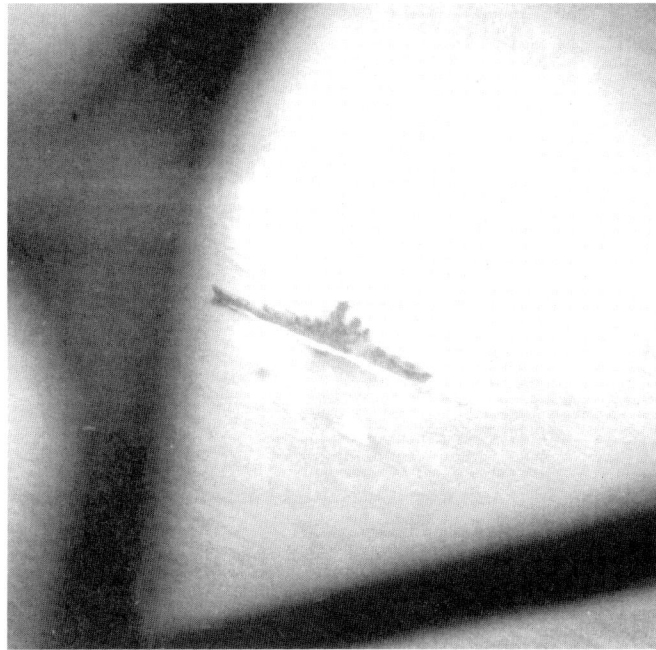

Seen from the rear cockpit of one of VB-14's Helldivers, heavy bombs explode astern and off the port side of the Yamato during the Battle of Leyte Gulf, 26 October 1944. National Archives, Washington DC

enemy fire. Meanwhile the Japanese ships turned 90 degrees onto a heading of 090 degrees. As the Helldivers made their initial approach, passing ahead of the now eastbound formation, the ships made another turn to starboard which continued for a full 270 degrees, so that by the time the dive-bombers rolled over from the south-west the fleet was again heading due north.

VB-14 made their approach from the north-east, circled round to the south side of the screen, then turned back in a 180-degree turn to roll over from 11,000ft

(3,350m) and dive generally to the northeast. This up-wind dive was necessary because the enemy fleet lay partly below an extensive cloud layer which covered the forward, starboard part of it. As the Helldivers approached, 'intense and accurate heavy AA was opened, mostly from cruisers, as planes passed east of the force. Augmented by medium and light weapons, it continued at extreme range during retirement.' Three SB2C-3s were hit by this, one being shot down and two of them severely damaged.

Also, as the Helldivers commenced their dive attacks, the enemy ships were still partially under cloud, and the rearmost aircraft of the attack, '... were unable to get as steep as they wanted to without losing sight of the target'. As it was, two planes were seriously handicapped by passing through cloud just before release point. By the time that Lieutenant (j.g.) Smith, last man in the second division, had pushed over, the Kongo class battleship which was his target was under cloud; so he dived instead on a heavy cruiser (wrongly identified as the Nachi) off to port which was in the clear, and his gunner claimed to have hit with at least one bomb amidships on this ship, causing a large fire. Two other pilots confirmed this hit and the subsequent blaze, and she was thought to have been seriously damaged.

Of those that attacked the Yamato, only one slight hit, or a very near miss, was claimed, Lieutenant (j.g.) Laz's bomb reported detonating close off the port side of the battleship's fantail. One pilot failed to release his main bomb at all, and Commander Blitch's bomb exploded in much the same place. Both Lieutenant Lewis and Lieutenant (j.g.) Doane missed short by 50 to 75ft (15 to 23m).

The Helldivers '... pulled out in heavy AA fire as they retired to the rendezvous area north-east of the fleet.' The dive-bombers had become quite well spread out because of having to jink in order to avoid flying over ships of the screen, although this could not always be avoided. The eight-tenths overcast which had so hindered the bombing now afforded some much-needed respite and protection as the Helldivers retired from the target area. 'The necessity for radical evasive manoeuvres and use of clouds prevented more than fragmentary observation of the results of the attack'. Two planes (Blitch and Doane), found destroyers lying across their retirement path, and strafed them vigorously, though Lieutenant (j.g.) Doane's

Berg's Helldiver, repeatedly burst into flames. 'About once every ten minutes a large blaze occurred, which would subside shortly afterwards. While it was a nerve-wracking business, he managed to make it back and get aboard safely.' Another testimony to both the skill of the pilot and the strength of the SB2C.

The next day VB-14 joined in a further heavy strike directed against the remnants of Kurita's force, which was now thought

meant an outward leg of 310 miles (500km) and a homeward journey of 290 miles (470km). They were led by Lieutenant Commander A. L. Downing, who also acted as strike leader for all the aircraft from McCain's task group. But almost immediately Lieutenant (j.g.) Skaggs had to land back aboard because of trouble with his propeller.

Downing set off for the estimated position of the enemy fleet and commenced a

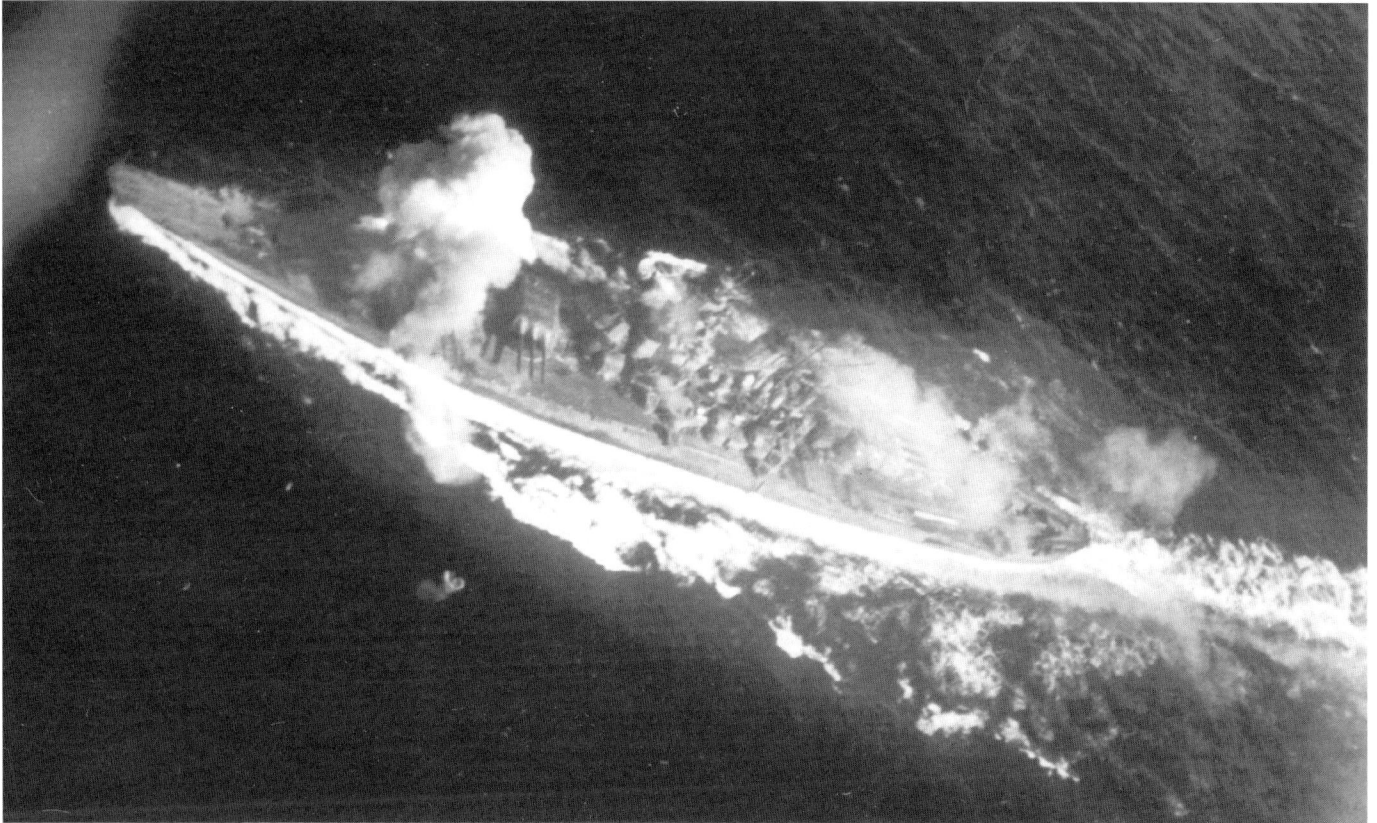

A direct hit scored on the fore turret of Yamato by Helldivers during the Battle of Leyte Gulf.
National Archives, Washington DC

Helldiver was hit by the destroyer he was strafing. Lieutenant H. J. Welker, flying Bu No. 18962, had led the second section of the first division against the *Yamato*, and was neither seen nor definitely heard from, after the attack; he was reported as probably shot down with the loss of both the pilot and his gunner, ARM2c N. A. Iorio.

The distance back, although only 280 miles (450km) due to the hard steaming of their task group to meet them, also took its inevitable toll of the attackers; for instance the damaged engine of Lieutenant (j.g.)

to be retiring west through the San Bernardino Straits. The search planes were already out looking for them when the VB-14 aircraft took off, with the enemy believed to be somewhere along the east coast of Mindoro. Four Helldivers lifted off at 06:08 on the 26th: three SB2C-3s all armed with a single 1,000lb SAP in the belly, and one 250lb GP bomb and a fuel tank on the wings; and one SBW-3 with one 1,000lb GP bomb in the bomb bay and a single 250lb bomb and fuel tank as wing-load, for again the strike

search, first covering the passage north of Mindoro. They located two groups of stragglers from the main bunch here, one a light cruiser or destroyer, the other a heavy cruiser.

Although Downing had hopes of finding the main fleet, he was already 300 miles from base, and it was by no means certain that it would be found before many planes in his group had reached their limit of endurance. Accordingly, he detached part of his striking force to hit each of these two ships, as they were located. On the second occasion he detached one division of *Hornet*'s VB12- as part of the attack force. Unfortunately these orders were misunderstood and all the *Hornet* dive-bombers joined in the attack. This left him with only the three *Wasp* Helldivers for the later attack on the main fleet. However, both ships attacked were severely damaged and subsequently sunk.

Their victims were the light cruiser *Noshiro* and the destroyer *Hayashimo*.

Soon after the second cruiser contact had been made, the search planes finally located the main body of Kurita's force, and their VHF reports were heard. The enemy were passing through the southern end of the Tablas Straits, heading south. Shortly afterwards, as the strike winged its way down the Straits from the north, the Japanese fleet was sighted off the north-west coast of Panay. By then it comprised three battleships, four heavy and one light cruiser, with seven to nine destroyers, steering 180 degrees at fifteen knots.

VB-14 passed west of this force on a parallel course, while Downing assigned targets to the various strike units. He then led VB-14 across and ahead of the destroyer screen, to assume a position on the enemy's port quarter at 12,000ft (3,650m) upwind and in the sun. As the three Helldivers crossed the screen, a heavy AA fire was opened on them, '... including at least one burst of white phosphorous, apparently containing shiny metal shrapnel of large size'. After they got up into the sun, however, this fire ceased.

The entire remaining strength of the strike force was assigned to attack the *Yamato*: as well as the three VB-14 Helldivers, there were eight Avenger torpedo bombers from *Wasp* and five from the *Cowpens* (CVL-25), with four escorting Hellcats which strafed the destroyer screen to facilitate their approach. The three Helldivers then rolled over at 11,000ft (3,350m) from a true bearing of about 060 degrees and obviously managed to achieve an element of surprise, as anti-aircraft fire was not re-opened on them until they were well into their dives, and the enemy warships did not turn to starboard until the attack was well under way. This gave the three Helldivers excellent bombing conditions, and they made a classic dive-bombing attack, diving almost directly down-wind and down the full length of the giant battleship. Their dive angles were steep, 75 degrees, and the results were, '... excellent'.

View of the tailplane of a VB-14 Helldiver, as near-miss bombs explode hard alongside *Yamato*.
National Archives, Washington DC

The anti-aircraft fire, if late in coming, was soon '... intense from all weapons. The *Yamato*'s secondary battery of 8in guns is apparently dual-purpose, as it was definitely firing, although it may have been directed against torpedo planes.'

Downing himself released at 1,800ft (550m), Ohm at 2,500ft (760m). Ensign Reardon continued his dive down to 1,500ft (460m) before release and made sure of a direct hit. He paid for this, '... with a large hole in his port wing, where the aileron is hinged, probably caused by 20mm AA'. Although it caused him some anxious moments during pullout, he got away successfully and returned safely. Carver, Downing's aircrewman, obtained, '... a remarkable ...' photograph of his pilot's direct hits forward, and another during retirement showing Reardon's hit. 'It is quite possible that Ohm, second man to dive, also hit, but there is no confirmation beyond the opinion of two VT pilots that 'all three VBs got hits.'

Retirement was made to the south-west, using radical evasive manoeuvres. As the last American aircraft left the area, the Japanese fleet, which had turned about 90 degrees to starboard, continued round to a 180 degree heading in a wide turn. The

Yamato was smoking from the hits she had taken, but her speed was not reduced.

After the join-up all aircraft set course for base – but they were to get one extra bonus from this outstanding strike. As they approached the southern tip of Burias Island, a lone Jake was sighted at 2,000ft (600m), about three miles (5km) off the beach at Nabasagan. When first sighted, the Jake was off Downing's port bow, and slightly below. Downing turned his Helldiver, overtook the dreaming enemy aircraft rapidly, at 190 IAS, and fired just thirty rounds of 20mm from an initial range of 300 yards. These hit in the wing root and cockpit of the target, and the Jake burst into flames and went down as its most surprised pilot bailed out. Ensign Reardon's rear seat man, ARM3c Landino, took a whole series of photographs of this event, another first for the Helldiver!

They all finally made it back aboard the *Wasp*, although they had cut it fine, and Reardon's aircraft, which could get no suction in his wing tank, landed with just 5 gallons (23litres) of fuel left in his tank!

Final Pursuit

Another strike was launched by the *Wasp* at 08:29, and VB-14, led by Lieutenant Walls, contributed eight SB2C-3s to it, six armed with one 1,000lb AP bomb and the other two with one 1,000lb GP bomb, each having a wing-load of one 250lb GP and a fuel tank. They were accompanied by six TBM Avengers and escorted by eight F6F-5 Hellcats. They were despatched to the western Visayans, confident of finding a target among the, '... many dispersed, fleeing or crippled enemy units ...' that had been sighted in that area throughout the early morning.

As the flight crossed the open water between Burias and Mindoro Islands, they were engaged by an unidentified ship, '... probably a destroyer ...'; but '... seeking bigger game ...' they ignored it, turning south-west and reached the vicinity of the Semirara Islands. Here, some ten miles south-west of Pucio Point on Panay Island, they located a large warship, dead in the water. Some ten miles north of her they sighted another, '... apparently intact ...' light cruiser steaming north-west at 15 to 20 knots. Radio discipline broke down, for while the strike leader was ordering part of his group to attack this vessel, the Avengers, worried about their fuel situation, '... were urging haste, and the necessity of attacking *something* and starting home. The ensuing confusion resulted in a misunderstanding of orders, and all planes of the strike attacked this supposed light cruiser, which pictures show to be a destroyer, probably a *Shiranui*.' (It was indeed this very destroyer.)

The enemy ship was approached from the north-west, the aircraft turning 90 degrees to dive downwind and across her as she proceeded north-west, and she took practically no evasive action whatsoever. Even so, the results of the bombing were. '... disappointing, partly due to the fact

she continued under way at only slightly reduced speed. (She sank the next day.) The return journey was uneventful.

One final strike was made by VB-14 that day, led again by Lieutenant Commander Downing and directed against much the same type of targets in the same area. At 12:35, eight SB2C-3s lifted off, armed with one 1,000lb AP and one 250lb GP wing bomb; together with five TBM Avengers, and top cover of eleven F6F Hellcat fighters, they steered for reported enemy ships in the vicinity of Mindoro Island, almost 300 miles out from the task group. This flight proceeded in the

south-west of Mangsoagui Point on Ilin Island. These ships were assigned to the Helldivers, who commenced to climb to attacking orbit. As they did so, enemy fighters attacked the torpedo bombers; but these were quickly disposed of by the Hellcats, although one F6F was hit by the destroyers' AA fire.

Having reached an altitude of 8,000ft (2,440m) the SB2C-3s levelled off as it was overcast at 8,500ft; they made their initial approach from the west, and the enemy destroyers, in V-formation, altered course from north-west to west. Downing led in until he had passed to the north of the

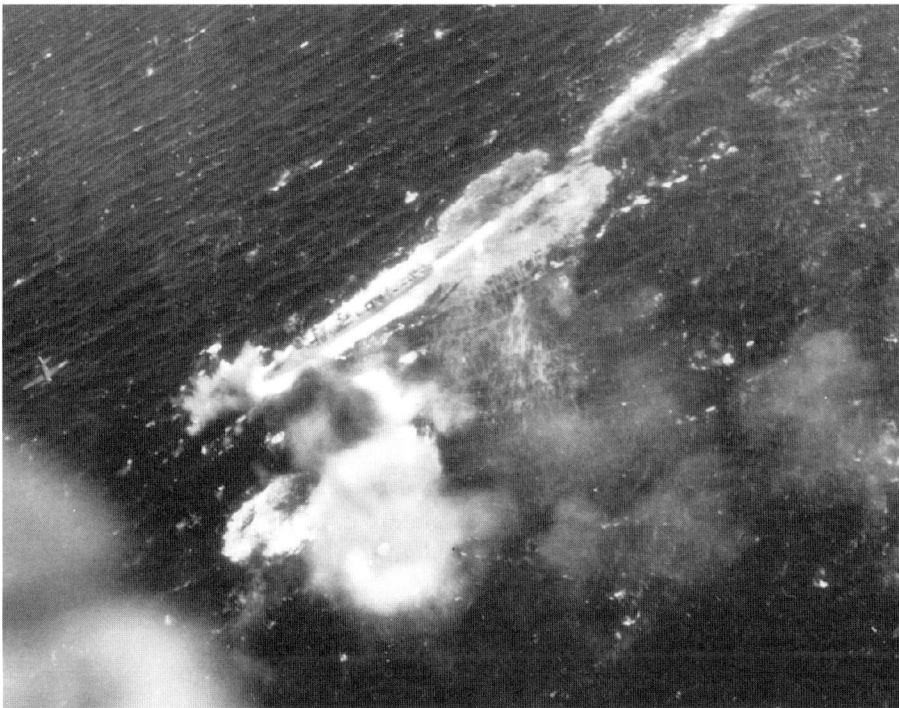

View from a VB-14 Helldiver of a hit on the bow of a Japanese destroyer, with near-miss bombs exploding off the port side and the splash of bombs astern of her wake, during the Battle of Leyte Gulf. National Archives, Washington DC

View astern of a VB-14 Helldiver as a Japanese destroyer takes a near-miss bomb just off her port bow. National Archives, Washington DC

that pilots thought they were attacking a light cruiser'. Three near misses were claimed, off the destroyer's bow, scored by Lieutenant Kane (port side), Lieutenant (j.g.) Forsgren (ahead) and Lieutenant (j.g.) Wisnyi (port side). Lieutenants (j.g.) Starkel and Haggerty made one run on her as she proceeded south through the Tablas Straits, but the results were unobserved. In return she, '... had put up a good volume of AA ...' but no Helldiver was hit. Although the ship's bow was blown off,

normal manner, with the Helldivers climbing to 9,000ft (2,750m) on the way out, though '... being forced by cloud conditions to make several deviations in course and altitude on the way'. They decided to investigate the straits northeast of Mindoro and had to come down to 3,000ft (900m) to make an effective search. Nothing was flushed to the north, but as they came down the Tablas Straits they sighted an enemy destroyer squadron of five (later amended to three) ships,

leading destroyer on an opposite course; then he rolled over to the right to dive down-wind, from 060 degrees. Although this led to a good approach dive, his 1,000lb bomb 'hung-up' on him – and that was not the end of VB-14's misfortune: 'Unfortunately, most of the other planes failed to follow his tactics, and instead of flying past their targets and turning back to dive downwind, they rolled over when Downing did, diving generally from the west, either upwind or crosswind on the other destroyers.'

A bomb splash can be seen and the explosion of another off the starboard side of a speeding Japanese destroyer, while the splashes of the twin 20mm cannon of the attacking Helldiver can be seen stitching their way across the sea towards her fantail. National Archives, Washington DC

'... although the one which had been hit was smoking, none was apparently having difficulty in maintaining speed.'

All in all the Helldiver had performed very well during the battle and had indeed proved to be the 'Fist of the Fleet' in these actions. One factor that helped them was the performance of their radio-radar equipment, which was made the subject of a special report in this action.[139] The overall performance was pronounced as follows:

Very good. In this period, thirteen planes were sent out on search missions. Contact reports were successfully made back to the parent ship from four of these planes, ranging in distance from 225 miles to 340 miles. These reports were made on the ATC transmitter, and from reports of our ship's communications department they were received as clear as a bell.

Some reports are still received as to hearing the YE homing signal on ZBX at distances of 150 miles or greater. In another instance a plane returning from a mission to a friendly island had some trouble locating the ship; the pilot finally contacted the CAP over the ship on VHF from a distance of 200 miles and was given the information necessary to return to the ship.

View aft from the rear seat of a Helldiver as a frantic turn is made by a Japanese destroyer during the Battle of Leyte Gulf. Two bombs have exploded off her starboard quarter as she heels round at high speed. National Archives, Washington DC

The destroyers were reported to be , '... probably *Shiranui* or *Takanami* class ...', and they now commenced turning in tight circles, '... so that the relation of the ship's axis to the direction of dive varied considerably.' As a result, there was poor shooting, only Lieutenant (j.g.) Elway obtaining a solitary hit on the port wing destroyer, his bomb striking her on her port bow. But none of the other AP bombs came close enough to cause any damage, being report-ed as falling within 20, 50 and 75ft (6, 15 and 23m), with damage 'doubtful'. Again the main bomb of Lieutenant (j.g.) Brown's aircraft, 'hung-up'. Lieutenant (j.g.) Berg and his aircrewman strafed the northern destroyer on pulling out from their dive, then turned to port to strafe the leading destroyer as they retired to the rendezvous point south-west of the target. When the attack was over, the destroyers resumed their north-westerly course, and

CHAPTER FIFTEEN

South China Sea to Iwo Jima

SB2C-5s of Marine Squadron VMSB-343, off Honolulu, Hawaii. Emil Buehler Naval Aviation Library, Pensacola

The Pacific war now entered its final phase, with the carriers leading the drive towards Tokyo Bay. By this time the Curtiss Helldiver was a tried and reliable weapon of great potency. There was no longer any need for it to 'prove' its worth, and with all trials and tribulations over, it only remained to hone it to perfection in the continuing light of battle experience.

The Arrival of the 'Dash 3'

After testing at Anacostia, SB2C-1 Mod. II No. 00008 was treated as an experimental test-bed for a more powerful engine. The Wright R-2600-8, rated at 1,700hp, was clearly insufficient for the mounting payload of 'The Beast' and alternatives had to be found. First tried was the Wright R-2600-20 engine. Rated at 1,900hp, this featured cast-aluminium cylinder heads and the Wright 'W' barrel fins. The addition of the new four-bladed Curtiss-Electric propeller, which was spinnerless, was also tried, and immediately became a ready identification feature. Although the improvement in performance was not spectacular (its top speed increased by 13mph

Specification – Model 84E XSB2C-3	
Engine type:	R-2600-20
Armament:	Two 20mm forward-firing fixed cannon; two 30mm rear-cover flexible machine guns; maximum bomb load 2,000lb (907kg); one 2,000lb torpedo
Dimensions:	Length 36ft 8in (11.17m); height 16ft 10½in (4.01m); wingspan 49ft 8⅝in (15.15m); wing area 422 sq ft (39.2 sq m)
Weights:	Maximum weight 16,800lb (7,620kg); empty weight 10,114lb (4,588kg)
Performance:	Maximum speed 294mph (473km/h) Take-off power 1,900hp (1,417kW) Range as dive-bomber 1,200 miles (1,931km) Climb height 1,750ft/min (534.4m/min) Service ceiling 25,000ft (7,620m)
Number completed: 1	
Serial numbers:	(ex-00008) (Bu. No.-Navy – Serial Army)
Contract:	Order dates; Jul 42 Delivery date; Sep 42

21km/h!) overall it was what was required; and range was also slightly improved to 1,200 miles (1,930km).

Other improvements were made – the fixed canopy was redesigned and given

extra strength with the elimination of the top rear windows on either side, while the rear canopy was similarly streamlined, being shortened and given an improved sliding mechanism. As the production run continued, the slotted dive brakes were worked into the line. On a few of these SB2C-3s, as they were designated, a wing pod housing the new AN/APS-4 airborne intercept radar equipment was fitted under the starboard wing, replacing the ASB sea-search equipment and eliminating the Yagi antenna. This search radar was made up of the transmitter-receiver pod, a junction box and a control box in the pilot's cockpit, and a pair of indicators, and the pod was completely transferable with a wing bomb when required. Aircraft thus modified were designated as the SB2C-3E. Both types began entering service with the fleet in 1944 and proved popular with their aircrew.

A total of 1,112 was built at Columbus, in the Bu. No. range 18599-19710. The SBW took on board these changes as we have seen and 413 of them became the SBW-3 (Bu. Nos. 21233–21645) and so of course did the Fairchild SBF-3s (Bu. Nos. 31686–31835). Meanwhile the war ground on.

New Targets

On 29 September 1944, at a top level conference in San Francisco, the decision was taken to invade the islands of Iwo Jima on 20 January 1945, and then Okinawa on 1 March 1945. Both these objectives were carried out, albeit at slightly later dates than at first envisaged; meanwhile further landings took place in the Philippines, at Mindoro in mid-December 1944, and at Luzon in January 1945. The Helldivers of Task Force 58 were fully engaged in supporting operations for these great advances, ranging the length and breadth of the South China Sea from the southwest Pacific to the southern homeland of Japan itself.

Carnage in the Philippines

On 27 October, General Douglas McArthur was forced to call for help from the fleet because the Army Air Force ashore had totally inadequate forces at Tacloban to do the job. Task Group 38.3 under Rear Admiral Sherman moved in to give fighter cover, and VB-15's SB2C-3s from the Essex (CV-9) hit a Japanese convoy, sinking the destroyers Fujinami and Shiranuhi. Two days later Task Group 38.2 carried out an attack against Japanese

Specification – Model 84E SB2C-3	
Engine type:	R-2600-20
Armament:	Two 20mm forward-firing fixed cannon; two 30mm rear-cover flexible machine guns ; maximum bomb load 2,000lb (907kg); one 2,000lb torpedo
Dimensions:	Length 36ft 8in (11.17m); height 16ft 10½in (4.01m); wingspan 49ft 8⅜in (15.15m); wing area 422 sq ft (39.2 sq m)
Weights:	Maximum weight 16,800lb (7,620kg); empty weight 10,114lb (4,588kg)
Performance:	Maximum speed 294mph (473km/h) Take-off power 1,900hp (1,417kW) Range as dive-bomber 1,200 miles (1,931km) Climb height 1,750ft/min (534.4m/min) Service ceiling 25,000ft (7,620m)
Number completed: 1,112	
Serial numbers:	18599/19710 (Bu. No.-Navy – Serial Army)
Contract:	Order date; 1943 Delivery date; 1943

Specification – CC & F SBW-3	
Engine type:	R-2600-20
Armament:	Two 20mm forward-firing fixed cannon; two 30mm rear-cover flexible machine guns; maximum bomb load 2,000lb (907kg); one 2,000lb torpedo
Dimensions:	Length 36ft 8in (11.17m); height 16ft 10½in (4.01m); wingspan 49ft 8⅜in (15.15m); wing area 422 sq ft (39.2 sq m)
Weights:	Maximum weight 16,800lb (7,620kg); empty weight 10,114lb (4,588kg)
Performance:	Maximum speed 294mph (473km/h) Take-off power 1,900hp (1,417kW) Range as dive-bomber 1,200 miles (1,931km) Climb height 1,750ft/min (534.4m/min) Service ceiling 25,000ft (7,620m)
Number completed: 413	
Serial numbers:	21233/21645 (Bu. No.-Navy – Serial Army)
Contract:	Order date; 1943 Delivery date; 1944

airfields around Manila, claiming the destruction of eighty-four Japanese aircraft for the loss of eleven American aircraft. All this time the fleet was the subject of severe kamikaze attacks which hit and damaged the carriers Intrepid, Franklin and Belleau Wood. The fleet then withdrew to replenish at Ulithi. They were back off Luzon at the beginning of November when Helldivers from the Wasp's VB-14, the, Hornet's VB-11, Intrepid's VB-8, Hancock's VB-7, Lexington's VB-19 and Ticondergoa's VB-80 made a mass attack on the Luzon airfields. Again results were spectacular,

more than 400 of the enemy being claimed destroyed for the loss of twenty-five carrier aircraft. Then the SB2C-3s of VB-19 helped claim a bigger scalp on 5 November when they caught the heavy cruiser Nachi in Manila Bay and hit her so badly that she sank; but the Japanese kamikazes took revenge by badly damaging the Lexington in reply.

Opposition was heavy in these attacks. Rear Admiral Edwin M. Wilson, then a Lieutenant with VB-11 flying from the Hornet on one such raid, describes his experiences thus:

Specification – Fairchild SBF-3

Engine type:	R-2600-20
Armament:	Two 20mm forward-firing fixed cannon; two 30mm rear-cover flexible machine guns; maximum bomb load 2,000lb (907kg); one 2,000lb torpedo
Dimensions:	Length 36ft 8in (11.17m); height 16ft 10½in (4.01m); wingspan 49ft 8⅝in (15.15m); wing area 422 sq ft (39.2 sq m)
Weights:	Maximum weight 16,800lb (7,620kg); empty weight 10,114lb (4,588kg)
Performance:	Maximum speed 294mph (473km/h) Take-off power 1,900hp (1,417kW) Range as dive-bomber 1,200 miles (1,931km) Climb height 1,750ft/min (534.4m/min) Service ceiling 25,000ft (7,620m)
Number completed:	150
Serial numbers:	31686/31835 (Bu. No.-Navy – Serial Army)
Contract:	Order date; 1943 Delivery date; 1944

SB2C-4s of VB-9 aboard the Yorktown **(CV-10) on 12 May 1945.** National Archives via National Museum of Naval Aviation, Pensacola

I never knew why the SB2C had such a large vertical stabilizer and rudder, as I once flew without them. On 6 November 1944 I led a strike on Clark Field on Luzon. The day before at Clark Field, I was on the deck, strafing, and they shot off my tailhook with small arms fire; so I pulled out higher this time. Just as I closed my dive brakes and pulled the stick back into my gut, blacking out, I heard a loud explosion. The one good thing about radio and engine noise was that it kept you from hearing anti-aircraft shells exploding. The 'word' was that if you heard an explosion, it would be last thing you would ever hear. So as soon as I finished my pullout and as my vision returned, I asked my gunner, Harry Jespersen, what happened. 'Mr Wilson,' he said, 'we have no tail.' Apparently, a 40mm shell had exploded, and blown the vertical stabilizer and rudder off at the fuselage.

Fortunately, I had pulled out in the direction of our task force, so I turned the lead over and kept the plane level. Figuring there must have been a purpose for that large vertical surface, I did not drop a wing or attempt a turn, and I set her down by a picket ship, the destroyer USS *Mansfield* (DD-728). I landed into a wave that broke over us, and immediately jumped out onto the wing to help 'Jes' get the raft out, which was in a tube between us, opening into the rear cockpit. He was still landing, with his arms over his face and head – we had no shoulder straps or hard helmets, so we had to protect our heads. When I tapped his shoulder, he was really startled; he must have thought it was St. Peter!

Standing on the wing, we inflated the raft and stepped aboard. The SB2C stayed afloat for about 45 seconds. There was a 40-knot sea, so we had a rough time getting aboard *Mansfield* – one minute we would be looking down on her, and the next looking up at her, but we finally got aboard. So my first 'sea command' as CO of a two-man raft was too short to go to my head.[140]

Aerial opposition was also encountered, the fighter strength of the Japanese being reinforced from time to time, despite everything. Not all the aerial fighting was one-sided however, the '2C' was sometimes able to give as good as it got – for example, Lieutenant Bill Colleran flying from the *Hancock* with VB-7 took that squadron's record during another strike over Clark Field by shooting down a Zero fighter at an altitude of 45ft (14m), and then nailing a second possible, that never became airborne.[141]

Continuing efforts by the Japanese to land troop reinforcements led to further action by the Helldivers on 11 November, when a Japanese troop convoy and its escort were attacked by 347 aircraft including SB2Cs from the *Hornet*, *Essex*, *Ticonderoga* and *Enterprise's* VB-20. A total of five troopships were sunk, as well as the destroyers *Hamanami*, *Naganami*, *Shimakaze* and *Wakatsuki* and the minesweeper *W-30* in this devastating strike – and it cost the American fleet just eleven aircraft.

The pressure was remorselessly maintained, and on 13 and 14 November the same Helldiver squadrons of Task Force 38, plus the SB2C-3s of VB-81, newly embarked aboard the *Wasp*, made equally effective strikes against the remaining

Japanese warships in Manila Bay. They sank the light cruiser *Kiso*, and destroyers *Akebono*, *Akishimo*, *Hatsuharu* and *Okinami* along with ten freighters, and damaged the destroyer *Ushio* and five other merchant ships. They returned to finish matters on 19 November, with further Helldiver strikes from *Intrepid's* VB-18 and *Hancock's* VB-7, which resulted in the sinking of a submarine-chaser and two freighters, and heavy damage to the cruiser *Isuzu*. A final raid was mounted on 25 November against the same anchorage, and this time the Helldivers helped to sink the cruisers *Kumano* and *Yasoshima*, the landing ships *T-6* and *T-10* and three freighters.

After another replenishment at Ulithi, Task Force 38 resumed the assault between 14 and 16 December, the Helldiver units involved being VB-3 (*Yorktown*), VB-81 (*Wasp*), VB-20 (*Lexington*), VB-7 (*Hancock*), VB-11 (*Hornet*) and VB-80 (*Ticonderoga*). Hitting airfields on Luzon, they claimed 170 Japanese aircraft destroyed for the loss of twenty-seven; however, a further thirty-eight, some of them Helldivers, were lost in landing accidents. This seems a high rate, but it was not uncommon.

Two SB2Cs from Marine Squadron VMSB-244 over Daveo, Mindanao, Philippines, on 12 June 1945. Cpl. E. Scarmellino via National Museum of Naval Aviation, Pensacola

An SB2C-4 of VB-12 in the landing circle above the Randolph (CV-15) in the south-west Pacific in 1945.
US Navy via National Museum of Naval Aviation, Pensacola

Transition from SBD to SB2C

One respected veteran pilot, Commander Harold L. Buell, has put on record his own feelings on the matter, proposing that perhaps one reason for such a high accident rate was too hasty deployment of not yet fully trained aircrew.[142] He recorded his reaction when his unit, VB-2, then at Hilo NAS, first saw the Helldiver:

My first reaction was that 'The Beast', as this Curtiss monster was called, would be trouble for

were already having trouble in an SBD, these advantages were far outweighed by the technical difficulties of flying this aircraft. I considered that to take the majority of VB-2 pilots out of the only combat aircraft they had flown, and one which they were only just starting to be able to use as a weapon, to put them into 'The Beast' and then launch them on a combat mission was imprudent. Yet this was to be the fate of VB-2. I can only conclude that we will pay a heavy price at sea for this rash decision made high up by someone who apparently has been forced into a very unpleasant choice due to other factors.

Further raids were planned against the Luzon airfields, but while *en route* to replenish, Task Force 38 ran into a typhoon, and as a direct result of this, several destroyers were lost because they capsized when their fuel ran out; it also caused irreparable damage to 146 of the fleet's aircraft in their hangars, many of them SB2C-3s. The planned strikes therefore had to be postponed, and the fleet returned to base where Helldiver numbers were soon made up again.

An SB2C of VB-7 from the Hancock **(CV-19) in January 1945.** Emil Buehler Naval Aviation Museum Library, Pensacola

the squadron. Compared to the steady, forgiving SBD, this aircraft required much more pilot ability to fly, both operationally and as a dive-bombing weapon. From a personal viewpoint, there were features about the SB2C that appealed to me from the start, for instannce it had a lot more firepower, a larger bomb load, it was a lot faster, both in a dive and cruising, it could really take punishment and still get home – this was a valuable asset. But for most of my companions in VB-2, and especially the least-experienced ones who

Typhoon Damage

One cause of severe losses of Helldivers at this time had nothing to do with inadequate training before combat – and it should be appreciated that US Navy pilots were receiving more than four times the hours of training as compared to their opposite numbers in the Imperial Japanese Navy's dive-bombers – nor to enemy action, but everything to Mother Nature.

Devastation in the South China Sea

The fast carrier forces of the US Navy now feared no enemy, whether land- or sea-based. They could roam wherever their fancy took them, almost inviolate to the best the Japanese could throw at them, and with the SB2C spearheading the attacks, they proceeded to do so. The offensive – Halsey's long-cherished Op-Plan 25-44 –

A formation of SB2C-4Es from VB-87 operating from the Randolph **(CV-15) off Trinidad in December 1944. A group of Avenger TBMs from VT-87 is in the background.** H. Paul Brehm via National Museum of Naval Aviation, Pensacola

An immaculate SB2C-4 of VA-73 over the Curtiss St. Louis plant in March 1945. Emil Buehler Naval Aviation Library, Pensacola

was resumed between 3 and 4 January 1945, when Task Force 38 carried out wide-ranging air strikes on airfields and other targets all over the island of Formosa, the Pescadores and the southern Ryukyu Islands. They sank the minesweeper *W-41* and destroyed over 100 enemy aircraft, at a cost of twenty-two American planes. One of these was that of the commanding officer of VB-18, Lieutenant Commander Cousins, who was killed on their strike on the Japanese Navy base at Formosa. He was succeeded in command by George Gesquire, and the battle continued.

On 6 and 7 January they turned their attentions to kamikaze airfields on Luzon, to pre-empt the Japanese reaction to the Lingayen Gulf operation, claiming seventy-five Japanese aircraft for the loss of twenty-eight of their own. They then struck north again at targets on Okinawa and Formosa on 9 January. Here they sank the destroyer *Hamakaze*, an escort ship and five merchant ships, while damaging nine small warships and three more freighters for the loss of ten aircraft. Next was an audacious foray deep into enemy waters: entering the South China Sea on 10 January, attacks were launched against targets up and down the Indo-China coast, with emphasis on Camranh Bay, as it was thought heavy units of the Japanese fleet might be seeking refuge there. Six Helldivers each from VB-7, VB-11 and VB-20 started to launch at 07:30 and, after first drawing a blank, they flushed a Japanese convoy of three valuable oil tankers and two freighters, with six escorts, off Phan Rang. While the '2Cs' of VB-7 attacked the warships, sinking four of them, VB-11 and VB-20's 'Beasts' hit the merchant ships and scored damaging hits on four of them; all the tankers and two cargo ships were finally sunk.

A second composite strike was 'weathered-out' but a third, launched at 12:40 had better luck and found another Japanese convoy off Qui Nhon. VB-11's Helldivers attacked the light cruiser *Kashii* and severely damaged her, and she was then finished off by a direct hit on her stern, a hit scored by Lieutenant (j.g.) R. A. Rhodes of VB-7. One of the smaller escorting warships was also destroyed by a '2C' from this squadron, Lieutenant (j.g.) P. L. Ruch hitting her fair and square and seeing her blow up with a satisfactory explosion. Yet a fourth strike was mounted that afternoon against the survivors of this

unfortunate convoy, some of which had beached themselves, and yet further damage was inflicted.

In total some 1,645 sorties were made against shipping targets in this rich area, resulting in the destruction of the Japanese cruiser *Kashii* and the Vichy-French light cruiser *Lamotte-Picquet,* both on 12 January, along with twelve lesser warships and twenty-nine freighters and oil tankers.

A further refuelling followed this foray, and the fleet returned the way they had

come, striking hard again at Formosa and the Pescadores and Japanese targets over the Chinese mainland in Fukien on 15 January 1945. Among their victims were the destroyers *Hatakaze* (12th), and *Tsuga* (15th), the landing ship *T-14* and two merchant ships. The fleet then turned and struck once more all along the South China coast at targets as diverse as Hong Kong harbour in the north, through Canton and to Hainan Island in the south. At Hong Kong, heavy flak met the attackers,

An SB2C-5 of VB-14 operating from the Intrepid **(CV-11) over Shanghai, China, in September 1945.** US Navy via National Museum of Naval Aviation, Pensacola

Composition of SB2C Units for January 1945, Carrier Raid

Task group	Carrier	Unit	Type
TG.38.1	*Yorktown* (CV-10)	VB-3	SB2C-3
	Wasp (CV-18)	VB-81	SB2C-3
TG.38.2	*Lexington* (CV-16)	VB-20	SB2C-3
	Hancock (CV-19)	VB-7	SB2C-3
	Hornet (CV-12)	VB-11	SB2C-3
TG.38.3	*Ticonderoga* (CV-14)	VB-80	SB2C-3

but four of VB-7's Helldivers penetrated it and planted a bomb on the tanker *Kamoi*. A second raid by three '2Cs' of VB-7 hit the Taikoo shipyards there, while a third, consisting of four VB-7 dive-bombers was launched at 13:02 – these met fierce resis-tance, but again selected a tanker as their target and set her ablaze. The final strike was made, with fourteen Helldivers from all three squadrons, with three of VB-7's dive-bombers hitting yet a third tanker; but this time the flak claimed the life of Ensign C. S. Snead when his SBW-3 (Bu. No. 21377) was hard hit over the target, although his rear-seat man parachuted to safety. Two freighters and three tankers were sunk at Hong Kong that day, and five small warships also hit.

An SB2C-3 of VB-3 operating from the Yorktown **(CV-10) in April 1945.** Dick Hill via National Museum of Naval Aviation, Pensacola

The launch! A Helldiver leaves the flight-deck of the Philippine Sea **(CV-47).** Jim Sullivan via National Museum of Naval Aviation, Pensacola

Another replenishment, another change of course and Formosa was again on the Helldiver's menu on 21 January, fourteen of them hitting Kiirun harbour where five more Japanese tankers and two freighters were sunk. In addition some 104 aircraft were claimed destroyed, and three destroyers – *Harukaze*, *Kahsi* and *Sugi* – and two landing ships were hit and damaged to varying degrees. On 22 January it was back for one more strike on Okinawa and the Ryukus, six VB-7 Helldivers joining a similar number from VB-11 to hit Yontan airfield. Finally the American fleet's amazing run of success ran out and Japanese counter-attacks and accidents brought damage to the carriers *Ticonderoga*, *Hancock*, *Langley* and *Cowpens*, shortly before the fleet took a well earned rest, arriving at Ulithi once more on 25 January. It had been a remarkable cruise with US Navy aircraft wiping out 300,000 tons of much needed Japanese mercantile tonnage and warship tonnage, and claiming the destruction of 615 aircraft; this was for the loss of 201 aircraft of all types, with 167 aircrew.

The Road to Iwo Jima

Between 10 February and 4 March Task Force 58, under the command of Vice-Admiral Marc Mitscher was at sea again, bringing death and destruction to the waning enemy and cutting another great swathe across the western Pacific. The fleet struck hard at the Japanese navy in its home bases in order to forestall any attempts to disrupt the planned landings on Iwo Jima, the final step before the invasion of homeland Japan itself, and in this the Helldiver was again its central strength.

Because of the need to carry extra fighter aircraft in order to deal with the threat of Japanese kamikaze pilots, a decision had been made to add VBF units to the fleet: these were F6G Hellcats equipped to carry a single 500lb bomb below the fuselage as fighter-bombers and formed into VFB units. They

lacked the range and punch of the Helldiver and the Avenger, but they served as makeshift strike aircraft while still giving the fleet extra protection. Towards the end of November 1944, orders had been issued to the fleet which had led to a reduction in the strength of both the VB and VT squadrons aboard each carrier, so they were down to just fifteen aircraft per Helldiver unit, and in some cases only twelve! This proved much too drastic, however, because

An A-25 Shrike towing a glider. Curtiss stated this was a test to prove the strength of the Shrike, but they could not conceal from their workforce that the Army Air Force had gone cold on their product; they had never embraced dive-bombing, especially in the higher echelons, and were seeking alternative employment for an unwanted aircraft. Curtiss-Wright via Stanley I. Vaughn Archive

Fine aerial view of a Shrike (18774) under test. Despite its mass-production at St. Louis, the USAAF rejected the whole dive-bomber concept in 1943 and the aircraft were surplus to requirements. Many ended up with the US Marine Corps. Curtiss-Wright via Stanley I. Vaughn Archive

the dive-bomber squadrons were being called on more and more to finish off the Japanese fleet at their anchorages, and this task required the heavier lifting capacity of the Helldiver. Another reorganization therefore shifted the balance back towards

the dive-bomber, the final ratio then being settled at fifty-six VF, twenty-four VB and twenty VT aircraft, while the light carriers were to carry thirty-six VF aircraft only.

The fleet sailed from Ulithi on 10 February with an initial strength of eleven fleet and four light carriers, six battleships, one battle-cruiser, fourteen heavy and light cruisers and sixty-five destroyers; the Helldiver units embarked were *Hornet's* VB-17 (SB2C-3), *Wasp's* VB-81 (SB2C-3), *Bennington's* VB-82 (SB2C-4), *Lexington's* VB-16 (SB2C-3), *Hancock's* VB-80 (SB2C-3), *Bunker Hill's* VB-84 (SB2C-4E), *Yorktown's* VB-3 (SB2C-4) and *Randolph's* VB-12 (SB2C-4).

Arrival of the 'Dash 4'

Improvements to the Helldiver continued with war experience, and the SB2C-3 was further modified, resulting in the introduction of the SB2C-4. Further 'beefing-up' of the cockpit area saw the deletion of all the small windows behind the pilot, giving yet further strength, and the other noticeable feature was that a prop-spinner was again re-introduced. Otherwise it was business as before, those planes fitted with the AN/APS radar pod becoming the SB2C-4E.

The introduction and increasing popularity of the rocket as a ground and ship attack weapon coincided with the widespread introduction of the 'Dash-4' Helldiver, and most were fitted with four zero-length rocket launchers under each wing for these.

The Dash-4 turned out to be the most widely produced variant of the Helldiver, and some 2,045 were built by Curtiss (Bu. Nos. 21646 to 21741; 64993 to 65286 and 82858 to 83127 inclusively); a further 100 air search types being built by Fairchild in Canada as the SBF-4E (Bu. Nos. 31836 to 31935). This model saw extensive combat service in the final months of the Pacific campaign, many surviving to be used post-war in both active and reserve units (*see* Appendices II–V).

Specification – Fairchild SBF-4E	
Engine type:	R-2600-20
Armament:	Two 20mm forward-firing fixed cannon; two 30mm rear-cover flexible machine guns; maximum bomb load 2,000lb (907kg); one 2,000lb torpedo
Dimensions:	Length 36ft 8in (11.17m); height 16ft 10½in (4.01m); wingspan 49ft 8⅝in (15.15m); wing area 422 sq ft (39.2 sq m)
Weights:	Maximum weight 16,800lb (7,620kg); empty weight 10,114lb (4,588kg)
Performance:	Maximum speed 294mph (473km/h) Take-off power 1,900hp (1,417kW) Range as dive-bomber 1,200 miles (1,931km) Climb height 1,750ft/min (534.4m/min) Service ceiling 29,100ft (8,870m)
Number completed:	100
Serial numbers:	31836/31935 (Bu. No.-Navy – Serial Army)
Contract:	Order date; 1943 Delivery date; 1944

Striking at Nippon

The first massive strikes against targets in the Tokyo Plain were launched by the air groups of Mitscher's fleet on 16 February 1945, from a position some 125 nautical miles south-east of the Japanese capital.

This showed the supreme self-confidence of the US Navy in its capability to look after itself anywhere. Bad weather conditions rather than enemy defenders proved the greatest handicap the Helldiver units had to face on this occasion, many of their prime targets – Japanese airfield and aircraft factories – being 'weathered out' and targets of opportunity having to be selected in lieu; but enormous damage was still done.

Next day the air groups went for Yokohama Navy Base, but few worthwhile warship targets were found – and for the loss of sixty aircraft, half of them due to accidents, only two small escort ships and a transport were sunk, which was not a good rate of exchange. Attacks were then switched by VB-3 and VB-12 to Hahajima and Chichijima. Between 20 and 23 February three of the air groups were used to support the landings on Iwo Jima itself; these had taken place on the 19th with little opposition initially, the '2Cs' hitting targets ashore there, and on the Bonin Islands, to good effect.

Nor did the Helldivers have much better hunting when, after concentrating and refuelling, they returned to harry the Japanese mainland once more. Strikes were launched from the west of Nanpo Shoto, and again the Helldivers left the decks in rough seas and flew into bad weather conditions on these missions. They struck at Okinawa on 1 March on their way back to base, sinking two small warships, but otherwise finding little else to occupy them.

Port profile of an SBW-3 (coded 585) at NAS Patuxent River testing station on 11 January 1945. US Navy

An SBF-3 (coded 778) at NAS Patuxent River, 12 January 1945. US Navy

Serial Numbers of the USAAF A/RA-25As		
Army serial no.	Type	Name
41-18774/18783	A-25A-1-CS	Shrike
41-18784/18823	A-25A-5-CS	Shrike
41-18824/18873	A-25A-10-CS	Shrike
42-79663/79672	A-25A-10-CS	Helldiver
42-79673/79732	A-25A-15-CS	Helldiver
42-79733/79972	A-25A-20-CS	Helldiver
42-79933/80132	A-25A-25-CS	Helldiver
42-80133/80462	A-25A-30-CS	Helldiver

Termination of the Shrike

With the general review of the Army Air Force's attack aircraft policy in June 1943, the general consensus had been reached that dive-bombers were more accurate than fighter-bombers; but whereas dive-bombers could not fight, fighter-bombers could dive, and the decision was made to reverse all the programmes of the previous three years and abandon dive-bombers as a distinctive type altogether.

The Air Force had only reluctantly adopted the dive-bomber, they had little experience of it, they lacked the will and dedication at higher levels to learn, and were very happy to get rid of it. All the types which they had had built in vast numbers and not used – including the A-24 Banshee (the army version of the Dauntless); the A-35 Vultee Vengeance (which they had taken over from large British orders) and the A-25 Shrike, renamed with its Navy name of Helldiver

– were redesignated as 'restricted use' (ie not suitable for combat duty) aircraft, the 410 Helldivers becoming RA-25As.[143] Those for which the Army did manage to find employment suffered the usual indignities of utilization as squadron 'hacks', target-tugs and instrument trainers, although a few served briefly on coastal patrol duties.

Marine Corps Helldivers

While the Navy Helldiver aircrew were carrying out these spectacular missions far to the north, other Helldivers were in action during the dogged land campaigns in the Philippine Islands themselves. Initially the US Marines had set up to close support marine air groups (MAGs) to assist the Army ashore, and these had proven most successful.[144] The Marines had also been arguing, ultimately successfully, for their own air group carriers to be assigned to them for the future invasions planned. These were a long time coming, but Helldiver units were established, and training commenced (see Appendix II).

Initially two squadrons, VMSB-334 and VMSB-342, had received a single Helldiver

(Above) **SB2C-3 'Rugged' of VB-80 at the invasion of Iwo Jima, 1945.** Charles R. Shuford

This photograph of Helldivers warming up aboard a carrier in the Pacific gives a good indication of the sheer size of the aircraft. The undercarriage can be seen to good effect, although the chocks seem rather inadequate! National Museum of Naval Aviation Library, Pensacola

each in November 1943, to begin familiarization flights. Although this was not taken further between May and June 1944, other Marine squadrons began to be fully equipped with between twenty-one and thirty aircraft each. The units were mainly equipped with Navy-type Helldivers and the 410 former Army A-25As that they had inherited (now renamed from Shrike to Helldiver and designated as the SB2C-1A) and reconverted at special modification centres – located at NAF Roosevelt Field, New York, Consolidated-Vultee, Allentown, Pennsylvania and Delta Airlines – to the Navy type electronics, the programme being completed by the end of 1944. Converted Helldivers were then transferred to the Navy Operational Training Command, to be used exclusively for training purposes. These aircraft – Bu. Nos. 75218 to 75588 (formerly 42-80053 to 42-80423) and Bu. Nos. 76780 to 76818 (formerly 42-804424 to 42-804462) – also included the residue of the 140 Australian A-25As.

They served briefly with VMSB132, -144, -234, -343, -344, -454, -464, -474 and -484 during that year, while VMSB-933 was equipped with SB2C-1Cs in November 1944, before VMSB-344 was disbanded and VMSB-132, -144, -234 and -454 became VMTB squadrons and re-equipped with the TBM. VMSB-464, -474 and -484 lingered on as training squadrons based at MCAS El Toro, California (*see* Appendix II).

The very last 'unhooked' Helldiver to come from the St Louis plant – these came as A/RA-25Cs and turned into Marine Corps SB2C-1As – joined VMSB-332 and was based at Oahu in 1944. This particular aircraft finally went to the pool at NAS Barber's Point where she lay unattended with scores of her sister Helldivers (both with hooks and without them!) until she was stricken in May 1945.

The first marine squadron to actually take the '2C' into combat was VMSB-231, the famed 'Ace of Spades' Squadron, then part of MAG-4 in the Marshall Islands. Led by Major Joseph White Jr. from Majure, they split into two divisions and attacked both ends of the main runway on Mille, and followed up to hit gun emplacements and revetments. too, with strafing attacks. Staff Sergeant Theron J. Rice was quoted as stating:

Two pilots, a bit over-anxious on their first combat mission, dived too low and returned with shrapnel from their own bombs lodged in their wing and fuselage. But the strike went off as scheduled, and pilots and gunners are more anxious than ever to become better acquainted with what they describe as a 'really sweet airplane'[145]

The squadron intelligence officer, Captain Charles E. Rogers, commented on the Marine flyers that, 'They like the steady way in which the plane dives, making it easy to stay on the target. They like the added firepower and the extra speed on straight and level flight.' The press renamed the squadron the 'Helldiving Gyrenes'.

Early 1945 and Helldiver 205 of CVG-84 comes up on the side lift of the USS Bunker Hill **(CV-17) in preparation for a fresh strike. In the background the battleship USS** South Dakota **(BB-57) keeps guard, with a destroyer of the screen beyond her.** US Navy via National Archives, Washington DC

LT. W.W. PERRY, JR.   LT. T.R. GAVIN   LT. E.F. GIBSON   LT. M.J. HEMBY   LT. W. GAUVEY, JR.   LT. T.C. THOMAS

BOMBING SQUADRON EIGHTY - FIVE

LT. CDR. A.L. MALTBY, JR.
COMMANDING OFFICER

LT. A.G. SYMONDS, JR.
EXECUTIVE OFFICER

LT.(JG) R.W. ELMORE   LT. (JG) G.C. SHOEMAKER

LT. (JG) H.E. EAGLESTON, JR.   LT. (JG) J.F. HUNGERFORD

ENS. R.W. JONES   LT. (JG) J.F. LACINA

ENS. H.W. FORSGREN   ENS. J.A. LOCKE

ENS. R.H. WENT   ENS. R.W. MANN   LT. O.S. HARGETT, JR.   LT. K.F. CALLAHAN   ENS. C.M. CRUMP   ENS. M.J. MITCHELL   ENS. W.J. DOERING

ENS. C.L. TRUMP   ENS. G.M. EVEN   ENS. S.G. PAYNE   ENS. M.L. SKINNER   ENS. J.J. SCHERTING

The men of Bombing Squadron Eight-five in 1944.
Maynard J. Mitchell

VMSB-244 was equipped with the SB2C-4s which replaced their SBDs on 19 May 1945. They made their first strike with the Helldiver on 2 June 1945 when they attacked Cabaguio on Mindinao Island with both bombs and rockets. Their sister unit, VMSB-245, was similarly converted in the same war zone, the whole of MAG-24 changing over to the Helldiver on 16 July 1945; but MAG-32 flew the Dauntless right until the end of the war before disbanding.

A taxiing accident at Columbus on 30 August 1943. American Aviation Historical Society Journal

CHAPTER SIXTEEN

Okinawa

Elaborate 'softening-up' moves preceded the actual invasion of Okinawa, set back to 1 April 1945, and as part of these Task Force 58 under Vice-Admiral Marc Mitscher sailed from Ulithi on 14 March with sixteen fleet and light carriers, eight fast battleships, two battle-cruisers, sixteen heavy and light cruisers and sixty-four destroyers to attack the fifty-five Japanese airfields on Kyushu. Helldivers acted as 'Fist of the Fleet' for heavy strikes made on 18 March, hitting hangars, barracks and workshops, and cratering runways; they met only light opposition, however, while aircraft targets on the ground were scanty, and many of those destroyed may well have been dummies.

Dusting the Emperor's Portrait

Next day the attacks were switched to the great Japanese naval base of Kure where lay many of the last survivors of the Imperial fleet: the carriers *Amagi*, *Katsuragi*, *Ryuho*, *Hosho*, *Kaiyo* and *Ikoma*, vast but impotent due to lack of aircraft and aircrew to man them; the still potentially dangerous battleships *Yamato*, *Hyuga* and *Haruna*; and the cruisers *Tone* and *Oyodo* – these were all among the warships targeted and hit by the Helldivers on that day ('2C' strength is given right).

After resting and refuelling, Task Force 58 sailed again for the waters around Okinawa with Helldivers embarked in nine fleet carriers to continue their pre-emptive strikes on 23 March, moreover on the following day they caught and attacked a Japanese convoy south of Kyushu, sinking eight freighters and an escort ship. They returned to strafing Okinawa's defences on the 25th before withdrawing to refuel and replenish once more. On 29 March they resumed the offensive launching strikes against targets on Okinawa itself, and on the northern Ryukyu islands and Kyushu again, and this pattern of strike and support was continued up to and beyond the day of the American landings.

VB-9 Helldiver from the carrier Lexington **CV-16 makes an unorthodox landing on 7 June 1945.** Emil Buehler Naval Aviation Library, Pensacola

SB2C Units at the Battle of Okinawa, April 1945				
Unit	Carrier	Commander	SB2C type	No.
VB-5	Franklin (CV-13)	Lt Cdr J. G. Sheridan	SB2C-4E	15
VB-6	Hancock (CV-19)	Lt Cdr G. P. Chase	SB2C-3 & 3E	5
			SB2C-4	3
			SBW-3 & 4E	4
VB-9	Yorktown (CV-10)	Lt T. F. Schneider	SB2C-4	15
VB-10	Intrepid (CV-11)	Lt Cdr R. B. Buchan	SB2C-4E	15
VB-12	Randolph (CV-15)	Lt Cdr R. A. Embree	SB2C-4E	15
VB-17	Hornet (CV-12)	Lt Cdr R. M. Ware	SB2C-3	9
			SB2C-4	2
			SBW-4	4
VB-82	Bennington (CV-20)	Lt Cdr H. Wood	SB2C-4E	15
VB-83	Essex (CV-9)	Lt Cdr D. R. Berry	SB2C-4	15
VB-84	Bunker Hill (CV-17)	Lt Cdr J.P. Conn	SB2C-4	2
			SB2C-4E	13
VB-86	Wasp (CV-18)	Lt Cdr P. R. Nopby	SB2C-4	15

On 1 April, Operation 'Iceberg' went in and initially met only slight resistance, the Japanese plan being to fight further inland from set and prepared positions of great strength.

Death Ride of a Giant

The desperate Japanese defenders had been using suicide pilots, in kamikaze attacks during the Philippines invasion, the practice became widespread, and it reached an awesome climax during the battle of Okinawa. It was here that the last of the great surface ships of the Imperial Navy, the *Yamato*, was destined to end her brief life in a courageous and brave but at the same time, forlorn and futile attempt at her own kamikaze attack. She sailed from Tokuyama Bay at 15:20 on 6 April, accompanied by the light cruiser *Yahagi* and a screen of eight destroyers, *Asashimo*, *Fuyutsuki*, *Hamakaze*, *Hatsushimo*, *Isokaze*, *Kasumi*, *Suzutsuki* and *Yukikaze*. The battleship, had only enough fuel in her tanks for a one-way trip to the Okinawa beach-head. First of all she was to act as a decoy to draw away Raymond A. Spruance's naval forces; should she by some chance survive, she was then to use her nine 18.1in guns against the American transport fleet and supply dumps ashore; and finally she had to beach herself as a permanent fortress there. The plan was a hopeless gamble and she never stood a chance. Her only slim hope was to evade detection for as long as possible so she could get within striking range of her juicy targets – but even this was doomed almost from the outset, because as the Japanese squadron left home waters, they were sighted in the Bungo Strait by two American submarines who sent a detailed description of the force to Task Force 58.

Vice Admiral Marc A. Mitscher prepared to strike with his full strength, eight fleet and four light carriers, and when at 08:32 the next morning Ensign Jack Lyons from the *Essex* (CV-9) sent a sighting report of the enemy, the first of 386 aircraft began to lift off those twelve flight decks to overwhelm the Vice Admiral Seiichi Ito's squadron by sheer weight of numbers. The first attack was despatched at 11:00, in which a force of 280 aircraft sank the *Yahagi* and the *Hamakaze*, and hit the *Yamato* with two bombs and one torpedo. In the second attack, launched at 15:00, 100 aircraft first sank the *Asashimo*, *Isokaze* and *Kasumi*, and hit the *Yamato* with a further three bombs and no less than nine more

torpedoes, finally sending her to the bottom with 2,498 of her crew. Only four destroyers survived this sortie – and the cost to the US Navy flyers was a mere ten aircraft, of which four were Helldivers. Let us examine the part played by the SB2C-4 and -4Es in this massacre.

The *Bunker Hill* (CV-17) contributed ten SB2C-4s to the first strike, which lifted off her decks at 10:15, led by Lieutenant Commander J. O. Conn. The Helldivers were all armed with either two 500lb bombs or a single 1,000lb bomb, and two 250lb wing-mounted bombs. They were accompanied by fourteen TBM-1 Avengers, fifteen F4U-1D Corsairs and two F6F Hellcats of CAG-84. Expectations were high among the SB2C-4 aircrew.

> Reports that the target would be warships of the Japanese fleet made this mission one that all pilots wanted to fly. There had been some juicy targets in Kure harbour about three weeks previously, but the current operation was the first for the squadron against enemy warships underway. Composition of the force was fairly well known. Rounding the southern end of Kyushu during the night, the enemy had been sighted by a scout and word was given to go for the kill.

The squadron duly rendezvoused with the rest of the group, which in turn joined up with other groups in the force and headed for a contact south-west of southern

Kyushu. They set a course to the north-west; crossing the Ryukus they encountered overcast skies at 7,000ft (2,130m) which kept the aircraft low, at 6,500ft (1,980m) altitude as they passed over Amami O Shima and headed out of the East China Sea. They picked up the tell-tale radar blips on their screens at a range of 32 miles (51km) and by the time they had closed the gap to 22 miles (35km) the enemy disposition was well defined and the strike leader was duly informed. Then the Japanese squadron appeared in the distance, an awe-inspiring sight for VB-84's fresh young pilots.

> Despite the knowledge of the whereabouts and composition of the Japanese force, it was still surprising to look over the sides of cockpits and actually see it below. Reports had been accurate, and there was the *Yamato* with her screen of one cruiser and eight destroyers. But even more surprising was the absence of opposition. The entire formation of more than 300 planes flew directly over the Jap force at 6,500 feet and did not draw a shot.

The Helldivers followed the lead of other groups, and circled north of the Japanese fleet, awaiting orders. After some time had elapsed no direct word had been received from the strike leader, and Conn, '… mindful of his plane's diminishing gas load …' decided to attack. At the time this decision was made, '… a destroyer was

SBW-4Es at Naval Air Station, Minneapolis in transit. E. J. Urana via National Museum of Naval Aviation, Pensacola

VB-82 Helldiver aboard the Bennington **(CV-20) in June 1945.** Naval Archives via Emil Buehler Naval Aviation Library, Pensacola

seen speeding below the squadron and it was designated as the target.' It appeared that this destroyer was one of the Japanese force's pickets, as it was several miles from the battleship and her screen; but it was admitted that, '... the exact position of the destroyer to the battleship was not known at that time because of cloud cover over the area.'

When VB-84 deployed for their attack they had to approach under a cloud base of only 3,500ft (1,067m). They went in from the starboard side of the destroyer's wake, and by the time they had made second runs against this ship her anti-aircraft fire had ceased and she was stopped dead in the water, although earlier she had been making flank speed, before the attack. 'White smoke continued to pour from the forward stack for several minutes ...' A short while afterward, three pilots and five gunners observed a terrific explosion aft of No. 3 turret: first there was a red-orange blast, then the fantail seemed to heave and

(Right) **A trio of SB2C-4/4Es from VB-83 over Okinawa in 1945.** Paul Madden via National Museum of Naval Aviation, Pensacola

Composition of the VB-84's Helldiver Strike

Aircraft	Pilot	Aircrewman
1st DIVISION		
1st Section		
201	Lt Cdr J. P. Conn	ACRM L. F. Dougherty
209	Lt (j.g.) R. W. Holladay	ARM2c G. E. Saylor
215	Lt (j.g.) H. R. Gordinier	ARM3c R. W. Maude
2nd Section		
212	Lt H. W. Wiley	ARM1c V. C. Olsen
205	Lt (j.g.) L. T. Hicks	ARM2c A. E. Sanchez
2nd DIVISION		
1st Section		
202	Lt (j.g.) W. B. Kornegay	ARM1c V. C. Rawlings
208	Lt (j.g.) F. A. Swearingen	ARM2c H. E. Pickering
210	Lt (j.g.) R. H. Elliott	ARM3c O. G. Rice
2nd Section		
211	Lt (j. g.) W. R. Lamb	ACRM W. C. Lineaweaver
204	Ensign G. E. Porter	ARM3c W. H. Dykes

shudder, and immediately afterwards black smoke billowed upwards. Observers believed they saw the vessel settling at the stern, but further observation was denied because of cloud cover, and the squadron returned to base.

Pounding the *Yamato*

Their sister squadron, VB-83, operating their SB2C-4Es from the *Essex* (CV-9) and led by Lieutenant Commander D. R. Berry, was part of Task Group 58.3, which included aircraft from the *Bunker Hill* (CV-17) and *Bataan* (CVL-29), with the commander of Air Group 83 as task group strike commander.[146] Commencing at 10:05 VB-83 launched twelve dive-bombers for this mission, eleven with two 1,000lb AP bombs and one with two 1,000lb SAP bombs up.

The combined air groups of Task Group 58.1 and 58.3 approached the Japanese squadron abeam of each other at about 6,000ft (1,800m), which was the ceiling under prevailing conditions. As VB-83 approached from the south-east the target was picked up by ACRM Eardley, the rear-seat man of the leading Helldiver at 30 miles (48km) range, as a single large return on his ASH gear. At 24 miles (39km) separate blips began to appear, and at 9 miles (14km) nine separate targets could clearly be seen. They visually sighted the enemy ships at a range of 4 miles (6km), although only part of the enemy squadron was visible through the broken clouds at 2,500ft (760m). VB-83 was at first told to orbit clear of the target and wait. From time to time the *Yamato* fired her main batteries at the aircraft, but these bursts were never close to VB-83. At around 12:50 the *Hornet* group attacked and then the *Essex* group followed them in. They watched the aircraft of 58.1 make their attacks, which left the cruiser dead in the water, listing to port and burning astern of the formation, with one destroyer protecting her. Then it was their turn.

They were assigned the *Yamato* herself, and led the way in for the 58.3 aircraft.

The co-ordinating of the attack within each air group and between the air groups was excellent. This was because it was impossible to maintain sight contact between the VB and VT leaders due to the weather conditions. The VTs were dropping their torpedoes when the last bombs were exploding. Thus, due to this co-ordination

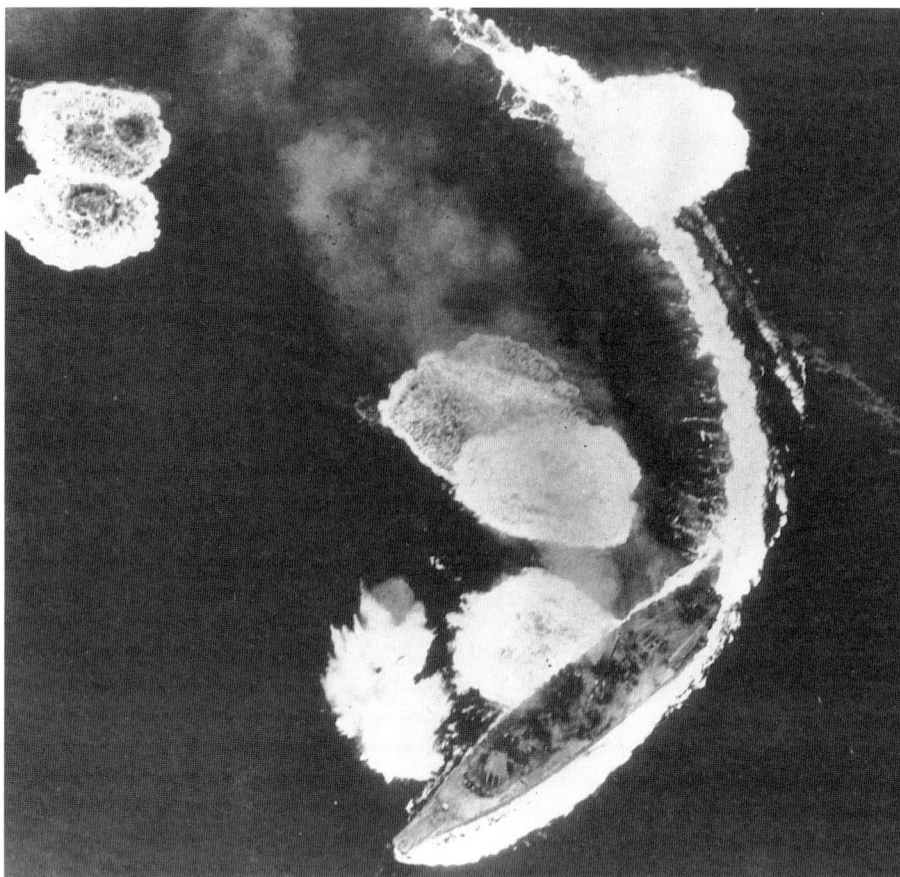

Death throes of a giant. The Japanese battleship Yamato **under attack by US Navy Helldivers of VB-83 on her last sortie, 7 April 1945.** US Navy via National Archives, Washington DC

and weather conditions that prevented the enemy from observing all the planes in the attack, our losses were very small.

Equally professional was the conduct of the Helldiver's opponents.

The manoeuvres of the enemy screening units were considered excellent. It looked more like one of our own task groups than the previous accounts of Jap ships going in all directions. The screening destroyers' AA fire was extremely accurate, as was found by the group commander at each approach near one, and especially when attempting to approach the *Agano* (sic) at 2,000 feet to get pictures. When gun flashes were observed, a diving turn was made and on two occasions the AA blast was felt but there were no holes in the plane.

The Japanese anti-aircraft fire was described by VB-83 as intense:

Heavy bursts were generally black, but about ten per cent were white phosphorous. A few dirty yellow bursts were observed, and there were a few that shot out flaming red balls about one inch in diameter. Throughout the attack the Jap ships maintained excellent formation discipline and kept closed up, even after having been hit. Fire was concentrated on attacking planes, rather than those retiring or preparing to attack. It is believed, from the above indications, that the commander of the Jap task group was an unusually capable officer.

No Helldiver was lost in this attack, but three were damaged by this fire, one was hit by a 25mm explosive shell in the port wing stub, a second took a 13.2mm ball through her port stabilizer and elevator, while the third had a heavy AA hit in the top part of the wing which blew all the skin off the upper surface of the wing between the main and after spars; this holed the wing tank, bulged the lower

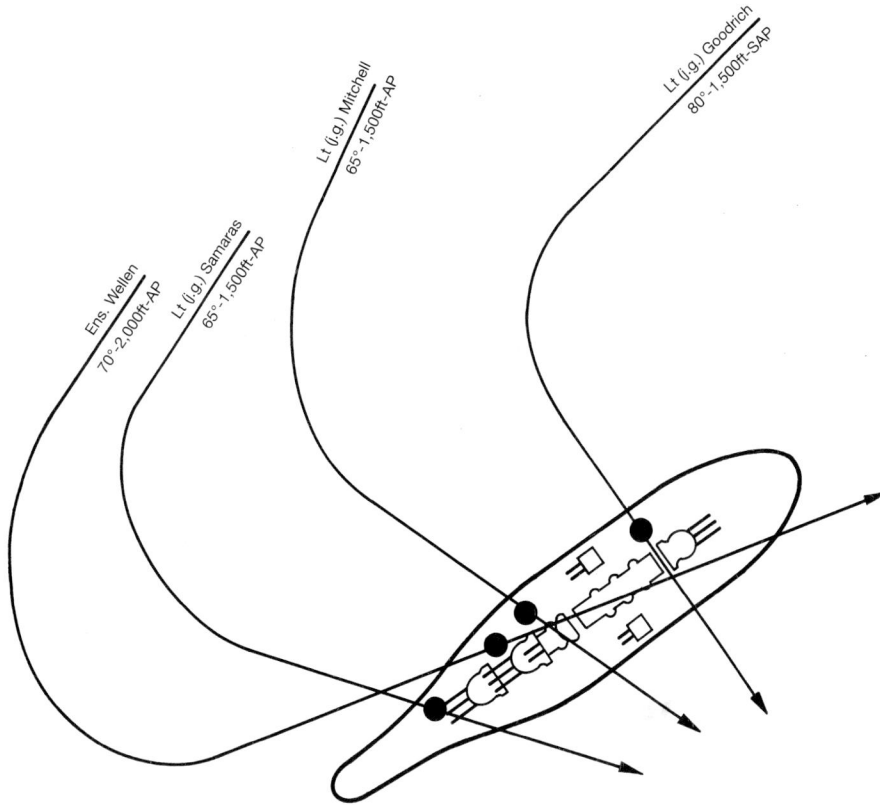

angle of dive, altitude of release, and type of bomb indicated below pilot's name

Bombing Squadron 83 attack on the Yamoto, **7 April.**

Another Helldiver's eye view of a Japanese Yamato-class battleship under attack. US National Archives, Washington DC

wing surface and started a fire below the engine.

Half a minute before VB-83's Helldivers commenced their first dives, the *Yamato* turned to starboard into the attack. Lieutenant Commander Berry led down from the starboard quarter, '… following planes rosebudding around to dive on the starboard bow'. The dive angles varied between 45 and 70 degrees, only one Helldiver using flaps. Soon hits started to thud into the leviathan. Lieutenant (j.g.) Mitchell planted a bomb amidships near the bridge tower structure, after a dive of 65 degrees without flaps and bomb release at 1,500ft (460m), with a pullout at 800–900ft. The tower structure was reported by one VT-83 gunner to have blown up a few minutes later. Ensign Samaras, with a similar dive, hit forward near no. 1 turret. There were two hits by the squadron, one slightly forward of the tower structure and one forward of the no. 3 turret. It was thought that these were made by Ensign Wellen and Lieutenant (j.g.) Goodrich respectively, both of whom made steep, no flap dives, Wellen releasing at 2,000–2,200ft (610–670m) and Goodrich at 1,500ft (460m). (*See* Fig. left.)

After completing their attack dives, all the Helldivers continued down low over the water, and throughout their retirement they were under continuous fire from destroyers of the screen on the side of the pullout. In the last sight VB-83 had of the mighty *Yamato* she was smoking slightly from their hits, but was still well under way.

Sinking the *Yahagi*

Flying from the *Yorktown* (CV-10) was VB-9, equipped with SB2C-4s and led by Lieutenant T. F. Schneider. VB-9 launched thirteen Helldivers at 10:50, eleven of which carried two 1,000lb SAP bombs, and two carried a pair of 1,000lb AP bombs. They formed part of the Special Shipping Strike #1 which comprised twenty F6F Hellcats and thirteen TBM Avengers from *Yorktown*, led by Lieutenant Commander H. N. Houck, along with similar contributions from the *Intrepid* (CV-11) and the *Langley* (CVL-27).

one light cruiser of the *Agano* class, two destroyers of the *Terutsuki (sic)* class and four destroyers of the *Takanami* class being noted. The cruiser was already dead in the water and surrounded by large oil slicks, with a destroyer close alongside either assisting her or evacuating her crew, evidence of the work of the earlier arrivals, although they could see, '... no topside damage of consequence on her'.

VB-9 veered to the northward to skirt the enemy out of gun range and to prepare for their attack, circling some six to ten miles north of the stationary cruiser waiting for the other groups' attacks and for

described as '... fairly accurate ...', although only minor hits were taken by two of the SB2C-4s at this time, one taking a 20mm shell through her starboard wing, while a second had shrapnel hits through the cowl into the oil cooler. To confuse radar-assisted gunnery they dropped 960 sleeves of CAFJ 10721 (282) 'Window' during the mission, to some effect.

Before the main sections could commence their attacks, one Helldiver, piloted by Lieutenant Harry W. Worley, broke his section of three aircraft while still abeam of the *Yahagi*, and they made their attack separately, coming in on the port

Helldivers score a near miss off the starboard side of the Japanese light cruiser Yahagi **and she is pulverized by the aircraft of the US fleet during the last fight of the** Yamato **squadron, 7th April 1945.**
US National Archives, Washington DC

Lieutenant Schneider formed his SB2C-4s into two groups of six and seven aircraft respectively, and headed towards the enemy in company with the other great mass of aircraft. Weather conditions were far from ideal with an overcast at 2,000ft (610m) right through the flight, forcing the Helldivers to stay down at 1,500ft (460m), and at an average speed of 145 knots IAS, on their course of 345 degrees. They reached the original sighting position at 13:05 and changed to a course of 270 degrees for the final interception. They ran into frequent rainsqualls and mist near the coast of Kyushu, '... and bombing conditions did not appear favourable'.[147]

At 13:25, through the rainsqualls, the Japanese force was sighted dead ahead, with one battleship of the *Yamato* class,

their own target assignment. During this time there was a heavy flak burst close to the Helldiver formation from the *Yamato's* main armament, reminding them that the enemy was by no means cowed by their vast numbers. At 13:45 the group commander ordered VB-9 to strike the light cruiser target, and Lieutenant Schneider led his Helldivers aloft into the clouds, spreading out in loose formation and proceeding south to swing into position for a fore-and-aft run. 'The 4,000ft ceiling prohibited the 70-degree split-flap dive the pilots had so arduously practised in view of such a target and necessitated glide bombing runs breaking through cloud cover.' All the time heavy, medium and light anti-aircraft fire was being delivered by both the cruiser and the destroyer which was

beam in glides at about a 35 degree angle and with no flaps. His wingmen, Ensigns Bell and Bowers, followed him down to low-level release, but no hits were scored on either ship by this section.

As Lieutenant Worley was making his recovery after the attack, a lone TBM Avenger, whose torpedo had failed to release on two earlier runs, radioed for strafing support so he could make a third attempt against the Japanese destroyer. Lieutenant Worley, no doubt sensing his section was in the best position to comply with this request, immediately pushed over into a strafing run, engaging the target with his 20mm cannon. He was followed by Ensign J. J. Bell, but his cannon jammed up and he was forced to break off, while Ensign W. W. Bowers had become separated and

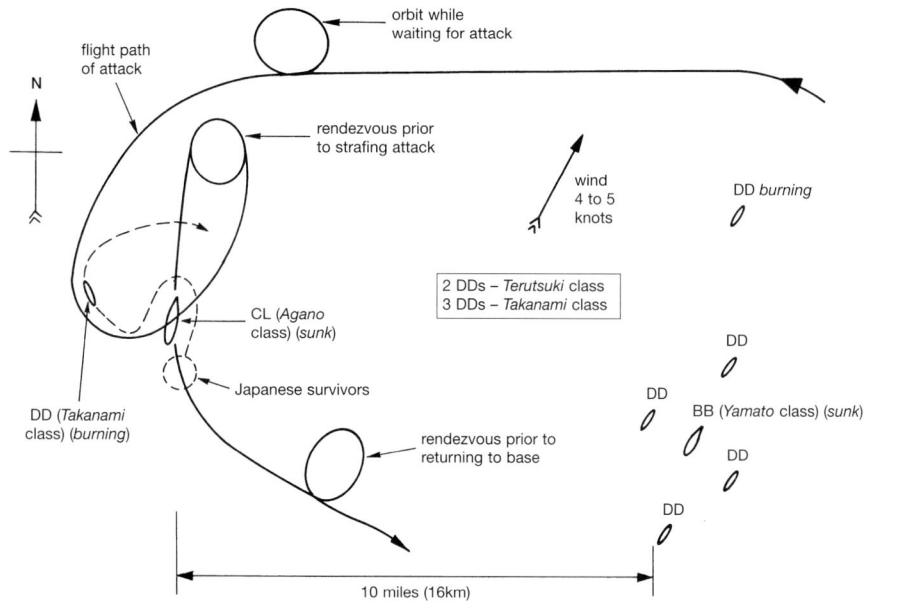

FLEET UNITS AS OBSERVED BY PILOTS OF VB-9

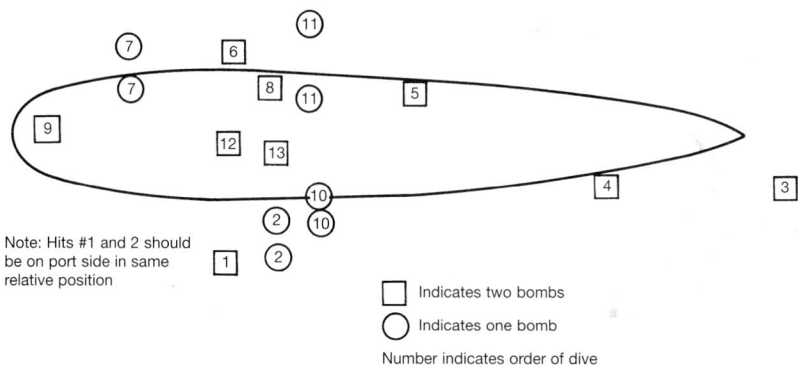

Note: Hits #1 and 2 should be on port side in same relative position

☐ Indicates two bombs
◯ Indicates one bomb

Number indicates order of dive

1. WORLEY, H. W., Lt	7. FRY, L. J., Lt
2. BELL, J. J., Ensign	8. VERRALL, R. L., Lt (j.g.)
3. BOWERS, W. W., Ensign	9. GREENWELL, J., Ensign
4. SCHNEIDER, T. F., Lt	10. DURIO, J. N., Lt
5. PLOSS, R. S., Lt (j.g.)	11. HANAWALT, W. R., Ensign
6. SHREFFLER, J. W., Lt (j.g.)	12. MARTIN, L., Lt (j.g.)
	13. SIGMAN, W. H., Ensign

Location of bomb hits by VB-9 on Yahagi.

(240 to 300m), followed by evasive jinking in a rapid climb back into cloud cover again (see Fig. left). 'The cruiser and DD were using every available gun to track the diving planes; however, due to the disposition of the enemy units, no planned defensive pattern was observed.'

Lieutenant (j.g.) R. S. Ploss, armed with armour-piercing bombs, followed Schneider down and witnessed his leader's bombs hitting near the bridge on the port side, '… but with very little discernible aftermath topside'. Lieutenant (j.g.) J. W. Shreffler's bombs missed off the port side amidships. The next section was led in by Lieutenant L. J. Fry who scored a hit, aft of amidships on the port side with one bomb. His wingmen both made their attacks from the Yahagi's port bow: Lieutenant (j.g.) R. L. Verrall's bombs landed near the ship's bridge, and Ensign J. Greenwell, scored hits on the cruiser's fantail.

The last section in was led by Lieutenant J. N. Durio, who attacked from the starboard quarter in a steep glide: he hit close aboard on that side with at least one of his bombs. Ensign W. R. Hanawalt planted one of his SAP bombs amidships on the port side, and Lieutenant (j.g.) L. Martin and Ensign W. H. Sigman placed their bombs aft and forward of amidships respectively.

After the first hits were made, explosions and fires were seen on the deck of the ship. Several F6Fs made their attacks simultaneously with the bombers, and the cruiser appeared to be covered by smoke and flame as witnessed by nearby observers. During the attack a lifeboat was lowered and evidently the ship was being abandoned.

The Helldivers made their rendezvous north of the blazing cruiser which, about ten minutes after the co-ordinated bombing and torpedo attack, was seen to roll over and disappear. They were not done yet, however:

Survivors and a 50ft motor launch remained in the general vicinity amid the debris, so the Helldivers returned to make a strafing attack, successfully sinking the last vestige of the cruiser's existence. [The report also observed that:] The use of the 20mm by an SB2C as a strafing weapon has proven very effective whenever used. The continued raids on Okinawa enabled pilots to become accurate with this weapon, and so they were able to destroy the 50ft motor launch without difficulty.

had proceeded to the rendezvous area after dropping his bombs. Worley therefore went in alone and was '… brought under intense fire from the DD [destroyer] and took the brunt of the AA. The returning torpedo pilot testifies that he, personally, was not brought under attack until he had released his torpedo and was beginning his retirement.'

Worley's sacrifice was total: hard hit, his Helldiver (Bu. No. 20277) was enveloped in flame and smoke, although he retained control long enough to pull out and break his glide. 'He then resumed a 60 degree dive towards the DD. Returning pilots

testify he was last seen heading directly at the DD, crashing into the water about ten feet off the destroyer's starboard bow. No survivors were seen to parachute or were seen in the water.' Both Lieutenant Worley and his gunner, ARM1c(T) Earl W. Ward, were killed in this gallant attack.

The other ten SB2C-4s were led into the main attack by Lieutenant Schneider, and broke through the clouds in 300-knot dives of between 40- and 50-degree angle, with no flaps, crossing the cruiser from bow to stern in individual bombing runs. Releases were made at about 1,500ft (460m), with pullouts occurring between 800 to 1,000ft

The Helldivers of the USS Bunker Hill **(CV-17) stand amid the smoke and fire as efforts continue to save their ship after the** Kamikaze **hits of 11 May 1945, off Okinawa.**
US Navy via National Archives, Washington DC

While VB-9 rejoined the rendezvous south-east of the scene of action and were circling, they saw the *Yamato* explode and roll over, and she finally disappeared at 14:25.

VB-85 over Okinawa

This was the last major threat from the Imperial Navy, and it had been disposed of decisively; then the Helldivers returned to the more mundane, but equally vital, tasks of assisting their army and marine colleagues ashore on Okinawa, where the fighting had settled into a long and very bloody slogging match. At the same time the fleet was beset by hundreds of kamikaze attackers, enough of which got through to cause damage to a large number of carriers. Replacement carriers and new air groups were continually arriving to take their place, however, and one such was VB-85, commanded by and embarked aboard the brand-new fleet carrier *Shangri-La* (CV-38) which joined the battered fleet at Ulithi on 20 April.[148] Four days later they were assigned to Rear Admiral A. W. Radford's Task Group 58.4, part of the First Carrier Task Force under Vice Admiral Marc A. Mitscher, in the Fifth Fleet, under Admiral Raymond A. Spruance. The group

by this time consisted of the carriers *Yorktown* (CV-10), *Independence* (CVL-22) and *Langley* (CVL-27).

The first combat mission flown by VB-85 took off at 13:30 on 25 April and was led by Lieutenant Commander Arthur L. Maltby Jr, the CO and Lieutenant Alfred G. Symonds Jr, the executive officer. Each Helldiver was armed with a single 1,000lb bomb and four rockets, under each wing, '... suitably inscribed in chalk with messages for the enemy such as "If you hear this one it will kill you".' Their target was Okino Daito Jima where there were important radar, radio and weather stations all vital to the steady flow of reinforcing guiding aircraft in from the mainland. Their task was to knock these out, and a direct hit on the radio station was scored by Lieutenant Merrill J. Hemby, with a damaging near miss on the radar site achieved by Ensign James A. Locke. They met only light anti-aircraft fire and no aircraft was hit or damaged, but the mission had to be cut short due to insufficient time and the rockets were jettisoned at sea.

On 26 April, VB-85's target zone was switched to the front line ashore on Okinawa Jima itself, where they were called upon to support the bitterly contested advance of the US Army against the

Japanese defensive line across the island. The squadron had devoted a large part of its initial combat training to air support exercises, so it had some idea of what to expect. Charles Crumb stated:

> These missions were characterized in general by good communications with 'Whiskey Base', the air support control unit which directed air operations in support of the 24th Army Corps. The gridded target charts were far superior to any used in training.[149]

The first day VB-85 carried out two such missions, with each Helldiver carrying a single 1,000lb GP bomb. On 27 April they carried a single 500lb GP bomb (with 8–11 second delay fuses) and four rockets. Two days later they switched back to cratering enemy runways with an attack on Wan field, Kikai Jima, and for this they again used 500lb GP bombs. The target was covered with five tenths cumulus clouds down to 2,500ft (760m) which ruled out precision dive-bombing: instead, the Helldivers made glide approaches. This of course was slower, and exposed the aircraft to enemy fire for a longer period, and VB-85 paid the price with their first combat loss when two SB2C-4Es were hit by flak, one of which, piloted by Lieutenant (j.g.) Robert W. Elmore, had one wing torn off by a direct hit; he was seen to spin into the sea just offshore and both he and his rear seat man, ARM3c Merrill H. Atwell, were lost. The second Helldiver, piloted by Ensign Roger H. Went, also took hits, one in the starboard wing, which blew a hole in it, and another which started a fire at the port wing root. Fortunately they managed to ditch safely in the water and were rescued from under the enemy's noses by a PBM, and both Went and his rear-seat man, ARM3c Milo J. Ireland, rejoined the squadron a few days later. The second strike was completely unopposed and without incident.

Not all the squadron's pilots were novices, of course – there was a leavening of combat-experienced men and by now VB-85 was working itself into a fully combat-experienced unit, the men having increasing confidence both in themselves and their aircraft. Charles Crump told me that:

> Several of our pilots had flown with other squadrons on prior tours of duty, and I believe, all of these were in the SBD. While they indicated that the SB2C was somewhat more difficult to fly because of its size and power, it was a much more rugged plane and they preferred it over the SBD.

They were also confident enough in their abilities to make on-the-spot changes to the Helldiver, as Charles Crump also related:

Cameras were mounted in the forward edge of the wings of some of the planes, generally about every second or third plane in the order of dive, pointing forward, taking 16mm movies. The 16mm camera was activated by the trigger for the fixed 20mm guns shooting forward. In analysing the movies, we found the camera was not being turned on soon enough to record all of the dives ahead. Therefore we modified the mechanism to start the camera so that it was activated by the master arming switch which was thrown shortly before the pushover for the dive. In this way we were able to record the sequence of bomb hits of several planes in advance of the camera. This was a substantial aid in assessing damage and identifying the bomb drops by the various planes.

Army support missions continued for the squadron on both 1 May (three strikes) and 3 May (two strikes), and on each mission the bomb load was a pair of 500lb GP bombs. The Japanese had a large labour force at their command, however, and no matter how many times the runways in the chain of islands north of Okinawa were cratered, sooner or later they were filled in ready to take transitory aircraft again; so back to Kikai Jima went Task Group 58 and VB-85, and took their place in a combined strike at this airfield once more. The wing-mounted pod containing their AN/APS-4 radar equipment was removed and two 250lb GP bombs were loaded, one on each wing, in addition to the normal two 500lb belly-loads. The enemy had resupplied his gunners, and the Helldivers were met by, '… moderate heavy, and meagre medium and light anti-aircraft fire …'. No aircraft was lost over the target, but that of Ensign William J. Doering took hits. The pilot nursed his Helldiver back to the carrier, but on receiving a 'wave off' his throttle failed to respond and he crashed into the sea alongside the ship. Fortunately the 'plane-guard' destroyer was quickly alongside and rescued both Doering and his gunner, ARM3c William W. Helterbrand.

Mission accomplished, VB-85 returned to air support missions on southern Okinawa itself for the period 6 to 11 May. The first attack on 6 May was against enemy AA positions on the ridges, and the ten-plane strike was armed with two 500lb bombs apiece, nineteen of which were dropped on the target. Two SB2C-4Es returned to hit these guns again, while a third attack was put in later by eleven Helldivers. In all three strikes the bombs were supplemented by strafing fire with both 20mm cannon and .30 cal machine guns and no reply was received from these sites after the first strike. After a short lull, attacks were continued on 9 May with an eight-plane mission against a supply dump, which caused serious damage, and a four-plane strike against Japanese troops. The next day enemy gun positions and troops were again the bombing and strafing targets for VB-85's Helldivers, flying

Long-range drop tanks litter the vulnerable decks of this Essex-**class carrier as 72 and 65, with wings folded, are towed by the flight-deck tractors to join the deck park aft. If a** Kamikaze **caught a carrier in this condition, the resulting fireball was horrendous.** US National Archives

so low that one was slightly damaged by its own bomb fragments. This period of intense combat ended on 11 May with two more eight-plane strikes against enemy artillery positions, pillboxes, caves and troops. In each case the bomb load was two 500lb GP bombs, but on the last strike of this period rockets were utilized to good effect as well.

After a short spell of rest and refurbishment, Task Group 58.4 again left Ulithi to resume their pounding, the carriers in the group now being *Shangri-La*, *Yorktown*, *Ticonderoga* (CV-14) and *Independence*:

> Co-ordinated attacks were made on the ships of the Group and on towed spars. In addition the squadron flew on three-plane anti-submarine patrol. Rendezvous at the replenishing area was made with the other units of Task Force 58 on 28 May. On that same day the Fifth Fleet became the Third Fleet under Admiral W. F. Halsey, USN, and Vice-Admiral J. S. McCain, USN, became Commander Second Carrier Task Force. Rear Admiral Radford remained as Commander Task Group 58.4, which took the new designation of 38.4. With Vice Admiral McCain embarked in the *Shangri-La* it became apparent that Air Group 85 would conduct its future operations under the watchful eye of the Task Force Commander.[150]

Missions resumed against shore targets on Okinawa once more, and eight sorties were flown between 29 May and 7 June, in each case with the usual two 500lb GP bomb loading, These attacks were made without loss, and were only interrupted by the appearance on 4 May of a typhoon with winds up to 54mph (87km/h), which caused aircraft to be battened down while the ships rode it out; but no losses were suffered, and strikes resumed again on 6 May.

The VB-85 dive-bombers went back into action again on 9 and 10 June, taking part in full squadron strikes against Minami Daito Jima with fifteen and fourteen aircraft respectively. Their armament was VT-fused 500lb GP and 260lb fragmentation bombs designed specifically to knock out the AA guns on the island. Tactics with this fuse called for a release point of about 4,000 to 5,000ft (1,220–1,525m). Although anti-aircraft fire encountered was only meagre to moderate, this high altitude of release proved a boon to flying personnel on later attacks in the Tokyo area.[151]

The fleet then departed for Leyte Gulf, bringing to an end the first phase of VB-85's work. It was typical of the outstanding contribution the Helldiver was making in this period of the war.

A montage of photographs of VB-85 operating from USS Shangri-La **(CV-38) in the Third Fleet, Pacific, 1945.** Charles M. Crump

CHAPTER SEVENTEEN

End of the Road: Japan

The continuing acceleration of the drive towards the invasion of mainland Japan itself, Operation 'Olympic', destined to take place in 1946, marked the climax of the 2C's part in the Pacific War and its final vindication. But development still continued back in the States as further wartime combat experience led to considerations of yet further modifications.

Concept of the XSB2C-6

Satisfactory as the Helldiver had gradually become, the original basic design flaw, that of longitudinal stability, remained. As one historian noted, '… ingenuity could not alter the laws of aerodynamics.'[152] While Curtiss, the US Navy and NACA continued to seek some simple method of mitigating this built-in limitation, it was clear that only by shifting the weight forwards could the '2C' be correctly balanced.

One idea was the shifting of the bomb shackles, and this was done in the final production model, the Dash-5, which we examine later. A more radical approach was decided upon when wind-tunnel testing had confirmed that the correct balance could only be achieved by moving the engine further forward and lengthening the tail. Opportunity was also taken, in the steady march of technology, to incorporate an improved powerplant at the same time, the thuggish 2,100hp Wright R-2600-22. This involved some nose redesigning to cope with this size of engine, and incorporating two air-scoops housed in prominent bulges on the upper cowling which made for a longer, flatter profile. Two SB2C-1Cs (Bu. Nos. 18620 and 18621) were thus modified into prototypes as the XSB2C-6.

Specification – Model 84H XSB2C-6	
Engine type:	R-2600-22
Dimensions:	Height 13ft 1½in (4.01m); wingspan 49ft 8⅜in (15.15m); wing area 422 sq ft (39.2 sq m)
Number completed:	2
Serial numbers:	(Ex-18620/21) (Bu. No.-Navy – Serial Army)
Contract:	Delivery date; Aug/Sept 44

Air/Sea Rescue Unit, Kaneohe, on 27 December 1945. Frank R. MacSorley Collection via National Museum of Naval Aviation, Pensacola

Other modifications included extending the vertical and horizontal tail surface, while at the same time narrowing its chord and adding a dorsal fin. The wing tips were 'squared off' and these combinations resulted in a much improved performance. Despite this fact there never was an SB2C-6, because the end of the war was clearly in view and other concepts had a higher priority. The plan was dropped and the two prototypes were utilized in various factory design experiments.

Concept of the XSB3C-1

This was Curtiss-Wright's planned replacement for the Helldiver, and it featured their massive 2,300hp R-3350-8 engine. The weight of this installation led to a redesign which featured a tricycle undercarriage (in much the same way as Douglas had modified one of its BT-1s the year before, and was to feature on its BTD destroyer dive-bomber design later in the war). With such a tremendous powerplant the new aircraft was expected to tote a bomb load of up to 4,000lb (1,814kg); it could also double-up as a torpedo bomber, being capable of lifting two of the standard US Navy air torpedoes of the day. The USAAC also expressed firm interest in this design, and allocated it the designation XA-40-CS.

A full-scale mock-up was built during 1941, and it looked very impressive; but things never progressed any further than that. It was not just aircraft design that decided the fate of a particular type or mark – other practicalities of war also came into play in the scheme of things. If the Navy had demanded the SB2C-6, then Curtiss would have had to close down the existing production lines at Columbus and to start all over again with retooling. This would have led to a repeat of the earlier confusion and delay, something the Navy did not want to go through again.

What finally scuppered the SBC3-1 for the Navy was fuel availability: the eventual planned introduction of the huge 2,300hp Wright R-3350-8 engine for this type which promised yet better performance, was only achieved by using grade 115/145 aviation fuel, and stocks of that grade just did not exist in sufficient quantities in the combat zone.[153] The US Army persevered with the concept for a while longer, but their subsequent policy to drop *all* specialized dive-bombers ultimately, consigned the XA-40-CS to oblivion also.

Warm-up: Wake Island again, June 1945

The last phase began with an attack by Task Group 12.4 – which contained the carriers *Hancock, Lexington* and *Cowpens*

Helldivers of VB-89 lined up aboard the *Antietam* **(CV-36).** National Archives via National Museum of Naval Aviation, Pensacola

commanded by Rear Admiral Jennings – on Wake Island, so often 'pasted' before and a deep symbol of America's determination in this war. On 20 June, VB-6's SB2C-4s from *Hancock* joined with *Lexington's* VB-94s in hitting established enemy targets on that island. This marked the combat debut of the 'Dash-5'.

As carriers arrived from the West Coast with their new air groups, these attacks on Wake were repeated, by SB2C-4s from *Wasp* (CV-18) on 18 July, and again on 6 August by SB2C-4s of VB10 from *Intrepid* (CV-11).

Arrival of the 'Dash-5'

The arrival of the SB2C-5 finally brought all wartime experience and years of modifications together in a package that suited most of its users. Three prototypes appeared as XSB2C-5s. Two Navy-type SB2C-4s (Bu. Nos. 65286 and Bu. No. 83127) were modified, as was a former SB2C-1C (Bu. No. 18308). These proved successful, and the Navy placed orders for 3,730 Dash-5s, of which some 970 were built by Curtiss-Wright, and eighty-five SBW-5s were built by Canadian Car & Foundry (Bu. Nos. 60211 to 60295). At the conclusion of the war, 2,500 SB2C-5s and 165 SBW-5s were cancelled outright.

The -5 could be distinguished by its increased fuel capacity – an extra 35 gallons could be carried in internal tanks; a frameless sliding canopy for the pilot, giving much clearer all-round vision; a spinnerless four-bladed, paddle-bladed propeller; and the fixing of the tailhook in its extended position. The ASB radar was also dropped from this mark but as the AN/APS-4 radar pod could always be fixed underwing this was not the retrograde step it first appears. Production of this mark, commenced as early as February 1945, and they started to reached the front-line squadrons in June, with VB-87 deployed aboard the *Ticonderoga* (CV-14), VB-93 aboard the *Boxer* (CV-21), VB-94 aboard the *Lexington* and VB-150 aboard the *Lake Champlain* (CV-39).

Specification – Model 84G XSB2C-5	
Engine type:	R-2600-20
Armament:	Two 20mm forward-firing fixed cannon; two 30mm rear-cover flexible machine guns; maximum bomb load 2,000lb (907kg); 1-Mk 13 torpedo
Dimensions:	Length 36ft 8in (11.17m); height 13ft 1½in (4.01m); wingspan 49ft 8⅜in (15.15m); wing area 422 sq ft (39.2 sq m)
Weights:	Maximum weight 15,918lb (722kg); empty weight 10,580lb (4,799kg)
Performance:	Maximum speed 260mph (418km/h) Take-off power 1,900hp (1,417kW) Range as dive-bomber 1,805 miles (2,905km) Climb height 10,000ft/8.9min (3,048m/8.9min) Service ceiling 26,400ft (8,047m)
Number completed:	1
Serial numbers:	(ex-65286) (Bu. No.-Navy – Serial Army)
Contract:	Order date; 1944 Delivery date; Feb 45

SB2C-5 firing rockets. J.C. Woodward

Main Assault Commences

On 1 July the full might of Task Force 58 set out from Leyte, under the command of Admiral Halsey, determined to finally avenge Pearl Harbour totally. In command of the carriers was another veteran, Rear Admiral 'Killer' McCain. The disposition of the Helldiver squadrons in the fleet is given below.

The fleet carried out oiling at sea throughout the 7 and 8 July, and then moved into position off Tokyo and began launchings on 10 July. In all, an awesome 1,022 naval aircraft were sent against enemy airfields on the Tokyo Plain. Typical of the Helldiver missions over Japan at this time, Lieutenant Commander Maltby led VB-85 which took off at 07:15 that morning, '... into crisp, clear air, with ceiling and visibility unlimited'. Their particular target was Japanese aircraft located in their revetments at Kasumigaura airfield. Crossing the coast north of Chosi, the '2Cs' headed west up the Tone river valley: arriving over the target, they pushed over from 12,000ft (3,660m), and dropped their VT fused fragmentation bombs at 4,500ft (1,370m) before retiring over Kasumigaura Lake. Despite heavy AA fire, none of the Helldivers was damaged and all got back safely. In the afternoon they mounted a similar strike at Konoike airfield on the east coast of Honshu, north of Chosi, again without loss. On neither occasion did they encounter any aerial opposition.

After this massive affirmation of US Navy power, the fleet headed north to a position east of the Tsugaru Strait where it replenished once more. Thick fog on 13 July, rather than any great Japanese attempts to dispute control of its own backyard, hampered operations for a while; but strikes

Specification – Model 84G SB2C-5	
Engine type:	R-2600-20
Armament:	Two 20mm forward-firing fixed cannon; two 30mm rear-cover flexible machine guns; maximum bomb load 2,000lb (907kg); 1-Mk 13 torpedo
Dimensions:	Length 36ft 8in (11.17m); height 13ft 1½in (4.01m); wingspan 49ft 8⅜in (15.15m); wing area 422 sq ft (39.2 sq m)
Weights:	Maximum weight 15,918lb (722kg); empty weight 10,580lb (4,799kg)
Performance:	Maximum speed 260mph (418km/h) Take-off power 1,900hp (1,417kW) Range as dive-bomber 1,805 miles (2,905km) Climb height 10,000ft/8.9min (3,048m/8.9min) Service ceiling 26,400ft (8,047m)
Number completed: 970	
Serial numbers:	83128/83751, 89120/89465 (Bu. No.-Navy – Serial Army)
Contract:	Order date; 1944 Delivery date; Feb/Aug 45

Specification – CC&F SBW-5	
Engine type:	R-2600-20
Armament:	Two 20mm forward-firing fixed cannon; two 30mm rear-cover flexible machine guns; maximum bomb load 2,000lb (907kg); 1-Mk 13 torpedo
Dimensions:	Length 36ft 8in (11.17m); height 13ft 1½in (4.01m); wingspan 49ft 8⅜in (15.15m); wing area 422 sq ft (39.2 sq m)
Weights:	Maximum weight 15,918lb (722kg); empty weight 10,580lb (4,799kg)
Performance:	Maximum speed 260mph (418km/h) Take-off power 1,900hp (1,417kW) Range as dive-bomber 1,805 miles (2,905km) Climb height 10,000ft/8.9min (3,048m/8.9min) Service ceiling 26,400ft (8,047m)
Number completed: 86	
Serial numbers:	60210/60295 (ex-SWB-IBs) (Bu. No.-Navy – Serial Army)
Contract:	Order date; 1944 Delivery date; Mar/May 45

SB2C Units at Start of the Final Drive on Japan, July 1945			
Task Group	Carrier	Unit	Type
TG. 38.1	Bennington (CV-10)	VB-1	SB2C-4
	Lexington (CV-16)	VB-94	SB2C-5
	Hancock (CV-19)	VB-6	SB2C-4
TG. 38.2	Ticonderoga (CV-14)	VB-87	SB2C-5
	Randolph (CV-15)	VB-16	SB2C-4E
	Essex (CV-9)	VB-83	SB2C-4
TG.38.3	Yorktown (CV-10)	VB-88	SB2C-4
	Shangri-La (CV-38)	VB-85	SB2C-4

against Northern Honshu and Southern Hokkaido and shipping in the Tsugaru Straits were resumed in still murky conditions on 14 July, and were repeated the following day.

On 14 July, only nine out of thirteen of VB-85's '2Cs' launched from the *Shangri-La* were able to rendezvous due to very low ceiling, and these were led by Lieutenant M. J. Hemby, USNR, against five freighters found north-east of Muroran. Five Helldivers carried out attacks against these ships, with Hemby and Ensign M. L. Skinner scoring direct hits. The other four aircraft, led by Lieutenant (j.g.) George G. Shoemaker, USNR, seeing that all the ships were under attack, sought targets of opportunity along the coast, they hit railway tracks, and strafed a train near Tomakomai. One of the Helldivers, piloted by Lieutenant (j.g.) Howard E. Eagleston Jr, was seen to pull up into the low overcast and vanish from view: he never made the rendezvous and both he and his rear-seat man, ARM1c Oliver B. Rasmussen, were listed as missing.

Further anti-shipping strikes were made in the afternoon as the weather lifted, with Lieutenant E. F. Gibson, USNR, leading an attack on freighters in Muroran harbour. They met 'moderate to intense' flak

Bad weather again hampered operations on 15 July, VB-85, for example, only getting away one strike that day. Muroran being closed in by cloud and fog, they pushed on south-west and struck at shipping and shore installations at Yokomo. One three-plane section, led by Lieutenant Thomas C. Thomas, with Lieutenant (j.g.) Lacina and Ensign J. J. Scherting, scored damaging hits on one freighter which left it '... burning furiously ...'; but other attacks were not so successful: Four '2Cs' hit docks and buildings in Yokumo town, another hit the railway at Mori.

The '2C's' had found only small fry, but had ventured out to sea and dealt with them in a proficient manner, the Japanese losing

On retirement from the Hokkaido area it was planned to attack the battleship *Nagato*, which had been located secured alongside a dock at the heavily defended Yokosuka Naval Base at the southern end of Tokyo Bay. Two strikes were planned for 18 July, but only one eventually got away due to bad weather. The Helldivers of VB-85 were each armed with a single 1,000lb G.P. for this afternoon mission, these weapons being fitted with the new water-discriminating fuse. Due to the action of these fuses, the aiming point prescribed in their operation order was the battleship's waterline on the port side, so that bombs dropping on, or short, would have a mining effect, while those hitting just over

An SBF-4E, flown by Ensign R. L. Jacobsen, making the 25,000th Helldiver landing aboard the Solomons (CVE-67) on 7 August 1945. National Museum of Naval Aviation, Pensacola

in their dives but, '... due to a combination of good luck plus an application of the standard evasive tactics of 'zig and zag and go like mad', no planes were damaged.' Gibson, Lieutenant William Gauvey and Lieutenant (j.g.) Hilmer W. Forsgren all scored hits on cargo ships, while Lieutenant William W. Perry and Lieutenant (j.g.) James F. Lacina severely damaged a gunboat and Lieutenant (j.g.) Richard W. Mann probably sank a small coaster.

49,000 tons of merchant shipping as well as six small escort ships, with a further 50,000 tons' worth of merchant shipping and another thirteen minor warships damaged. Showing complete contempt for the enemy American and British battleships and cruisers sailed up and down the Japanese coast carrying out bombardments of important factory sites and sweeping the East China and Yellow Seas as far as the Yangtse estuary at Shanghai, without challenge.

would damage the ship's superstructure.

Air Group 85 made landfall at the western end of Sagami Wan and set course to approach the target from the south-west. The Helldivers followed the Hellcat fighter-bombers and Avenger glide bombers, both equipped with VT fused fragmentation bombs to obliterate the ship's AA guns; but even so, the defending fire which rose up from the battleship to meet the '2Cs' was described as intense: 'Not only

A SB2C of VB-85 from Shangri-La **(CV-38) in August 1945.** US Navy via National Museum of Naval Aviation, Pensacola

were the many guns of the naval base, the Yokosuka Naval Air Station and surrounding areas active, but a flak ship, destroyers and other ships in the harbour were also firing. The *Nagato* herself appeared to be firing all secondary guns until these were silenced by several thousand pound bomb hits on the ship.'[154] There was no aerial opposition encountered.

The dive-bombers scored four hits on the prescribed aiming point, and four bombs detonated aboard the battleship itself. Her superstructure could be seen to be badly smashed in the resulting photographs, but the underwater damage could not be ascertained. In return, five '2Cs' took hits, including Maltby's in which an explosive shell severed all his control cables on the port side and took away his tail-hook; despite this, he got his aircraft back and made a safe deck landing. Lieutenant Gauvey took a burst on his way down, but found he still had full control – though his rear-seat man, ARM3c Alfred P. Bonosconi, was not so fortunate and received lethal wounds from this burst, despite which he managed to observe and

A Curtiss SB2C-4 of Fasron One. Emil Buehler Naval Aviation Library, Pensacola

report a direct hit. However he died of his wounds on the flight back to the carrier. Lieutenant (j.g.) Hungerford's dive-bomber was hit while retiring from the target, and

his rear seat man, ARM1c George E. Kaufmann received severe fragmentation wounds in his arms and legs, and fractures of his left leg.

146

A typhoon warning was received on 19 July and the fleet withdrew for another replenishment, it was not until 24 July that attacks were again resumed. This time the remnants of the Imperial fleet were hit, holed up in their bases around the inland sea at Kure and Kobe harbours. The blows struck were hard ones, and the battleships *Haruna*, *Hyuga* and *Ise*, as well as the heavy cruiser *Aoba* and the training ship *Iwate*, were all hit and settled on the bottom – a fitting retribution for the Pearl Harbor attack. The carrier *Amagi* and the light cruiser *Oyodo* were both sunk outright, as were fifteen merchant ships, while

was made the target of live dive-bombing and severely damaged.

This was not all accomplished without loss, of course, and VB-85 recalled that these final attacks were, '… very successful but costly …'. For instance on that particular day VB-87, fresh out from training in Hawaii with its new aircraft, lost eight out of thirteen '2Cs' while sinking the *Hyuga*, or forced to ditch through lack of fuel on their return to the fleet. Nor was this the end of the carnage from the air, because after another refuelling, the Allies were back again on 28 July. Further blows were made on many of the above ships: more signifi-

25mm light weapons so she was quite a 'bristly' target. She was first attacked at Shikoku on the morning of 24 July by the thirteen Helldivers of VB-85, led by Lieutenant Commander Arthur L. Maltby. They scored three direct hits from the dives of Lieutenant Edward F. Gibson, Lieutenant (j.g.) Richard W. Mann and Ensign Glen M. Even, with another five very near misses, all proved from photographs taken at the time. Richard Mann graphically described his experiences thus:

We pushed over into our dives at approximately 13,000 feet. Our target was the Japanese

An **SB2C-5 of VB-150 from the** Lake Champlain **(CV-39) in June 1945.** US Navy via National Museum of Naval Aviation, Pensacola

the aircraft carriers *Hosho*, *Katsuragi*, *Kaiyo* and *Ryuho*, the light cruiser *Kitakami* and destroyers *Yotsuki*, *Hagi*, *Kaba* and *Tasubaki*, and three lesser warships were badly damaged. As a final telling and symbolic blow, the old Japanese battleship *Settsu*, which for years had been used as a radio-controlled target ship for Japanese dive-bombing practice – much like the American *Utah* and the British *Centurion* had been in their respective countries –

cantly the heavy cruiser *Tone*, the training ship *Izumo*, the destroyer *Nashi* and submarine *I-104* were sunk, along with eight auxiliary vessels, while nine smaller warships, another submarine and the large submarine depot ship *Komahashi* were badly damaged.

The *Tone* was a veteran ship with a good war record and a reputation as a crack AA ship. As well as her main armament of eight 8in guns, and eight 5in heavy AA guns, she mounted no less than fifty-seven

heavy cruiser *Tone*. I could hear the flak crackling all around me on the way down in an almost vertical dive. At 1,500 feet I released my bomb load, consisting of a 1,000lb G.P. and a 260lb fragmentation bomb – both of which scored direct hits.

On pullout some Jap gunner scored a direct hit on my left wing which burned so badly I was forced to ditch in Kure Harbour. Both Robert F. Hanna, ARN2c, my radioman-gunner, and I were fortunate in getting out of the plane with only minor injuries. My hand was broken and badly cut and Hanna received a scalp wound, a cut on the arm and a strained back.'[155]

From above his colleagues saw him make '… a hard water landing …'. No survivors were seen to get out and he was posted missing in action, but in fact both Mann and his rear-seat man survived, and after some very bad times as POWs, were to rejoin their comrades after the war.

Having already been badly damaged in VB-85's earlier attack, the *Tone* was located anchored in Eta Shima Bay; her final destruction was allocated to Air Group 87 aboard the *Ticonderoga*. The twelve SB2C-4Es of VB-87 took off at 07:45 that morning, each laden with one 1,000lb SAP and two 260lb fragmentation bombs, and accompanied by the TBM-3E Avengers. They were escorted by the F6F Hellcats of the Group, and set course for Japan, although one '2C' had to abort the mission early on with engine problems.

Over the city of Hiroshima itself they were met with light flak, but the eleven Helldivers commenced their dives from 11,500ft 3,500m), releasing high, at around 3,500ft (1,000m), and claiming four direct hits on the cruiser, three amidships and one on her starboard quarter. The ship settled on the bottom, finally out of the war for good. The price was one '2C' that had been badly hit in return. This aircraft was piloted by Lieutenant (j.g.) Raymond Porter, with ARM3c Normand Brissette as his rear-seat man: at the rendezvous point his engine was seen to be smoking badly and it soon dropped out of formation, although Porter managed to ditch her safely in the Iyo Nada area of the Inland Sea. His squadron mates saw both men escape from the wreck and enter their life-raft. Alas, there was to be no happy ending, because having survived the hit and the water-landing, they were among the American prisoners incarcerated in the Chigoku Military HQ in Hiroshima City when the first atomic bomb was detonated. Porter died instantly in the blast, Normand survived the resulting firestorm, but died on 19 August from acute radiation poisoning.[156]

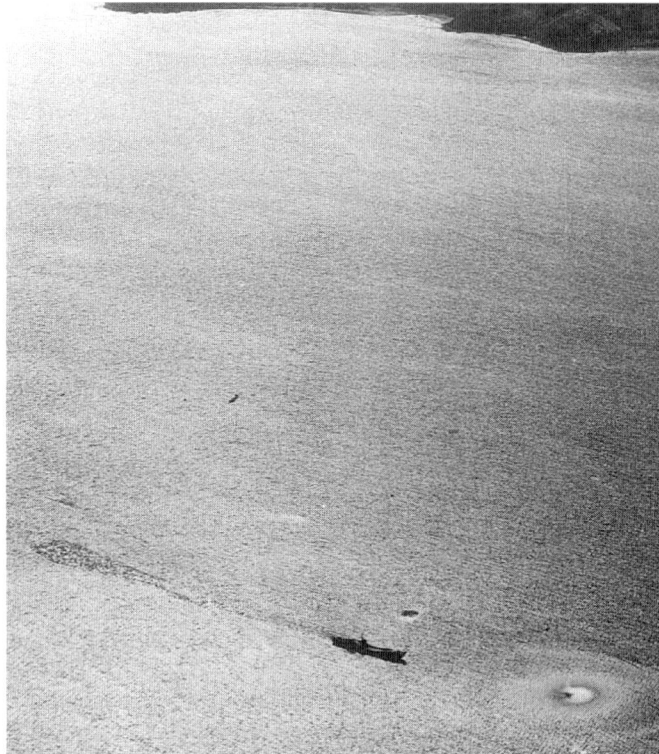

US Navy SB2C-4 Helldivers attacking a Japanese Navy landing craft off mainland Japan, April 1945. Official US Navy Photo

End of the Japanese Battle Fleet

On the afternoon of 24 July the attacks on the remnants of the Imperial Fleet continued unrelentingly. There is little doubt that the remaining Japanese big ships, had they been able to make the choice, would have emulated the last sortie of the *Yamato* and sailed to give 'do-or-die' battle against the Allied battleships off their shores; but they were powerless to do so, lacking the fuel for even brief suicidal sorties. They just had to sit and take it, and they proved tough and resilient opponents to the bitter end. A strike was sent off against Kure by VB-85, which after crossing Shikoku, found heavy cloud and haze over the Inland Sea, causing them to lose cohesion. They made a circuit over Kure and then began their attack, '… with the planes strung out in what could hardly be called a formation …' Nonetheless the '2Cs' were effective, one four-plane section targeting the light cruiser *Oyodo* and with Lieutenant A. G. Symonds Jr,

USNR, and Lieutenant Hemby both scoring direct hits on this ship, with three other very near misses. It was enough to finish her off, and she duly rolled over onto her side.

Five of the Helldivers took on the battleship *Haruna* and attacked her through heavy defending fire. A direct hit was scored by Lieutenant Hemby, with three near misses close alongside. As they flew over Kure through thick flak on the way out, Gibson's rear-seat man, ARM2c Charles H. Linsz, was taking photographs with an F-8 camera of the battleships *Haruna, Hyuga* and *Ise,* for future attacks, these pictures being the first close-ups of the two converted battleship-carriers' flight-decks.

Casualties were taken in this attack: Lieutenant Alfred G. Symonds was reported as crashing in flames near the target, and he and his rear-seat man, ARM1c Edward Hicks, were posted missing in action. Others had close shaves, Lieutenant (j.g.) Rexford W. Jones took a 40mm explosive shell in the cockpit over Kure harbour which injured him and demolished his instrument panel, destroyed the generator control box and cut the hydraulic line so that fluid sprayed in his face. Before his radio gave out Lieutenant Hemby managed to give him instructions for joining up, and then led him over the mountains. With Lieutenant Gauvey and Ensign C. L. Trump as wingmen, Hemby, unable to locate the rescue submarine because it had submerged, directed Jones towards the nearest picket destroyer and he ditched alongside her safely, both he and his aircrewman, ARM2c Paul R. Isenberg, were safely picked up. This done, the three Helldivers now found themselves very low on fuel, and flying in darkness. While Gauvey got down safely on *Shangri-La,* Trump had to go aboard the *Yorktown,* the first carrier he could find in the darkness. Hemby also got to his home base, but just as he was '… coming up the groove …' his engine cut on him as he reached the ramp, and those watching with baited breath on the

A fine aerial view of an SB2C-5 bearing the squadron designation V-35, circa 1945. National Museum of Naval Aviation Library, Pensacola

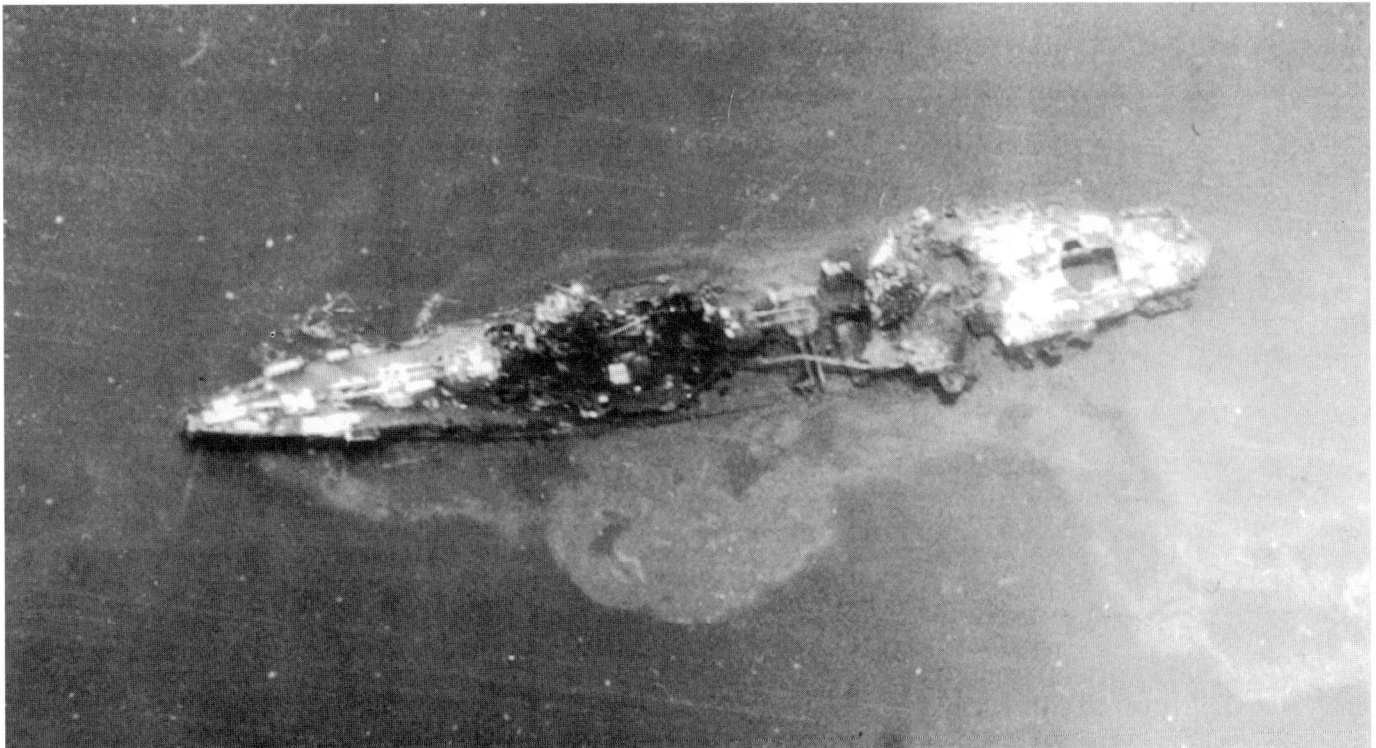

The Japanese battleship Ise on the seabed at Kure after being sunk by the combined attentions of the Helldivers and Avengers of task Force 38 on 19 March 1945. Note the flight deck aft with bomb damaged hangar open and the oil slicks surrounding the ship, 18 July 1945. US National Archives, Washington DC

carrier saw his Helldiver disappear below the flight deck. Word was being passed round that he had gone into the water when he was seen again, flying up the starboard side of the *Shangri-La* below the level of the hangar deck! He had shifted from a dry tank to an 'empty' one, and with the mixture control in automatic lean, came safely aboard at his next approach. Very quick thinking indeed!

A strike on 25 July was aborted due to bad weather, VB-85 jettisoning bombs off the southern coast of Shikoku. On 28 July VB-85 returned to the Inland Sea with the battleship *Haruna* as their assigned target. She was moored in an inlet on the eastern shore of Eta Shima, some three miles west of the Kure naval base. The '2Cs' made their approach from the south-east at 14,000ft (4,270m), peeled off to the right and dived on the target in a first-rate attack. Four direct hits were made by Lieutenant Commander Maltby, Lieutenant Hemby, Lieutenant (j.g.) Hungerford and Lieutenant (j.g.) William J. Doering. In addition, photographs showed three very near misses close aboard the ship and two

other near misses. But again, a heavy penalty was exacted: as the Helldivers retired they had to fly over Nishinomi Shima on which the Japanese had many AA guns sited, and as they did so three aircraft were damaged by flak. Lieutenant (j.g.) Shoemaker took an explosive shell into the wing-fold hydraulic lines and the resulting fire burnt until the fluid was exhausted. Lieutenant Thomas P. Gavin's aircraft was hit in the outboard section of the wing with only minor damage. The rear-seat man of Ensign S. G. Payne's '2C', ARM2c James L McCarthy, had a tracer pass *through* his .30 calibre ammunition belt, less than two feet away from him, which exploded several rounds – incredibly he suffered no injuries from this. All three aircraft got back safely.

The squadron was sent back to repeat the attack that same afternoon, but could only mount a seven-plane instead of a planned fifteen plane strike, due to damage to aircraft. The cut-back in the number of Helldivers carried caused this, as there were now insufficient 'spare' aircraft to immediately make up for losses. Led by

Lieutenant Gibson, these seven dive-bombers were attached to the *Yorktown's* VB-88, which had the strike group leader. VB-88 led in first against the *Haruna*, and VB-85 followed them down against the same target, diving from north to south; but they found that '… again, anti-aircraft gunners, who by this time must have had a great deal of practice on retiring planes, put up an intense and accurate fire.' One direct hit with a 1,000lb bomb was made by Lieutenant Gibson, and there were very near misses as well.

Losses were again severe. Lieutenant Gibson's aircraft was caught in a vicious cross-fire as he turned to cross the island of Nishinomi Shima, and his '2C' was observed to burst into flame in the area of the front cockpit and crash heavily into the water, killing both the pilot and Linsz. The same gunners hit Lieutenant (j.g.) Maynard J. Mitchell's Helldiver, seriously wounding his rear-seat man, ARM2c William C. Pinkerton. The latter recalls events thus:[157]

The *Haruna* was at a berth in Kure harbour. Our squadron scored two hits that I saw, but I don't

The Japanese battleship *Haruna* **was claimed as 'sunk' more times by the Flying Fortress crews of the USAAF between 1942 and 1945 than the British carrier Ark Royal was claimed sunk by the Luftwaffe between 1939 and 1941, but they didn't even scratch her in all that time. Here she is seen overwhelmed by direct hits and near misses from the SB2C Helldivers of Task Force 38 at her anchorage at Kure harbour, her final resting place, 28 July 1945.** US National Archives, Washington DC

know who scored them. The flak was very heavy and intense. It was also very colourful, because the Japanese used various smoke colours in the AA shells so they could see where they were shooting. Also they had white phosphorous shells which streamed among coloured smoke, – very pretty!

After releasing our bombs Mitch applied full power and we got out of there as fast as we could. We joined up with Lieutenant Gibson and gunner Chas. Linsz. We were leaving the area at tree-top and wave-top height when we crossed some small islands and a point of land. I noticed tracer shells coming towards us, so I picked up my mike and said to Mitch, 'Start jinking!' [ie take evasive action] – and that's as far as I got because at that moment a shell burst in my compartment at a point adjacent to my left ankle. Both my legs had multiple wounds, and my left arm and my right hand were also hit. I was bleeding profusely, and really hurting bad. I didn't know the engine had been hit, or that we were going to have to ditch.

Our intercom was also knocked out by the AA fire. At the same time we were hit I saw Lieutenant Gibson and Linsz hit the water violently, and they were surely killed. Things were bad in my compartment: I was semi-conscious at times and fully conscious at others, and the way I was bleeding I knew I had to apply tourniquets or I would bleed to death. I managed to put tourniquets on both legs and my left arm. I was in the process of trying to give myself a morphine shot to stop the pain, but found I couldn't do it as my hands couldn't hold the syrette [needle].

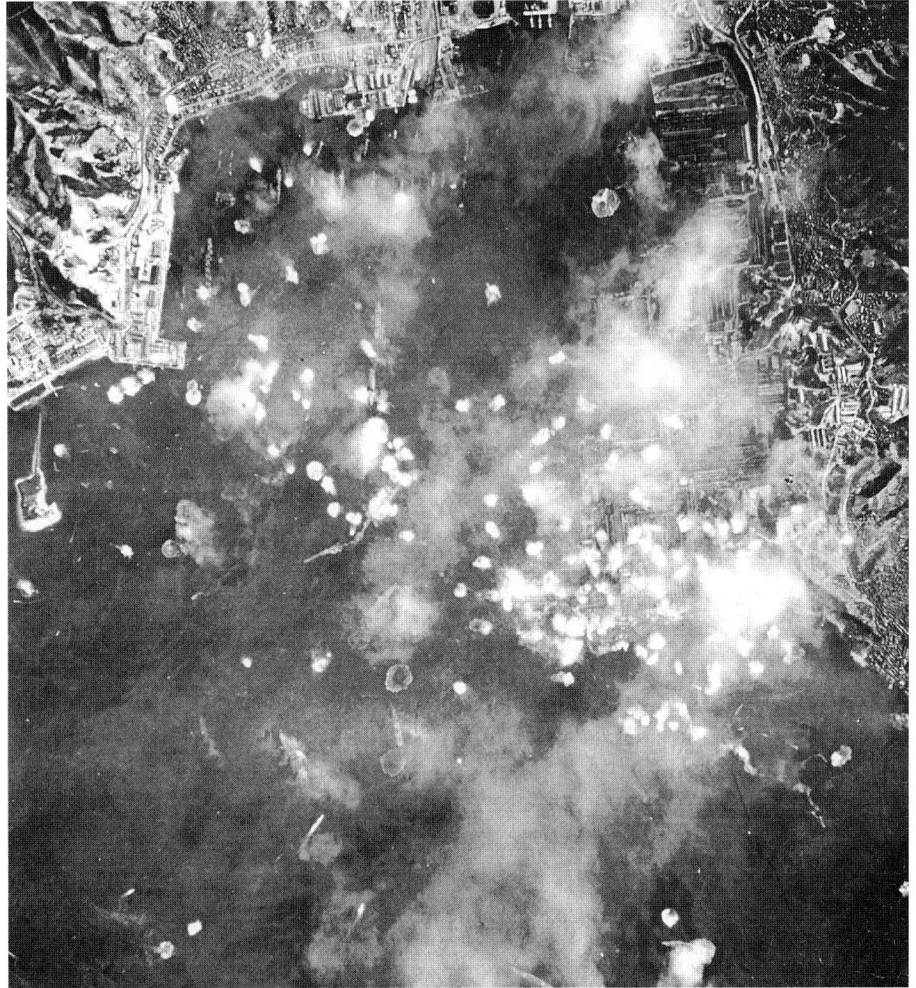

(Above) **Curtiss SB2C Helldivers finishing off the remains of the Imperial Japanese fleet in Kure Harbour, July 1945.** Commander J. Alton Chinn, USN, Rtd

The final strikes go in. SB2C Helldivers from the USS Randolph **(CV-15) make low-level bombing runs against a Japanese coaster off Nokkaido, Japan, 24 July 1945.** US National Archives, College Park

The oil pressure started to fall, and Maynard turned his aircraft south-west in order to gain open water before ditching. At the extreme western tip of Shikoku, separating Iyo Nada and Bungo Suido, the Wright finally packed up on him and he was forced to ditch. William Pinkerton recalled this moment vividly:

It was at this point that I knew we were going to make a water-landing, then we were in the water! But everything was *wrong* for a water landing! My cockpit canopy was closed, and it should have been open, so with my broken hand I turned the canopy handle and opened it, a tough job even with two good hands. I was still attached to my parachute by four really hard to-open-buckles, with water coming in fast – so I opened those buckles! Also, when making a water-landing I was supposed to be fastened by my lap belt and shoulder straps, but mine were off as I was giving myself first aid, and, at the moment of contact with the water, I was bent over trying to reach a morphine syrette that had fallen from my fingers to the cockpit floor. So, everything was wrong. What was right was the will to live and the adrenaline rush that goes with it.

While his comrades circled him, Maynard inflated the raft and got the badly wounded Pinkerton into it before the Helldiver sank. Applying tourniquets, he kept his gunner alive until they were rescued by an amphibious rescue plane and flown to Okinawa. Mitchell returned to the squadron on 19 August.

Lieutenant Gauvey's 1,000lb bomb failed to release in his dive, due to material failure in the bomb-rack. Moreover, in pulling out of his dive he registered 9.5gs which resulted in wrinkling of the fuselage and wing skins, and 'mushing' to a dangerously low altitude. Having used up several of his nine lives getting back control, he then almost forfeited the rest when he was hit by a 12.7mm projectile – this entered the lower part of the engine cowling, severed a hydraulic line, pierced the left leg of Gauvey's flying suit leaving a 6in gash, but *without even scratching his leg*, and passed out through the port side of the cockpit near the throttle. Still alive after all this, Gauvey faced another problem, because with the loss of the hydraulic fluid, he feared that if he opened his bomb-bay doors to jettison his 'hung-up' cargo, he would be unable to close them again. He elected the dangerous option of trying to land back aboard with bomb still up, and

FIRST TOKYO STRIKES

BRIEFING ON TOKYO AIR FIELDS

"SHOE" & "MURPH" READY FOR ACTION

KASUMIGAURA AIR FIELD

RETIREMENT

"KONOIKE" A/F. TARGET FOR SECOND STRIKE

CLOUDY ISN'T IT!

FUJI YAMA AT 13,000 FT.

A montage of photographs of VB-85 in action against targets on the Tokyo Plain, July 1945. Charles M. Crump

advised the carrier, who gave him the necessary clearance and cleared the flight deck. Fortunately he made it without any further incident, but it must have been the most eventful few hours of his young life!

For the Helldiver pilots of Task Force 58 the battle was now becoming mere target practice, and worthwhile ship targets were

becoming harder and harder to locate. Kobe, Nagoya and Maizuru were attacked on 30 July and four small warships and 11,000 tons of merchant shipping were sunk, while the destroyer *Yukikaze*, submarines *I-202* and *I-153*, submarine depot ship *Chogei* and three small warships were damaged.

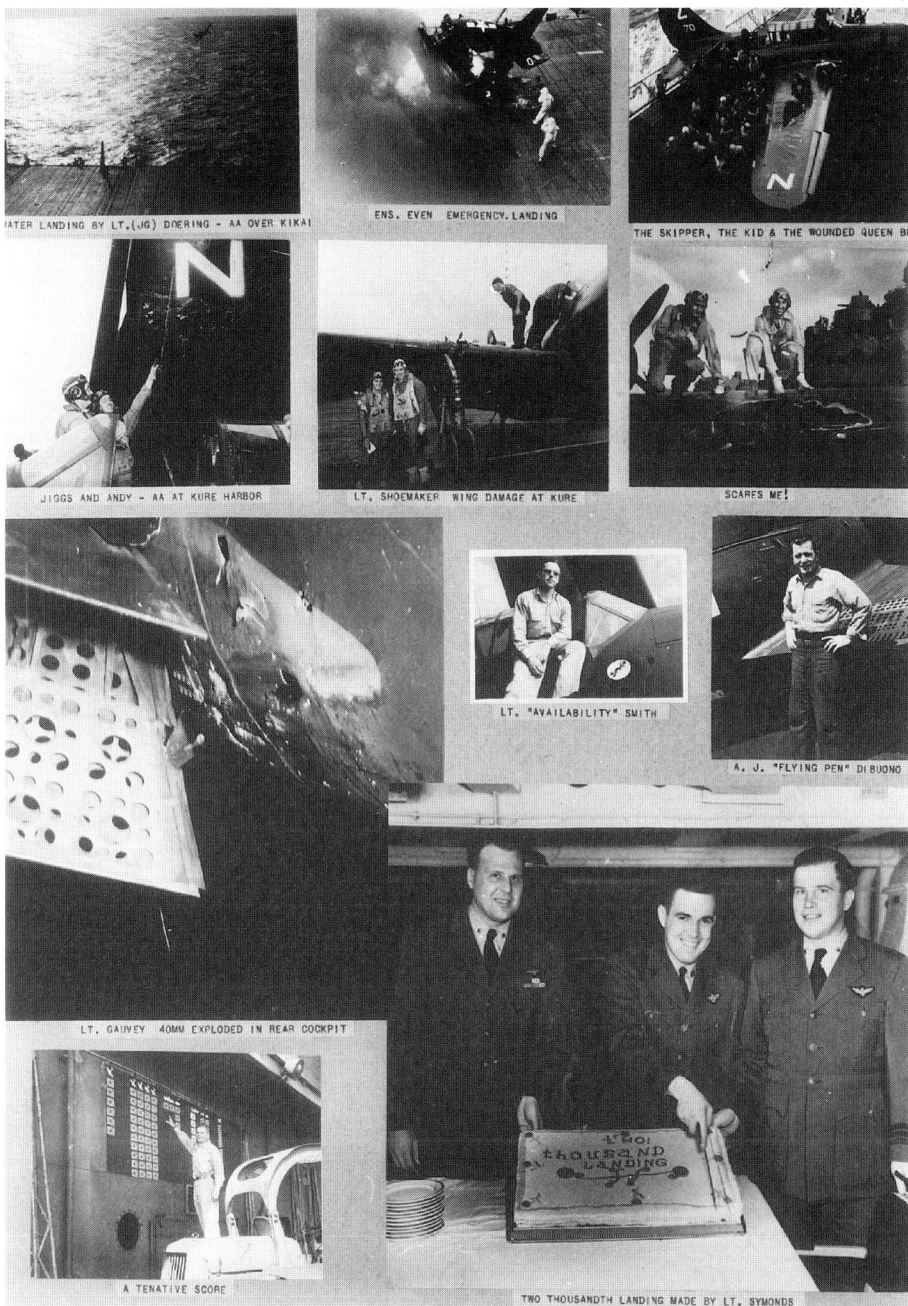

A montage of photographs of VB-85 operating from the USS Shangri-La (CV-38) in the Third Fleet in the Pacific, 1945. Charles M. Crump

at immortality, however, though fortunately no American warships were hit.

Transition from War to Peace

It was almost an anti-climax when the Japanese did finally surrender. It was unexpected, and most Helldiver aircrew had resigned themselves for many more months of combat and hard fighting before they saw their families again. Euphoria greeted the announcement: the war was done, and the Helldiver had played a full and worthy part in it, not least in the final, almost complete annihilation of the Imperial Japanese Navy. This did not mean the Helldivers were idle, however, as this extract from VB-85's unofficial history recounts:

As the Task Force retired to await formulation of plans for the surrender and occupation of Japan, Air Group 85 expected to be relieved: but as the days passed it soon became apparent that the Dragons would fly over the empire again. Beginning on 24 August the Squadron began a series of special photographic and publicity reconnaissance flights for the task force Commander, to obtain information on Japanese airfields and prisoner-of-war camps. To those on the first few flights it seemed unbelievable that they could be flying safely at tree-top level over the same territory which only a few days or weeks before had been bristling with anti-aircraft weapons. The guns were there, but with covers on them; the Japanese fighter planes were there, neatly lined up on the airfields, but with their propellers removed; and the Japanese people were there, but they either waved to the airmen or completely ignored them. As prisoner-of war-camps were located by their large 'POW' letters, so packages of food, cigarettes, candy and so on were dropped to the grateful inmates who wrote on the roofs of buildings such things as 'Thanks', 'Home' and 'Candy'.[158]

At the massive fly-past over Sagamai Bay on 2 September, after the official surrender ceremony, Helldivers from the following squadrons took part: VB-1 (*Bennington*); VB-6 (*Hancock*); VB-10 (*Intrepid*); VB-16 (*Randolph*); VB-85 (*Shangri-La*); VB-86 (*Wasp*); VB-87 (*Ticonderoga*); VB-88 (*Yorktown*); VB-89 (*Antietam*) and VB-94, (*Lexington*).

Last Days of Combat – August 1945

The fleet withdrew again to replenish between 1 and 8 August, and this period saw the dropping of the atomic bombs on Hiroshima and Nagasaki. The enemy continued to procrastinate, however, and Task Force 38 resumed the offensive again on 9 August with strikes against Northern Honshu and Hokkaido, sinking five small enemy warships, damaging two others and destroying some 250 enemy aircraft on airfields

there. On the following day they returned to the same area to complete the work.

On 11 and 12 August they refuelled and then further air strikes were launched on 13 August, the '2Cs' once more returning to the Tokyo plain, where another 250 Japanese aircraft were destroyed, and twenty or more knocked down in the air. The final strikes were made in the same area on 15 August, although many were cancelled with the formal Japanese acceptance of surrender terms. This did not stop kamikazes from making one final attempt

Maynard J. Mitchell was born on 31 October 1924, in Chicago, Illinois, at the height of the prohibition era. At an early age his family moved to the western suburbs of that city and the young Maynard attended public schools until his graduation from J. Sterling Morton High School, Cicero, Illinois, in June 1942.

Like any patriotic young American at that time, he wasted no time in enlisting in the Naval Air Corps as soon as he reached his eighteenth birthday, and he was called to begin his naval pilot's training on 10 December of that same year. His first training was done at El Paso, Texas, and it was known as the 'wash-out' session because it sorted out the hopeless cases – for Maynard it ended up as approximately 35 minutes of flight time in a Piper V-3 airplane, and he soloed after 4½ flight training hours, which he told me '… was about normal'.[159]

After this he spent two months in pre-flight training instruction at Athens, Georgia, followed by two months or so in Memphis, Tennessee, flying the two-winged 'Yellow Peril' as the biplane trainers were known. From there he went to Corpus Christi, Texas, for advanced and instrument flight instruction, flying the North American SNJ 'Texan' trainer and the Vultee SNV Valiant, known as the 'Vibrator', along with 'link' training. In January 1944, Maynard finally received his commission as an ensign, along with his naval aviation designation.

His next move was to Deland, Florida, to join his air group, VB-85. Here he and his fellow young pilots flew the Douglas SBD and eventually they received their quota of SB2Cs.

'You perhaps read about the former president of the United States, George Bush; well, he was a torpedo bomber pilot, and it has been claimed that he was the youngest-ever Navy pilot. However, the skipper of our dive-bomber squadron, Lieutenant Commander Art Maltby, claimed that I was the youngest, for whatever that means!'

The group was stationed at Otis Field, Maine, for advanced combat training manoeuvres before they joined their carrier, the *Shangri-La* (CV-38) in Newport News, Virginia, and went on their shake-down cruise. They eventually traversed the Panama Canal, and then sailed on to Hawaii in early 1945. Then it was on to the combat zone, and starting in April and May, they began to take part in active battle, flying close support missions for the Marines on Okinawa. They then moved on to the first carrier attacks in the Tokyo area, an offensive that eventually led to Maynard's plane being shot down on 28 July 1945.

'Dick Mann and I were very close,' Maynard said, 'and we roamed together for much of our time in our air group – I always look forward to seeing him at our reunions. During a span of three or four days, our three plane section were all shot down – Lieutenant 'Hoot' Gibson and his gunner/radar man were killed, and Dick Mann went down and was taken prisoner.'

When Maynard was flying his SB2C-4E (No. 20902), carrying a 1,000lb GP bomb in an attack against the Japanese battleship *Haruna* at Kure harbour, near Tokyo, on 28 July 1945; his regular gunner/radar man William Pinkerton was with him for the trip. Here he describes the experience:

'As we pulled out from our bomb run we were headed almost directly at a Japanese gun emplacement. Unfortunately, even though I fired our 20mm cannons, I was unable to put it out of commission, and as a result of their fire we sustained two hits, one in the engine area and another directly behind my cockpit and almost between the legs of William Pinkerton.

Due to the engine area being hit we lost all oil pressure, and it was evident that because of that, and possibly other engine failures, we did not have much flying time left in the Beast. I headed for open water south-west of the target area in the Sea of Japan, with the intention of ditching. Bailing out was not an option as I was reasonably sure that Pinkerton was in no condition to withstand such an ordeal, and so I thought our best chance was ditching.

This marvellous machine handled like a dream during the ditching manoeuvre. By the time we reached a point a short distance off a peninsula jutting off the island of Shokoku, we had lost all power. I maintained adequate air speed and glided down to the water. It was in our favour that there were very light winds at the time, probably less than five knots, and no apparent swells, and this, along with the physical shape of the SB2C with its smooth cigar-shaped underside (with wheels up of course), allowed us to make an unbelievably smooth ride on top of the water until gravity and friction slowed us down; finally the plane nosed over briefly and then settled back in a floating posture.

I don't know how long the Beast floated before heading for the 'deep six', but I can tell you that I made some of the fastest moves ever! Even while the plane was nosing over I was on my way out of the cockpit and onto the wing! It was only then that I realized the extent of the injuries to Pinkerton. I reached in his cockpit and

Lieutenant Maynard J. Mitchell, USN, of VB-85 seated in his Helldiver in 1944.
Maynard J. Mitchell

removed the two-man raft, inflated it, got him out of the cockpit, and then pushed away from the aircraft so that we would not be struck by the tail assembly as this great machine made its last dive!

I then helped Pinkerton into the raft and gave him what first aid I could under the circumstances. After only about ninety minutes we were picked up by a Navy sea plane and taken to a hospital ship at Okinawa. Certainly the handling characteristics and durability of the SB2C played a major role in this successful outcome – though really Pinkerton survived only through his will to live: as we sat in the raft, I rated his chance of survival as 'slim' to 'none'. He had extensive injuries to his legs and without the quick action by the rescue plane, he would not have lived.'

Maynard paid moving tribute to the crew of the PBM that rescued them: 'These rescue crews were fearless in their efforts to find and rescue many airmen during the conflict – one only has to look at what this one crew did in saving the six airmen with Pinky and me. This rescue was made within sight of enemy territory and with the possibility of facing enemy aircraft, yet there was no hesitation on their part.'[160]

This was Maynard's most traumatic landing, but not his last ditching with an SB2C. After the war was over the group transferred to a small jeep carrier for the journey home: 'During my flying career I prided myself on handling emergency situations, of which I had more than a few, along with good pilot procedures. However, during the transfer I did not follow the instructions of the landing signal officer, a cardinal sin, and as a result I dumped one of the beautiful Beasts into the ocean!'

After his discharge from the Navy on 10 December 1945, the following year Maynard entered Purdoe University; after three years' hard work he accepted a sales position at a company that manufactured Industrial Products and Aircraft Components. For the next forty years he was involved in various sales positions, in 1969 finally joining a Kansas City firm which manufactured products for the veterinary industry. He retired, as sales manager, in 1989.

'In 1959,' Maynard told me, 'I was fortunate enough to meet Pauline Donahue, who became my wife in 1961. We will celebrate our thirty-sixth wedding anniversary on 30 December 1997. Pauline travelled with me on many occasions throughout the United States and Canada; she also came on a trip to Tokyo (which we thought was terrible), to Hong Kong (which was great) and to London (which was a delight!).'

In retirement, Maynard spends much of his time trying to improve his golf handicap, and in periodic visits to friends and relatives throughout the United States. When he is able he attends the reunion of the VB-85 veterans, where memories of stirring times and gallant deeds are shared.

'In all sincerity, it was a great experience to pilot the Beast from our initial checkout in the aircraft in Deland, Florida, to my last flight when we transferred from the *Shangri-La* to a jeep carrier at war's end,' Maynard told me.

CHAPTER EIGHTEEN

Flying the Helldiver:
the American Experience

Almost all the combat flying done during World War II in the Helldiver was in the Pacific and was conducted by the men of the United States Navy and Marine Corps. It was to be the same with the much improved models such as the SB2C-5, most of which saw their service in the post-war fleet. It is important therefore to record the eyewitness accounts of these aircrew in order to achieve a balanced picture of 'The Beast', because many of those who most criticize it, did not fly it.

'It was one of the more difficult planes to land and take off from a carrier,' Roger Went of VB-85 told me.[161] 'But despite these problems, many of us when offered a chance to join a fighter-bomber squadron of F4Us declined, preferring to stay with the Beast.' One of the qualities that endeared it to many Helldiver pilots was its toughness, and Roger has several memories which exemplify this:

I recall a practice dive one day that points out the rugged qualities of the plane. I had not been having a good day, missing the practice target regularly with the practice bombs. On my last dive I was determined to score a bull's-eye, but in my desire to do so I dived too low – so much so that in pulling out over the Delaware Bay I was so close to the water that my gunner called me on the intercom and said, 'You're leaving quite a wake!' I was glad there were no ships in front of me, because I don't know whether I could have gotten over them.

Upon landing back at the base, one of the chief mechanics came up to me and said, 'What did you do? The wings are wrinkled!' So they were, and when I looked back into the cockpit I discovered I had pulled 11Gs!

Another incident was during my training on the '2C'. On a training flight over New York City I developed an oil leak and had to fly under the rest of my formation, looking up at them to maintain my position. I didn't realize that the flight leader had decided to fly low over the Statue of Liberty; so imagine my surprise when I glanced through the windshield, which was

'Jap Hoodoos'. A wartime composite of the squadron badges of some of the SB2C Helldiver units flying against targets in mainland Japan in 1945. Commander J. Alton Chinn, USN, Rtd

covered with oil, and saw a big dark blur in front of me! Quickly turning to port I avoided a collision that would have made me thoroughly infamous – and very dead.

The Helldiver always got me home, except for the time I was shot down in action – yet even then the three-foot hole in the starboard wing and the two-foot hole in the port wing didn't bring the SB2C down – that decision was mine. A fire had developed under the fuselage someplace, and unable to blow it out and not knowing the source, I decided to ditch. Both my gunner and I got out without a scratch, which was amazing because I was forced to land about twenty knots faster than in a normal carrier landing due to the holes. When landing on the water you realize that, at speed, it is hard.'

Edward J. McCarten renewed his acquaintance with 'The Beast' after a two-year combat gap, on 11 January 1945. At that time he was flying as an instructor in dive-bombing and combat tactics, at NAS Cecil Field, Jacksonville, Florida, along with several other VB-5 veterans, and his first '2C' flight was in an SBW-3. The idea was that pilots were exposed to the Helldiver before assignment to a squadron – a good notion, but one that came, 'rather late in the war'.[162] He recalled that: 'The training was conducted with the plan that teams of six pilots would leave Cecil Field and report as a team to squadrons. In fact I

The last moments of a SB2C-5 of VB-4 alongside the Tarawa **(CV-40) in 1946.** US Navy via National Museum of Naval Aviation, Pensacola

never did learn how well the system was followed, or how well it worked.'

Another who flew the '2C' after the war was Captain John Shea, Jr. 'I had no combat service in the SB2C; however, I did instruct in 'the Great Iron Bird' in 1946 in the Advanced Training Command. Believe it or not, we used the SB2C for the first fighter-bomber instruction at NAS Dayton Beach and NAS Deland, Florida. The syllabus called for a mix of fighter training and bomber training, and this included all VF gunnery runs, air combat and so on. It was an experience, dog-fighting in the SB2C!'[163]

The SB2C and the SBD Compared

Commander Charles R. Shuford, US Navy, rtd had a wide experience of Navy dive-bombers, and he gave me this verdict:[164]

I flew the SBD, SB2C, F4U, AM and AD (Spad) all as dive-bombers. I joined VB-80 in early 1945 as a replacement pilot after the *Ticonderoga* was hit by kamikazes. The air group switched to the USS *Hancock* (CV-19). Prior to joining the group I had been doing utility flying out of Pearl Harbor and Guam, and all of this was in the 'Beast'. VB-80 reformed in April 1945, and went to the Pacific in the fall after the end of the war.

As you know, many pilots did not like the SB2C, and preferred the SBD. There were a lot of fatalities during training prior to entering combat. The electric prop wasn't too popular, and the hydraulic system was a plumber's nightmare!

I had good experiences myself, and had only two instances that are worth mentioning in

which I encountered problems. While flying ASW for the *Saratoga* (CV-3) out of Hawaii, I made a carrier landing and my tail-hook bounced over all except the last wire and I hit the barrier. This required an engine change, which was done overnight and I flew away the next morning. The 'Beast' did have some tail-hook problems, besides others.

While practising dives on a target boat off the coast of California in 1946 with VB-80, my engine swallowed a valve on one cylinder causing a vibration, and an explosion that blew the cowling off. I was able to return to base without complete engine failure, however, and make an

emergency landing. This brings to my mind the following poem dedicated to the 'Skid':

Needle and Ball
Here's to the kid that flew in a skid,
Long may he live, I say,
When the gas is low and the gauge says 'NO',
And the carrier is miles away …

Then it's a watery fate for the kid in the crate,
and Davy Jones makes another haul,
So, if you want to get back to jump in the sack
…
Keep your eye on the needle and ball!"

Another very experienced dive-bomber pilot is Benjamin G. Preston, who flew all types in his career, When he first went to the fleet he flew the SB2U2 in Bombing Squadron 3; then in February 1941 he was transferred to Bombing Squadron 5 aboard the old *Yorktown* (CV-5) which was then still flying BT-1 dive-bombers. When that ship was transferred to the Atlantic from Pearl Harbor in April of that year, she was engaged in escorting convoys to England and Russia, and the neutrality patrol protecting shipping to England from German surface raiders. Benjamin recalled that his squadron first received the SBD in July 1941,[165] and commented: 'After flying the BT-1 whose hydraulic systems were operated by a hand-hydraulic pump, the SBD was well received by all the pilots and crews.' After *Yorktown* transferred back to the Pacific directly after Pearl Harbor, Benjamin took part in operations at the Marshal and Gilbert Islands, New Guinea, and in the battles of the Coral Sea and Midway, Guadalcanal and Tulagi.

The Marianas battle in June 1944 was probably the last time SBDs were used off carriers. There were probably some shore-based squadrons that still flew them for a while. The squadron I was in received SB2C aircraft in July 1944, and conducted strikes for the rest of the year from the USS *Intrepid*. Strikes were made on Palau, the Philippines, Formosa and Okinawa, including the Battle of Leyte Gulf. We made several strikes against the Jap battle fleet on that occasion.

With such a wealth of experience his viewpoint is well worth quoting:

I liked both the SBD and the SB2C, although they flew very differently. The SBD was easy to fly and the SB2C was a heavier, stiffer plane. However, it was faster, it had rocket rails, two 20mm guns and it could carry two .50 calibre

An SB2C of VA-1A landing aboard the Tarawa (CV-40) in 1947. US Navy via National Museum of Naval Aviation, Pensacola

Over she goes, an SB2C (Bu. No. 89205) of VB-4 (which later became VA-1) takes a wrong turn from the Tarawa (CV-40) in November 1946. US Navy via National Museum of Naval Aviation, Pensacola

gun packages under the wings for strafing and a bomb bay to carry the bombs. It also had an excellent surface search radar.

Deck Control Viewpoint

What of the men who had to keep 'The Beast' flying in combat, on whom the air-crew totally relied to have their aircraft up and running for each and every mission at the peak of its performance, as their lives depended upon it? Chief Warrant Officer John Patrick Piercy, US, Navy told me his views.[166] He was the Chief Machinist's Mate at the time of the Tarawa strikes, and knew the SB2C as well as anyone aboard:

My flight-deck duties consisted of checking the aircraft for 'ready to launch' condition as they came up to the catapult on the flight-deck launch mark. This mark could vary a great deal, depending on the aircraft's bomb or fuel load. I had to check that the wings were spread properly and the locking pins were in place; that the flaps were down and the bomb-bay doors closed. After certifying to these safety features, I handed on the aircraft to the 'launch officer' for the final checks before the launch. These consisted of the engines being turned to full power, and in the event of a catapult launch, the pilot had to place his head against the head-rest to prevent damage to his person when the aircraft accelerated from zero to 100mph in just seconds!

Any person who worked on the flight deck during war time was subjected to the most hazardous duty of any naval person. Sometimes an aircraft returning from a mission would not have dropped its ordnance, and when it was arrested these bombs would come loose and come up the deck at us. It was the same with rockets and machine-gun fire. Some men even walked into spinning propellers or were run over by taxiing aircraft.

I believe the Helldiver was an exceptionally good aircraft for the job for which it was designed. It could carry a large bomb load internally, plus eight 5in rockets under the wings. Quite a package! And it was tough.

One thing I remember in action is of a Helldiver returning from a mission with just one bullet hole in the fuselage. This solitary bullet had, by a freak chance, killed the radioman, ARM2c Joseph Applefeld, instantly. He was the only Jewish boy in the squadron, and an outstanding young man.

Another thing that comes to mind is while we were attacking Tarawa in the Gilbert Islands during the Marine invasion there. The squadron used up all its general-purpose bombs and had to

Two views of an SB2C-5 of VB-4 en route from Groton, Connecticut to Rochester, New York in December 1945. Mac McCullum via National Museum of Naval Aviation, Pensacola

typical strike approach was a division of six SB2Cs in two sections of three aircraft in 'V' formation. The first section was 'lead' and the second section was free to move to left or right of 'lead' according to conditions. [*See* Fig. right.]

Often, the dive-bomber force would consist of two or three divisions along with TBM Avengers and fighters for ship targets. Journey to the target was usually at 12–15,000ft, depending on the weather. About fifteen to twenty miles from the target we would move down 5 degrees or so, maintain the same power setting, pick up speed to the zone above the target, then commence to break at very short intervals, establishing the dive angle directly and pounce on the target as if sliding down a very narrow 5-degree cone.

From the enemy viewpoint, his AA weapons could not track us effectively, and could only rely on proximity fusing for making hits on us.

Done properly, each attacker was at 70 degrees, but following a slightly different angular path with respect to the target – simply, we didn't all come down the same path, nor did we pull out in the same path.

Our training was translated into combat practice, that is, formation, target run-in, and actual dive at 70 degrees at target; later on, in 1945, VBs adopted the four-plane division (two two-plane sections) for ease of manoeuvre and flexibility, similar to the fighters' 'thatch weave'.

The Recollections of Ron Hinrichs

I met Ron Hinrichs at the San Diego Aerospace Museum. He had trained on the SB2C-3 when a young ensign, and gave me copies of his logbooks for the period. In May 1944, he flew on twelve days with up to three flights a day, the maximum of which was of 2.3 hours' duration. The Helldivers he piloted during that same period were from a wide range of -3s[168] and he was also a passenger to Ensign Wright in a solitary SB2C-1(01101). June followed a more intense pattern, with twenty-two days' flying, again mainly with the -3, including two deck landings aboard the escort carrier *Card* (CVE-11) and four aboard the *Tripoli* (CVE-64) which were being utilized for training carriers at that time. The latter involved Hinrichs in a barrier crash with Helldiver No. 18859, from which he walked away.

He left San Diego on 5 July where he continued his training with the SB2C-1C,[169] making twenty-two flights in August and seven more from NAS Hilo.

use depth charges, which are normally used for anti-submarine warfare. However, these explosives worked quite well, since the enemy was in underground bunkers and other places that the regular ordnance could not penetrate; the blast effect of the charges killed many of the enemy inside their dug-outs quite effectively.

The 'Beast' did not provide any particular problems that I can recall at this time, except, since it was bigger and heavier than the Douglas SBD Dauntless, the deck run was considerably longer and the catapult launches utilized more steam power. The hydraulic system was considerably more extensive with folding wings, bomb-bay doors and bomb displacement gear, gun chargers and so on and so forth.

A VB-80 Pilot's Experiences: Captain Chuck Downey

Captain Chuck Downey flew combat missions in the Helldiver with VB-80 and gave me this very valuable insight into the tactics employed at that time.[167]

Many of us flew the SBD Dauntless before the SB2C, so diving the Beast at 65 to 70 degrees was the 'drill'. Anything less was too shallow and the bombs would fall short. Over 70 degrees, you, as a pilot, are also way over in the straps (about 85 to 90 degrees), very uncomfortable and not likely to retain accurate control and obtain a satisfactory hit and recovery. A

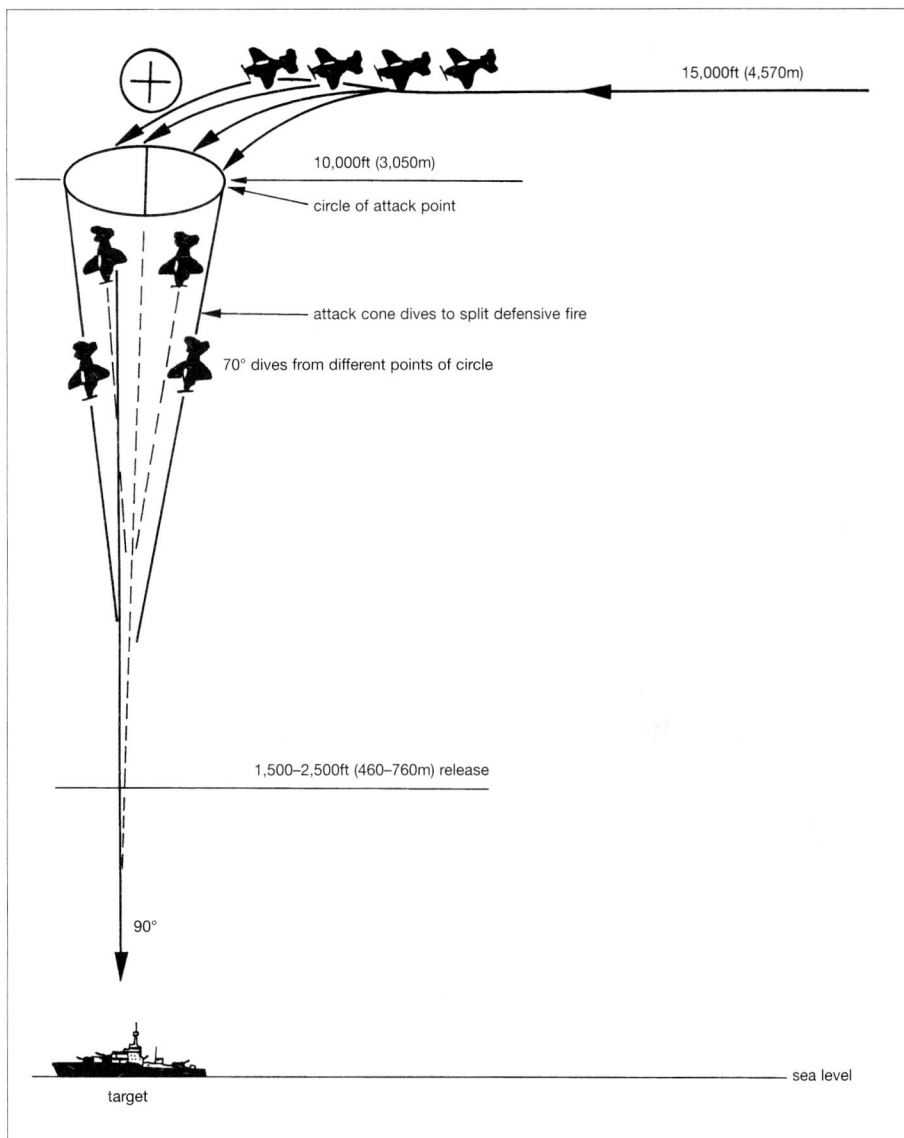

Methods of attack used by VB-80.

Diagram labels:
- 15,000ft (4,570m)
- 10,000ft (3,050m)
- circle of attack point
- attack cone dives to split defensive fire
- 70° dives from different points of circle
- 1,500–2,500ft (460–760m) release
- 90°
- sea level
- target

complement on board to combat the kamikaze threat. However, in January 1945 I flew as top cover for a raid to Cape Saint Jacques, French Indo-China. dive-bombers from VB-3 (SB2Cs) sank four destroyers in a classical attack and VT-4 torpedo planes sank three tankers. This was the only time I observed the dive-bomber in combat.

Observations by Commander Mann

'I would like to point out a couple of things about a dive-bomber that are often overlooked,' Commander Richard W. Mann, USN, rtd, told me recently.[171] He listed them as follows:

First, even though we dived at a 70 degree angle, the plane itself is at 90 degrees to the surface. Even though we are at 90 degrees we are still getting lift from the wings and this causes the angle of dive to go to 70 degrees. A pilot would have to be on his back, so to speak, in order for the angle of dive to be more than 70 degrees, making it impossible to control the plane.

Second: on pullout from a dive we would pull from 6Gs to 8Gs, which, as you can imagine, puts a severe strain on the wings – in effect a 15,000lb plane suddenly weighs about 100,000lb. Yet, when we returned to the carrier and landed, we pulled a lever and the wings folded. To me, that is a remarkable engineering feat.

And third, in most war moves a dive-bomber executes his dive and immediately pulls right back up – but this is suicide because you are losing speed as you climb, and you would be a dead duck. Any dive-bomber pilot (that's still around) will tell you that the best procedure is to release your bomb load at 1,500ft and execute your pullout so as to be straight and level right on the water or land and depart the target area at your maximum speed right on the deck.

Commander Mann's memories of the Helldiver – '... our beloved Beast' – are fond ones. He told me:[172]

The SB2C was an outstanding dive-bomber from roll-in at high altitude to pullout at 1,500ft. It was extremely accurate and capable of absorbing much punishment. Take-offs and landings were critical. We always seemed to be overloaded on take-off and would settle at the bow. Coming aboard, we had a critical range of speed to maintain: too fast, you missed the wires; too slow and you lost control. It was a narrow envelope and it took skill to make a successful landing.

Ron had several interesting observations to make concerning the aircraft:[170]

The SB2C was known as 'The Beast', but where this name came from I do not know. The plane was easy to fly, but it did have several faults which may have led to this nickname. The propeller was an electrically controlled one, and had the bad and dangerous trick of slipping out of pitch and going completely flat; while VB-4 was aboard the Essex and Bunker Hill, four of the planes went into the drink on take-off because of this. The take-off torque was very noticeable and required 15 degrees of rudder tab to offset this. Another problem was the tendency of the tail hook to bounce and thus skip a wire on a

carrier landing. This occasioned my one and only barrier crash.

It was a much faster plane than the SBD, and of course was faster in a dive. We used the same diving tactics in both planes. We would approach the target as close as possible, roll over on our back and pull the ship down the target in as near to a 90-degree dive as possible. Then it was a matter of rolling the plane around to keep on target. Our usual approach altitude was 10,000ft with a pullout around 2,000 to 1,500ft; we pulled higher Gs during a pullout in the SB2C, but not anywhere near a blackout condition.

I wasn't on any strikes with VB-4, and that squadron was transferred off to Guam at the end of 1944 in order to increase the fighter plane

(*Above*) **On the side elevator of the** Bunker Hill **(CV-17) is an SB2C-5 (Bu. No. 83359) of VB-95 in October 1945.** Dave Lucabaugh via National Museum of Naval Aviation, Pensacola

Loading torpedoes onto an SB2C of VB-2 for trials. Naval Photographic Center, Naval District Washington DC via National Museum of Naval Aviation, Pensacola

The moment of truth! An SB2C-5 of VA-1A disappears over the side of the Tarawa **(CV-40) in February 1947.**
US Navy via National Museum of Naval Aviation, Pensacola

VMSB-244 in China

For some units the ending of World War II has a fuzzy blurring into the Cold War era. Take for example Marine Corps Helldiver outfit VMSB-244. One day it was fighting the Japanese on Mindanao, a few days later the young American aircrew found themselves in the middle of a Civil War in China. Let Tom Ross, one of those Marine Corps flyers, tell the story:

We had the SB2C-4s and -5s, Navy versions with full tailhook and folding wings in VSMSB-

A little accident-damage in the tail of the 'Big Tailed Beast', but there was still plenty left to get her home!
This lucky Helldiver is from VA-9A aboard the Philippine Sea **(CV-47) in 1948.** Emil Buehler Naval Aviation
Library, Pensacola

244. We were sent to Tsingtao and a Corsair squadron went to Peking, both to assist in formalizing the Japanese surrender in China – at least, that was the cover story. I believe our real mission was to discourage Mao and his Red Legions from undertaking any concerted action against Chiang Kai-Chek and his Nationalists.

The flight from Malabanag, Mindanao, to Tsingtao, China was a very hairy experience. We flew through the tail end of a typhoon on the way to Laog in northern Luzon and had to land in Manila. We lost two planes and four crew on that leg. Then, when we finally arrived at Okinawa, our final stop before China, we were hit with a monster typhoon that caused a tremendous amount of damage. We stayed in the wind and the rain all night long to keep our SB2Cs turned into the raging and shifting wind. The arrival in Tsingtao seemed a let down after all the previous exitement. It was, however, a totally different experience because the Nationalist and Communist factions started shooting at each other the day after the Japanese surrendered, and we were, more often than not, right in the middle of it!

We flew patrols between Tsingtao and Peking to discourage the rebels from cutting the tracks or the telegraph wires. We had ammo for our two 20mm wing cannon, but the belts were not fed into the guns, so we couldn't do anything to stop any mischief even if we saw it! The Commies quickly realised this and acted with impunity and expended great amounts of their ammo trying to shoot us down! Most frustrating, that combined with our absolutely horrible living conditions (in barracks built by and for the much smaller Japanese) made us all eager to get the hell out of China! I departed a month before six of our aircraft, flying in formation, crashed into a mountain while trying to find home-plate in a blinding snow storm. All twelve aircrew were killed.

Experiences of Post-war Helldiver Pilots

One of the very best and most detailed accounts of post-war flying of the 2C was given to me by Lieutenant Commander

Okey C. Roush, US Navy, rtd. His experiences deserve retelling in full:

I turned age seventeen in 1943, and as soon as I graduated from High School I immediately enlisted in the US Navy. Two and a half years later that I was designated a naval aviator, so I had to wait a long time before I got to fly the Curtiss-Wright SB2C Helldiver! By the time I did begin to fly it in November 1946, I had spent my earlier Navy education in ground

SB2C-5 flown by Ensign Okey C. Roush (inset). Okey C. Roush

school, and being taught to fly in the Stearman N2S biplane; then advanced and instrument training in the North American Aviation SNJ Texan, known in the Air Force as the AT-6; and pre-operational training in the Douglas SBD dive-bomber. The SBDs had already been replaced in the fleet by SB2Cs and were being used in the training command.

Next we were returned to the training command headquarters in Pensacola, Florida, and did field carrier landing practices in the SNJ in prepa-

ration for qualification in a newly introduced feature of our training, which was a mandatory ten SNJ carrier landings. Obviously the SNJ trainer was not designed as a carrier aircraft, but a tailhook was attached externally under the tail of the SNJ, and after taking off from the carrier, the trainee would lower the jerry-rigged tail-hook with a clothes-line rope attached to it!

Then came operational training in the Grumman [General Motors manufactured] TBM Avenger torpedo bomber, along with the required field carrier landing practices and the mandatory six carrier landings. Since the war was over, it took five months to complete the TBM operational phase and carrier landings. And at no time did we learn that only six months earlier a flight of five TBMs had been lost at sea from our base at the Naval Air Station, Fort Lauderdale, Florida, and have never been found!

My next assignment was my first introduction to the Helldiver, an SB2C-4, while assigned to the Carrier Aircraft Service Unit at Naval Air Station, Norfolk, Virginia, for familiarization. The next two flights were in an SBW-4, an exact duplicate of the Curtiss-Wright Helldiver except the W was made by the Canadian Car and Foundry under contract to Curtiss-Wright. Then, back to the SB2C-4 for two more familiarization flights, for a total of about nine hours of flight time in the Helldiver.

I must say it was quite a thrill to feel all that power up front after spending so much time earlier in less impressive planes. As I recall, the TBM had the same engine as both the SB2Cs and the SBWs, the Wright R-2600 Cyclone, but somehow it didn't *feel* the same. Maybe it was the four-bladed propeller that made the difference, but there *was* a distinct difference. I later learned that the Helldiver SB2Cs dash 3, dash 4, and dash 5, all had the four-bladed propeller.

After several more hops in both Helldiver types, orders came through for me to report to Squadron VA-1B (formerly VB-74), assigned to the aircraft carrier USS *Midway* (CV-41) which was then in port at Norfolk, Virginia, with the air group training out of Naval Air Station Oceana at Virginia Beach. From December 1946, through to July 1947, we underwent field- and sea-based tactical training and bombing practice, and field carrier landing practice. These sessions were interspersed with short day-cruises for bombing and tactical training, as well as one cruise to the Caribbean where the *Midway* worked out of the Port of the Naval Station at Guantanimo Bay, Cuba. Personal experiences during this time span were my first carrier landing in the Helldiver in January 1947, followed by fourteen more carrier landings and one catapult shot.

This was the era of the last of the true pilot take-offs and landings aboard the carriers. The Helldiver and other propeller driven planes of its day were perfectly capable of taking off the decks with a full load without the assistance of a catapult, unlike the jets that came along later. Practice catapult shots were made so that pilots would be familiarized with the technique in the event that deck crowding left insufficient room for a take-off run.

It was still the era of the straight deck carriers, too; the 'Meatball' landing signal system hadn't yet been introduced, and all landings were made by making 180 degree approaches and taking the directions of the landing signal officer (LSO), as soon as the landing plane's pilot got him in sight. And it was either to land, or to take a wave-off; the luxury didn't exist then as it does now, that if a wire isn't caught the pilot can add power and take off the angled deck to go around and try again! If you got a cut from the LSO and missed a wire, you ended up being stopped by the steel wire barriers rigged about midway up the flight deck!

Another interesting thing we did at Guantanimo Bay was that the ship's captain purposely sailed into the bay where he knew the ship couldn't turn itself around in order to get out. The pilots were all briefed when departure time came, and as many planes as possible were

placed forward of the island facing the port side of the ship, and an equal number were placed aft of the island facing starboard. Upon radio command all engines were started and were run at a predetermined power setting; the ship weighed anchor, and would then begin to rotate slowly on her own axis until she had turned the 180 degrees necessary to point her out to sea. No one was sure that a ship of this size, 45,000 tons plus, could be turned in this fashion, but operation 'Pinwheel' was quite a success!

Two views of XBT2C-1, No. 1. Stanley I. Vaughn Archive

In July 1947, VA-1B was the first squadron to receive the new Douglas AD-1 Skyraider, and my flight time for the next few months until my release from active duty alternated between the SB2C and the AD-1, and several other types. Following my release from active duty in November 1947, my next flights in the Helldiver began as a naval reservist at Naval Air Station, Columbus, Ohio, in the early months of 1948. This was the home of the Helldiver, since the Curtiss-Wright plant in Columbus

was built specifically to manufacture the Helldiver aircraft. My flights there began with the SBW-5 and then alternated with the SB2C-5, as well as other types, throughout the remainder of that year. In December 1948 I flew seven Helldiver flights for acceptance and ferry; and my last Helldiver flight was in May 1949, to ferry our last SB2C-5 to Corpus Christi, Texas, for overhaul, after which I returned by commercial air. I had previously ferried a number of other Helldivers to Corpus Christi for overhaul, but usually returned in a newly overhauled one.

It is a pleasure to report that in all the above-mentioned flights in the Helldiver, and other types as well, I was never involved in any accident or major incident of any kind. This was easier to accomplish in the Helldiver than many others, however, as it was a completely predictable airplane. The extra power it seemed to have over the others gave the pilot a feeling of confidence; and the wide stance of the main landing gear and the large effective area of the rudder made take-offs and landings easy as far as directional control was concerned.

So far I have not mentioned that the Helldiver was practically universally known as the 'Beast', for fear the reader may misinterpret that the name originated because the plane was a beast to fly, which was not the case. Actually the nickname came about from the phonetic pronunciation of the symbol 'SB2C'. Another nickname was 'SB Ducey'.

Speaking of nicknames, another interesting one came about among 'Helldiver drivers' having to do with the cockpit arrangement of the secondary controls and switches in the cockpits of early Helldiver models. The stick and rudder pedals, throttle quadrant and instrument panels were in the usual locations, but the flap and dive brake handles, landing gear, radio switches, and numerous other necessary component controls were lined up on both sides of the cockpit in no logical fashion. This caused the pilots to refer disaffectionately to the manufacturer as the 'Curtiss novelty factory'! This shortcoming was corrected when the SB2C-5 came out, with well organized control panels on both sides of the front cockpit, and with a more logical and easy-to-reach arrangement.

Otherwise, as far as pilot comfort in the cockpit was concerned, they had done a good job of

Helldiver ramp strike on the wooden deck of the USS Shangri-La **(CV-38).** US Navy

fitting the plane to the pilot. The seat was easily adjustable fore and aft as well as vertically, and my friends who were as tall as six feet five inches, and others as short as five feet five inches, had no trouble in accommodating themselves comfortably. All controls were easily within reach (better in the dash 5s), and visibility was good in all directions except straight ahead in the three point attitude. There was no air-conditioning, but the heaters were quite adequate on the coldest days. The canopy was easy to operate, and wind in the cockpit was not intolerable with the canopy open, whether cruising or diving. Canopy positions had no effect on the trim of the airplane.

When fully developed, the Helldiver performed well in its role as a dive-bomber. Although my experiences were never during wartime, the airplane could be trimmed perfectly under any load and flight conditions, including in the 70 degree prescribed dives for bomb delivery. This perfectly trimmed condition allowed the pilot to make minor corrections as

appropriate without having to 'fly' the airplane all the way down. Although I was flying 'tail-end Charley' in one eighteen-plane formation from the USS Midway for a demonstration for a visiting admiral, our squadron skipper told me I had the best hit on the target of any in the entire group. I believe this kind of accuracy was made possible by the design and functioning of the split dive flap assembly incorporated on the Helldiver, and the inherent controllability of the airplane.

The Helldiver also exhibited good aircraft carrier compatibility with regard to shipboard operations. At the time I was aboard, the USS Midway still had a straight deck. With full practice bomb load, the Helldiver needed a relatively short take-off run, and settled only slightly as it cleared the bow of the ship.

But take-offs are easy! What is difficult is getting back on board. With the 180-degree approach, I was especially impressed with the controllability of the Helldiver at slow speeds. Holding about a 90-knot airspeed from the

180-degrees position until picked up on the LSO's paddles, one could get the trim tabs and throttle setting correct and the check list completed. After that, if you got a 'slow' signal you just needed to give it a little forward stick to speed up slightly, or if 'fast', a little back stick would slow you down. Similarly, if low, a little added throttle brought you up a little, and if high a slight cut in throttle would return you to the correct level. The Helldiver responded exactly as we were told it would.

One small difficulty on the carrier landing approach was that the engine and cowl of the plane were fairly big up front of the pilot. In the last few seconds before the LSO's cut was to be expected, there was a danger of losing sight of the LSO under the nose of the plane unless you sat up as high as possible in the cockpit and leaned towards the left side of the plane. If you lost sight of him you had to take a self-initiated wave-off, as it was too late by that time, and things were happening too fast to try to fly back over to the right where you could again see the LSO.

164

Once the cut was received and executed, the Helldiver's nose fell through the horizon on its own, allowing the plane to enter a gentle, but positive, glide towards the deck. All the pilot had to do was keep the plane directed down the centreline of the flight deck, and when the appropriate time came, to break the glide

unequal cable lengths steered his Helldiver even more to the left, off the flight deck, over the catwalk, and over the side! Once again the trusty Helldiver performed its duty well, as the hook didn't break, held onto the plane, and the plane dangled against the side of the ship while the captain stopped the ship. By that time my

This concludes my recollections regarding what I consider to be a good old friend, the Helldiver. I could never have asked for a more steadfast and reliable aircraft to have spent 250 or so flight hours with. Moreover, I imagine there are many others like myself who regard the Helldiver with the same fond memories as I do.

Trio of SB2C-3s over Oxnard, California, in 1944. S-12, S-10 and S-14 are from VB-80. Charles R. Shuford

gently, thus allowing the tail-hook to engage the next nearest cable of the carrier's arresting gear, and the gentle Helldiver came to a stop after the three-point landing.

Staying in the centre of the deck was very important. One of my squadron mates drifted left of centre just before touch-down, and caught the cable in the left one-third. The

friend had shinned up the rope lowered to him from the flight deck and was safe. The Helldiver was no longer visible; only the hook could be seen sticking above the waves. Finally, with no means of retrieving the plane, the cable had to be cut, and the Helldiver then slipped off the loose end of the cable and went on to Davy Jones' locker.

The Helldiver Experiences of Ron McMasters

Ron McMasters was another pilot who gave me a very detailed description of his Helldiver experiences in the fleet at this time.[173] He first explained briefly his initial training: 'I received my wings in 1946 so I

(Above) **An SB2C-3 over Oxnard, California, in 1944. S-12, from VB-80, with Ensign Charles R. Shuford as pilot.** Charles R. Shuford

Ensign Charles R. Shuford and Airman W. L. Wonderly of VB-80 aboard the USS Hancock **(CV-19), 1945.** Charles R. Shuford

didn't see combat in the aircraft. I did fly the SB2C-4E in operational training at Jacksonville, Florida and then went to the fleet squadron VA-7A. I flew the SB2C-5 in VA-7A and spent most of 1947 at sea in manoeuvres, cruising the Mediterranean and North Atlantic. I was then released from active duty and flew the SB2C-5 in the Reserve until July 1948'. He then went on to describe some of the characteristics of the Helldiver thus:

The SB2C-4E was the first operational aircraft I flew. The cockpit was not well arranged, with the wheel and flap levers on the right so the pilot had to switch both hands to operate them. This was corrected in the SB2C-5. The SB2C-4E was not very smooth in a 70 degree dive, our standard dive angle for dive-bombing; with dive brakes open it was vibrating and very difficult to hold on target. The SB2C-5 helped this situation with a selection of 'rocket flaps' which closed the outer portion of the dive flaps. The stick had a 50lb counterweight, so it was not easy to roll back on target with a wind shift.

The elevator control cables had a tendency to stretch, which sometimes made the pullout scary sometimes! Trim tab was used to assist the pull-out. The hydraulic system was a nightmare, and there were 240 changes made by Curtiss to try to correct the problems. The wing flaps on one side of the aircraft had a tendency to collapse on landing approach, and at slow speed it was impossible to prevent the aircraft from rolling – if the pilot realized quickly what was happening he could close flaps, go to altitude, bleed them or land with no flaps. During operational training we lost one pilot due to this condition, and another was killed in the Mediterranean while landing.

Landing aboard the carrier was interesting at times! The fuselage was wide, which made it easy to lose sight of the LSO, and this necessitated the approach to be in a turn to the ramp. The wide landing gear made the actual touchdown comfortable, however.

Take-off required some technique. Once the tail was off the deck and speed was picked up to take-off, the tail had to be lowered to help the plane 'fly-off'. This was particularly the case if you were carrying armament.

The aircraft was fine if everything went well. Cruise speed was 165 knots, which was not an improvement over the SBD Dauntless. As you know, the SB2C was nicknamed 'The Beast' and as a new ensign I thought it was an airplane which a pilot had to fly continually. The early models had problems of the tail coming off, and solid flaps, which made it difficult to control in a dive.

Our dive angles when at sea were 70 degrees with rocket flaps, which made the aircraft more stable. When diving at sea our attacks were always against the home carrier or on the escorting cruisers. Also our approach was always in a nose-down speed run, rolling into the dive from a slight bank to the right or left. The reason for this bank was to create a 'fan-out' of the diving aircraft so that one or more of us would have a fore-to-aft release to the targeted ship. We always practised this type of run every time we flew in the Caribbean, Mediterranean or North Atlantic; however, I do not recall ever being assigned any Soviet ships as specific combat targets should the 'Cold War' have turned hot!

'Seven Tons of Nuts and Bolts'

Not *all* those who flew the Helldiver, liked the experience very much. As Colin Shadell confirmed to me:[174]

The SB2C was not a particularly popular plane among the pilots who flew it. We all referred to it as the 'Beast'. I had around 500 hours and forty

The spectacular disintegration of an SB2C-5 of VA-34K aboard the carrier USS Kearsarge **(CV-33) in November 1948, while conducting exercises in the Atlantic.** Emil Buehler Naval Aviation Library, Pensacola

A Canadian Car & Foundry-built SBW-4E (coded 664) with underwing rocket fittings and open bomb bay doors. Ken Johnson via National Aviation Museum of Canada, Ottawa

The Curtiss SB2C-4E (97) of Ron McMasters during an operational training flight from NAS Jacksonville, Florida in 1946. Ron McMasters

shoulder harness unlocked, I could just barely reach the magneto switch. The landing gear actuation lever was on the floorboards on the right side of the pilot's seat. On take-off, raising the gear required taking the right hand off the control stick, taking the left hand off the throttle to grab the stick, and raising the gear with the right hand. Very often it was hard to get the friction lock on the throttle quadrant tight enough, so the throttle would creep back during this gear-up operation. This all made take-offs very exciting – particularly on a carrier.

The cowl flaps were operated by a manual hand-crank, low on the forward console and also hard to reach. The Mark 8 gunsight was stored on the floorboards on the right side of the seat and was supposed to remain there at all times when not in use, but especially during take-off and landing, to remove it from its storage place and install it in position in the windshield while flying in formation was quite an accomplishment!

Fortunately, when the SB2C-5 came out someone had figured out that some cockpit changes were in order. The overall size of the cockpit was smaller, and the magneto switch was right at your fingertips on the left console. Wheels, flaps and cowl flaps were also more sensibly located, so that all in all, it was a much easier plane to fly, especially in formation.

The best description of the Beast I have ever come across was when I once heard one of my friends discussing it and say, 'There I was, going straight down with seven tons of nuts and bolts strapped to my ass!' Most apt!

Final Torpedo Trials

As late as 1945 the Helldiver was involved with carrier trials as a torpedo bomber. The unit concerned was VT-97 which, with its companion VB-97, had been commissioned at Wildwood in November 1944, and later transferred to Grosse Ile. The main function of these two squadrons was as a training pool for replacement pilots for the Pacific combat squadrons; VB-17 was finally disbanded in April 1946. VT-17 did much work on developing the SB2C for successful torpedo dropping, which helped the post-war decision to combine both dive and torpedo bombing into the new 'Attack' (VA) designation. Stanley Vaughn II told me about these trials:[175]

In 1945, at Oak Harbour, Whitby Island, Washington Navy Base the field engineer was using SB2Cs to drop torpedoes for testing, using flight crews officially supposed to be on 'rest and

carrier landings in the SB2C3, -4, 4E and 5 and the SBW. The SBW was built by the Canadian Car and Foundry Works. The SB2C-4E was equipped with a search radar for submarine patrol.

During World War II the plane always carried a rear-seat radio-gunner, but after the war about the only time we had a radioman along was so he could get his flight pay and the machine guns had all been removed.

The best things I can say about the 'Beast' are that it had a very reliable Wright R-2600 engine, and that it was a very stable platform for accurate

dive-bombing. In a near-vertical dive with the dive brakes out, the manoeuvrability and target acquisition was excellent. I personally believed that the size of this aircraft deserved a more powerful engine, especially for carrier work.

My main complaint, however, had to do with the cockpit configuration. Whoever designed that cockpit must have been over six feet tall or had never flown an airplane – or both! I'm 5ft 6in short and in that barn-like cockpit almost everything was out of my reach (this criticism does not include the SB2C-5 which I will touch on later). With my seat all the way down and

recreation' for the flights. The hope was for a dual-purpose machine, and certainly in a dive the Helldiver was fast enough to drop the torpedo and then get away at 300 knots. However, the general consensus among the dive-bomber crews was that they just did not want to drop torpedoes, they thought it was too darn dangerous!

Once a '2C' Man – Always a '2C' Man!

Ralph Charles of Somerset, Ohio, is as old as the century, but that did not stop him going out at the age of 97 and buying himself an Aeronca which he named 'Blue Boy II'. This was in spite of the fact that he had not flown for fifty years, having made a promise to his late wife, Leona, that he would not do so. However, she passed away three years ago, and now Ralph thinks it is all right for him to get airborne once more.

Ralph once worked at Orville Wright's original aircraft factory, and he built no fewer than seven aircraft of his own, without the benefit of blueprints. He also worked as a test pilot at the Port Columbus plant of Curtiss-Wright from 1942 onwards, flying both the Helldiver and the Seagull. Once he made a 'wheel-gear up'

belly landing in a Helldiver and walked away from it. He had been patiently circling the field waiting for a smaller aircraft to take off so that he could land. The pilot of the other machine seemed oblivious to the tower's clearance commands, and in the repeated circuits, Ralph lost his concentration; when the other pilot did finally take off, Ralph went straight in without

the benefit of lowering the SB2C's undercarriage! He recalled to journalist Mike Hardin, 'It's embarrassing as hell, but I did it!'[176] When, during the subsequent Navy investigation into the cause of the crash, they asked him the question, 'When did he realize that he didn't have the landing gear down?', Ralph promptly replied, 'When the prop got shorter!'

A tight echelon formation of Curtiss SB2C-4Es during operational training out of Jacksonville, Florida, 1946. Ron McMasters

An SB2C-5, coded 307, flown by Ensign Colin Shadell, the Assistant Flight Officer of VA-5A. Note that the slats or airfoils on the leading edges of the wings, which extended when the wheels were lowered for landing, supposedly to give more stability at speeds near stalling for carrier approaches, were one feature omitted on the SB2C-5. Colin Shadell

CHAPTER NINETEEN

Post-war Operations

Despite the endless criticism of later historians, the Helldiver was one of only two US Navy aircraft of World War II to extend its useful service life well into the post-war era, and indeed, it was not finally phased out of military service in the USA until June 1949. Abroad it continued in service even longer and nor were its actual live combat days over, by a long chalk. As the Confederate Air Force point out:

> The Helldiver's performance has been maligned by many critics, frequently without regard to the facts. Some comparisons to other Navy aircraft are enlightening. The SB2C-4 had a higher cruising speed and greater range (without drop tanks) than the TBM Avenger, and a significantly higher top speed. It easily outperformed the SBD Dauntless in every category except range. Its cruising speed was only two mph slower than the F6F Hellcat. Only the F4U Corsair, among contemporary carrier-based aircraft, had a significantly superior speed. The Corsair could carry the same load as the Helldiver, but over a much shorter range. Of aircraft designed in the same, immediately pre-war period, only the Corsair outlasted in front-line Navy service.[177]

An SB2C-5 of VB-9A from the Philippine Sea **(CV-47) in 1948.** John M. Moore via National Museum of Naval Aviation, Pensacola

Post-war US Navy Operations

In the much-reduced post-war Navy the Helldiver was retained when other war-built types, other than the Corsair, were discarded. The old idea of Admiral Halsey to utilize the Helldiver in the combined role became a reality and the Navy adopted the Army-style attack designation for its new units. This led to some renumbering and reassigning of Helldiver squadrons in the immediate post-war period. VB-17 for example, had reformed at Alameda in August 1945, ready for its third tour of combat duty. When the air group became a carrier battle group (CVBG) and was transferred to Brunswick, VT-17 became one of the first to trade in its TBM Avengers and re-equip with the SB2C, so that their assigned carrier, the *Coral Sea* (CVB-43), had an autonomous

attack squadron, VB-17 and VT-17 becoming VA5B and VA6B respectively.

Therefore VT-17,(Commander Rubin H. Konig), which was using SB2C-4Es from March 1946 onwards, was redesignated as Attack Squadron Six B (VA-6B) on 15 November 1946, when it began using the SB2C-5. It re-equipped with the Douglas AD-1 on 23 September 1947 and was redesignated as Attack Squadron Sixty-Five (VA-65) on 27 July 1948. Similarly our old friends VB-17 were re-designated as VA-5B on 15 November 1946; and then became VA-64 on 27 July 1948.[178] To illustrate the closing service of the Helldiver – how it took over from the Avenger, and then in its turn, how it was gradually phased-out and its own successor, the Douglas AD-1 Skyraider,[179] introduced – we can look at two Navy units in the period 1946–48.

Peacetime Work in a Helldiver Squadron, VB-6B 1945–47

VT-17 had reformed on 21 August 1945 at NAS Alameda, California under Lieutenant Commander William N. Janes, but began moving to NAAS Fallon, Nevada. On 11 October, Janes became the executive officer and Lieutenant Commander Rubin H. Konig took over command, being promoted to full commander a month later. The squadron was redesignated as VA-6B and carried out routine training with rockets, gunnery, bombing and navigation exercises; in between there were several aircraft ferrying hops between Hollister, Livermore, Alameda and Fallon. These activities were severely curtailed during December and January 1946, however, due to bad weather.

On 1 February 1946, the squadron was transferred from ComAirPac to ComAirLant, and moved to their new base at NAS Brunswick, Maine. In the period January/February 1946, most of the pilots were busy checking in SB2C-5s for ferry to NAS Memphis, Tennessee from San Diego, and it was during these check-outs, on 30 January 1946, that Ensign Robert Bion Kinney was killed while flying an SB2C-5 (Bu. No. 83225). the aircraft entered into a spin, approximately five

During March 1946, SB2C-4Es were received by the squadron, and by the end of June, VB-6B had an aircraft strength of twenty-six SB2C-4Es and one TBM-3E. Time was spent getting used to the Helldiver, and then on 28 June 1946, the air group departed to Cleveland, Ohio, to take part in the air races. This led to a series of accidents: for instance, on making his landing at Cleveland, Lieutenant (j.g.) Joseph W. Nelson, piloting an SBW-4E (Bu. No. 60207), at the end of his landing

Then on 19 July, Ensign Alexander McNeill Jr, flying an SB2C-4 (Bu. No. 83078) on a landing approach back to NAS Brunswick, was forced to make an emergency landing due to loss of power caused by a faulty propeller governor. The pilot was not injured, but ARM3c. C. B. Young in the rear seat suffered slight abrasions to his left arm.

The squadron took departure for NAS, Seattle, Washington on the 12 and 13 August, with twenty aircraft departing for

SB2C-5s of VB-20 parked at Corpus Christi Naval Air Station, late 1946.
US Navy via National Museum of Naval Aviation, Pensacola

miles north of NAS Fallon, and he could be seen struggling to regain control. He managed to effect two or three partial recoveries, but then the Helldiver went into the ground and both the pilot and aircraft were '... burned beyond recognition'.

run, taxied into the tail of an SB2C-4E (Bu. No. 83059) piloted by Ensign Mahlon H. King, when the latter was forced to stop at a converging runway to await the take-off of an airliner. King's plane was a write-off, and Nelson's required a new engine.

that base each day; but on the 15 August orders were received to change their home port from Seattle to NAS Norfolk, Virginia. VB-6B was caught with its Helldivers spread all over the country at Minn, Minnesota, Glenview, Illinois and Rapid

field'. He was not hurt but the Helldiver was written off completely. It was later found that no less than nineteen faulty spark plugs were the contributing cause of his sudden loss of power.

Between October and December 1947, VB-6B spent its time in carrier qualifications aboard the escort carrier *Salerno Bay* (CVE- 110) and the giant *Franklin D. Roosevelt* (CVB-42), finally exchanging their SB2C-4 Helldivers for SB2C-5s in November. This training continued during January 1947, with field carrier landings and in developing their break-up and rendezvous technique. On 23 January 1947 they embarked aboard the carrier *Valley Forge* (CV-45), which then sailed for Guantanamo Bay, Cuba. The only accident was that of Ensign John Warlick who, while effecting a normal landing in his new SB2C-5 (Bu. No. 89406), turned off to the left of the duty runway to clear himself of planes coming in to land behind him, and in so doing, collided with an F6F Hellcat. However, during the actual shakedown cruise of the air group that February, VB-6B found itself in trouble on three more occasions.

On 5 February, while landing aboard in SB2C-5 Bu. No. 83378, Ensign A. C. Shaw bounced and crashed into the barrier; fortunately he was all right and his Helldiver was repairable. Not so lucky was SB2C-5 Bu. No. 89405, which was a total write-off when Ensign F. C. Richards ground-looped while practising field carrier landings at NAAS Leeward Point, Guantanamo – although again, the pilot walked clear of the wreck. Despite these mishaps, the air group itself came in for praise at the annual inspection by Vice Admiral Bogan and Rear Admiral Holden on 17 March. The carrier herself was in not such a good state, however, and had to terminate her cruise two weeks early, with one steering control engine out of order and the other in bad shape.

VA-6B was then based back at Norfolk, still under Commander R. H. Konig, for an uneventful period. The SB2C-5's days were numbered, however, because he reported[180] that their time during the period July to September 1947 '... was spent mainly in dive-bombing and rocket training, but with all pilots looking forward to the change from SB2C-5 aircraft to the AD-1. Near the end of the period all pilots attended the AD-1 mobile training unit. All pilots in the squadron who are expected to participate in the "shake down" cruise of

Four SB2C-5 Helldivers of VA-1A (formerly VB-4) tip over into an attack dive in 1948.
US Navy via National Museum of Naval Aviation, Pensacola

City, North Dakota. Commencing on 16 August the Helldivers began to reconcentrate at Norfolk, and by the 20 August all had arrived safely at their new base.

The next accident took place on 9 September, when Lieutenant (j.g.) Vernon N. Winquist's aircraft crashed immediately after taking off from the south-west course of East Field, due to a sudden engine failure. By '... skilful handling of his aircraft ...', Winquist '... managed to avoid numerous obstacles and landed in a dirt

the USS *Coral Sea*, were familiarized in the AD-1 aircraft and qualified in FCLP.'

This might explain why they were not too careful with their SB2C-5s. On 7 July, Lieutenant Richard D. Greer, USN, managed to hit a gasoline truck while taxiing to the parking area after completing a routine training flight. Fortunately both the Helldiver (Bu. No. 89464) and the truck suffered only slight damage, and Greer none

gallon fuselage tank since he had taken-off from NAS Norfolk, and that he had completely forgotten to switch tanks before running out of gas in this tank!

On 1 October 1947, the squadron reported aboard the carrier *Coral Sea* (CVB-43) for commissioning exercises. On that date it had one SNJ-6 aircraft, eight AD-1 Skyraiders and nineteen SB2C-5s on its strength. Over the next three months all

Everett W. Herman, USN, flying Bu. No. 89434, stalled and ground-looped on take off, following a momentary cutting out of the engine. A small fire resulted but this was quickly extinguished by the crash crew, and neither Herman nor his rear-seat man, ACOM Lawrence L. Mahone, was hurt; but the Helldiver was another write-off.

The final casualty was SB2C-5 Bu. No. 89360, flown by Ensign Frederick H.

Helldivers at The Rock. The Gibraltar landing strip plays host to an SB2C of VA-11A from Valley Forge **(CV-45) during a courtesy visit in 1948. There are Grumman TBF Avengers in the background from the same ship.** Emil Buehler Naval Aviation Museum Library, Pensacola

at all, except maybe to his pride! Then on 11 August, while practising field-carrier landings, Ensign Miles L. Lacey, piloting an SB2C-5 (Bu. No. 89439) crashed at the down-wind end of runway no. 6 at Fentress Field. Lacey was not injured either, although he managed to cause 'class A' damage to his Helldiver. It later transpired that Lacey had been flying on his fifty-five

the SB2C-5s were transferred out, and AD-1s steadily received in their place, until a full complement of two dozen AD-1s and one SNJ-6 was achieved. On 31 October, command of the squadron was handed over to Lieutenant Commander Glen B. Butler.

Accidents continued, with two AD-1s and two SB2C-5s being written off. On the afternoon of 17 October, Lieutenant (j.g.)

Mann, USN, who ditched his Helldiver in Pokomoke Sound near bombing target no. 25 on 12 November 1947. His engine had cut out at 1,000ft (300m), but Mann was able to turn cross-wind before making a water landing in rough seas. Again, neither the pilot nor his companion, AOM1 M. D. Gerke, was injured and both swam ashore safely as their aircraft sank.

SB2C-5s of VA-5A aboard Shangri-La **(CV-38) in 1947.** Harold C. Gustafson via National Museum of Naval Aviation, Pensacola

An SB2C-5 (Bu. No. 83245) of NARU, Dallas, DA 103, in 1948. US Navy National Museum of Naval Aviation, Pensacola

Transition to the AD-1 Skyraider, VA-5B

It was the much the same story for their sister squadron. In April 1947, VA-5B was based at NAS Norfolk, Virginia, and working from East Field there. They were fully employed in the routine training work of Navy Reserve flyers, with emphasis being placed on bombing and the use of rockets. They suffered two accidents at

underside of the fuselage and the speed ring were damaged in the belly landing, but nobody was hurt, only shaken. Then on 29 May another SB2C-5 was part of a formation taxiing along the runway, when one (Bu. No. 83594), piloted by Ensign Harley T. Briggs, USNR, was forced to apply both brakes, '… to prevent colliding with the plane ahead, resulting in the nosing up of his plane with consequential bending of the propeller blades'.

the wires and went heavily into the crash barrier. In the bland words of the official report: 'Damage to the plane was such as to recommend it for a strike.' Nobody was hurt. On 29 July, Ensign Park L. M. Gourno, USNR, ground-looped his in SB2C-5 (Bu. No. 83522) while making a field-carrier landing practice. He got away with it, but once more, his aircraft was written off.

In fact the greatest potential danger to the squadron at this time was the forecast

SB2C-5s over F4U-4s (units VT-75/VF-75) over the Franklin D. Roosevelt **(CV-42) in May 1946.** Emil Buehler
Naval Aviation Library, Pensacola

this time. On 2 April 1947, Ensign Charles E. Snyder, USNR, made a normal take-off piloting his SB2C-5 (Bu. No. 83745); when the Helldiver was approximately fifteen feet off the ground, however, her engine began to miss and the aircraft, '… settled back on the mat'[181]. The damage was light, the propeller was bent and the

Rather more potentially dangerous were two accidents that took place later that summer. On 9 July the squadron was carrying out carrier qualifications, with pilots making their landings aboard the escort carrier Sicily (CVE-118) off the coast. Ensign Richard F. Johnston, USN (T) came aboard the carrier in his SB2C-5 (Bu. No. 83174), missed

that a hurricane was working its way up the eastern coast of Florida and would hit East Field with some strength. As a consequence, all flyable aircraft were evacuated on 15 September 1947, flying to airfields further north; VA-5B took its aircraft to Rome, New York, where they arrived the next day, after an overnight stop at Floyd

By 1 April 1948 VA-5B was as a fully AD-1 Skyraider squadron, and in this capacity embarked aboard the carrier *Coral Sea* (CV-43) for her shake-down cruise.

Experimental and Civilian Service

In contrast to the more docile SBD/A-25 which lingered on in a variety of roles after it was retired from the services, the 'Beast' proved too much of a handful for most civilian flyers, and survivors are comparatively rare.

US Coastguard Experiments

Two SB2C-5 Helldivers were included in a number of types which were evaluated post-war by the USCG, to improve the efficiency of the air-sea rescue service. These Helldivers were equipped with life-rafts and ration packs as part of a high-speed initial rescue service.[182] The idea was to expedite rescue operations, and between 1945 and 1947 a number of experimental sorties were carried out with these two aircraft, SB2C-5s obtained from the US Navy. The idea was for the Helldivers to be 'first-out' to the scene of the disaster, dropping life-rafts and rations to keep the survivors going until the arrival of the slower float planes or surface ships which would effect the actual rescue, but the idea was never adopted in practice, however.

NACA Experimental Helldiver

Another SB2C-5 (Bu. No. 83135) was retained for several years by NACA for tests and trials. Painted deep blue with TEST in large white letters along each side of the fuselage, and with the distinctive NACA logo on the tail, this machine was seen with what appeared to be a second pilot's cockpit built aft of the normal one, and at a greater height. It was fitted with an experimental sighting device at one time.

Flying Circus!

Probably the most bizarre peacetime use of the 'Beast' was as part of the Wilson King All-Star Air Show, which toured the USA in the late 1940s. The show had commandeered a former A-25A (No. 42-67896) – in its new role it was painted red overall

An **SB2C-4E of VT-75 taking off from the** Franklin D. Roosevelt **(CV-42) in January 1946.**
US Navy via National Museum of Naval Aviation, Pensacola

Bennett Field, New York City. In the event the hurricane changed course, crossing Florida and the Gulf to smash into New Orleans instead. All the aircraft were therefore recalled on the 17 September.

It was during this time, April to July 1947, that the squadron gradually switched-over from SB2C-5 to AD-1 aircraft, as shown by the figures (*see* box below).

VA-5B transfer from SB2C-5 to AD-1 April–July 1947					
Aircraft transferred	No.	Date	Aircraft received	No.	Date
SB2C-5	1	4-4-47	AD-1	4	2-5-47
SB2C-5	2	17-4-47	AD-1	9	9-5-47
SB2C-5	4	22-4-47	AD-1	2	23-5-47
SB2C-5	10	8-5-47	AD-1	2	6-6-47
SB2C-5	2	23-5-47	AD-1	1	9-6-47
SB2C-5	2	6-6-47	AD-1	1	10-6-47
SB2C-5	3	9-6-47	AD-1	1	27-6-47
SB2C-5	1	31-7-47	AD-1	2	31-7-47
SNJ-4*	1	4-8-47	AD-1	6	15-8-47
AD-1	1	22-9-47	AD-1	3	21-8-47
AD-1	2	23-9-47			
AD-1	4	26-9-47			
*For instrument training					

A tight fit for the SB2C-5s of VB-4 aboard the Tarawa **(CV-40) in 1946.**
Commander Harold W. Calhoun via National Museum of Naval Aviation, Pensacola

with a white stripe down the centre of the cowling and fuselage, and white flashes on both wings and tail surfaces plus a yellow star on the cowling. It was christened 'Bernadine', and while its ultimate fate is unknown, it was certainly one Helldiver that ended its days with a bang and not a whimper!

Foreign Service 1945–1959

Thus passed the 'Son-of-a-Bitch, 2nd Class' from US Navy front-line service, although many lingered on in the Reserves. Others were sent to be 'mothballed' at NAF Litchfield Park, where they were put into storage; a large number of these were subsequently refurbished and gained a new lease of life flying with America's allies during the turbulent postwar era.

Greece 1949–1953

Greece was one of the first small nations to take delivery of the Helldiver. After an agonizing debate in the US, Congress had finally authorized help to such nations who were under threat from Communist expansion and insurgency, and as a result forty-two SB2C-5s were delivered in 1949.

Initially, two SB2C-5s were delivered on 2 July 1949 and these were used mainly for pilot training and familiarization at Hellenic Air Force base Larisa. This duo was followed by a further twenty Helldivers which were delivered in August 1949, and used mainly in the same role. Finally another batch of twenty SB2C-5s arrived, and these formed the 336th Light Bombarding Squadron, also at Larisa.

The Helldivers replaced the unsuitable Supermarine Spitfire then being flown by 336 Mira in the Greek civil war which had raged from 1946. The Helldivers proved far more suitable to their task, that of tactical reconnaissance and strikes against Communist guerrilla forces operating with Soviet and Yugoslavian backing in the Greek mountains to overthrow the democratic Hellenic government and take that nation behind the Iron Curtain.

Serial Numbers and Known Squadron Codes of Hellenic SB2C-5 Helldivers

Serial number	Squadron code	Serial number	Squadron code
3186		3724	
3224		3350	17
3260		9134	
3264		9137	
3296		9163	
3313		9193	15
3321		9209	
3328		9224	
3329	9	9250	11
3330		9301	
3331		9350	
3353	8	9386	6
3442		9390	
3480	2	9396	
3719	10	9398	
3721		9401	
3723		9404	
		9453	

Commanders of Hellenic Air Force Helldiver Squadron 1949–1952

Commanding Officer	Appointed	Relieved
Squadron Leader E. Athanasopoulos	15 May 1949	15 July 1950
Squadron Leader P. Fragogiannis	15 July 1950	7 October 1950
Lieutenant K. Nikolopoulos	7 October 1950	15 July 1951
Major G. Lagodimos	15 July 1951	15 October 1952
Major G. Tsitsoglou	15 October 1952	20 October 1953

Pilots of the 336th Light Bombarding Squadron, Hellenic Air Force, with their SB2C-5 Helldiver, showing the squadron insignia. Hellenic Air Force Official

(Below) **An SB2C-5 of the Hellenic Air Force, No. 9453.** Hellenic Air Force Official

An SB2C-5 Helldiver (coded 7) of the 336th Light Bombarding Squadron Hellenic Air Force, based at Larisa airfield during the Greek Civil War of 1946–49. The aircraft is equipped with underwing machine-gun pods for these operations. Hellenic Air Force

(Below) Head-on view of an SB2C-5 Helldiver of the 336th Light Bombarding Squadron, Hellenic Air Force, based at Larisa airfield during the Greek Civil War 1946–49. Hellenic Air Force

(Bottom) Rare photograph of an SB2C-5 Helldiver of the 336th Light Bombarding Squadron, Hellenic Air Force, based at Larisa airfield during the Greek Civil War of 1946–49. Hellenic Air Force

The SB2C-5s gave good service until they were finally replaced by F-84Gs as late as 1953. A few Helldivers remained in service even after that date, however, and were used for air photography purposes as late as 1957.[183]

Italy 1950–1959

Although the *Aeronautica Militare Italiano* was not allowed to operate bomber aircraft under the terms of the 1943 armistice with the Allies and subsequent agreements with the four victorious powers, the Communist threat was very real in Italy and for this reason the USA supplied two dozen SB2C-5 Helldivers in 1950 under the same Mutual Defence Air Program (MDAP) as was used to equip Greece and Portugal. These machines (serial numbers 80015 to 80038 inclusive) were shipped to Italy aboard the escort carrier *Mindoro* (CVE-120) in September 1950, and unloaded at the port of Brindisi. They were redesignated as S2C-5s (the offending letter 'B' for bomber being deleted to conform to the letter, if not the spirit of the law! It was also reported that their bomb-bay doors were welded shut for the same reason.) They were used to form two squadrons of anti-submarine patrol aircraft, *Gruppi Antisommergibill* 86 and 87 respectively, which was part of the Air Force-controlled *Aviazione per la Marina*. They took these twenty-four Helldivers formally 'on charge' on 16 September.

In December 1950 the first Air Force unit, the 86 *Gruppo Antisom* was set up with two squadrons, 162a *Squadriglie* and

Italian Navy SB2C-5 Helldiver, coded 45, of 86 Gruppo, flying over the battleship Andrea Doria **at Taranto.** Ufficio Storico, Roma

Two Italian SB2C-5s (Nos. 101 and 102) undergoing assembly. Ufficio Storico, Roma

164a *Squadriglie*, but only eleven pilots for the twenty-four aircraft. They were based at Grottaglie airfield, initially under the command of Capitaine Vodret and then, from 1 March 1951, under the command of Lieutenant Colonel Pelosi. Familiarization flying commenced, and continued into 1951; with the first fatal accident being the loss of Lieutenant Guerrisi on 11 January 1951. Practice bombing was conducted with both 500lb bombs and torpedoes against a variety of naval warship targets, and for the designed anti-submarine role, a large part of the training programme involved the practice and use of their AN/APS-4 search radar for surfaced submarines, and the carrying out of both depth-charge and rocket attack against such targets. They also took part in NATO exercises operating from Hal Far airfield at Malta, another 'first'.

Meanwhile the Italian Navy had been attempting to form its own Naval Air Arm but had been thwarted by the Air Force who did not understand specialized naval needs any more in 1950 than they had in 1930, despite the lessons of a lost war in between. Navy pilots were sent out to the USA to train, and a sympathetic US Navy made a gift of two SB2C-5s which were painted in unofficial Navy colours with the Italian Navy badge on the tailplane and 'Hall-type' anchors superimposed on the national insignia on fuselage and wings. The two Helldivers were formally handed over to the Italian Navy at NAS

Italian Navy Officers Trained in the USA for the SB2C-5	
Rank	Name
T. V. Oss	Carlo Jorio
T. V. Oss	Anton Vittorio Cottini
T. V. Oss	Mario Volpi
T. V. Oss	Mario Albanese
T. V. Oss	Luigi Piamonete
T. V. Oss	Danilo Guyon
T. V. Oss	Donato Iorio
S. T. V.	Fiorenzo Rosso
S. T. V.	Guido Sessa
G. M.	Filippo Cali
G. M.	Italo Saaccardo
G. M.	Vittorio Valente
G. M.	Veniero Di Marzio

Cabaniss Field, Corpus Christi, Texas, in mid-1952 and were given the *Marina Militare Italiana* serials nos. MMI 101 and MMI 102. These two Helldivers were the only ones supplied to Italy which had the tail-hooks fitted; all the others had them removed, being shore-based under Air Force control, and Italy having no aircraft carriers of her own.

Those Navy pilots that graduated were kept 'in limbo' until the dispute between the Navy and the Air Force had cooled down a little, and it was not until November that the first two Navy pilots, Valente and Volpe, flew to the two Navy Helldivers to Norfolk, Virginia, for embarkation and the voyage to Italy. They were shipped out aboard the

They were hustled out of their machines, and charged with violating national air space; flying unidentified aircraft, flying unregistered aircraft, flying without valid nationally recognized licence and failure to comply with control tower landing instructions, the latter when one of the pilots was told by the tower to land on two separate air strips at the same time! Both Helldivers were treated as American machines and stored in a hangar.

On 4 August 1953, after eight months the two Navy Helldivers were confiscated by the Air Force, who thus successfully nipped in the bud the birth of a true Italian Navy Air Arm! They were assigned the serial numbers MM. 4698 and MM.

submarine force, and the 87 Gruppo was formed from 164a *Squadriglie* at Catania-Fontanarossa air base in Sicily in November with ten machines. In the interim, 86 *Gruppo* had received a new squadron to replace 164a, when 163a *Squadriglie* was formed on 15 July 1952, the group having twenty-six Helldivers in total.

The SB2C-5s all served well in this unusual role until, from about February 1953 onwards, they were replaced by the more suitable long-ranged Lockheed PV-2 Neptune aircraft, designed for the job. The SB2C-5s then began to be relegated to second-line duties, one going to the Guidonia Experimental Centre as a glider tug, and others similarly converted, with

Italian Helldiver No. 101 seen here at NAS Cabaniss Field, Texas in September 1951, awaiting delivery to the Italian Navy via the USS Midway. Launched on 12 December from Midway's flight deck the two Helldivers were treated as bitter enemies by the Italian Air Force authorities and on arrival at Capodichino, Naples their pilots were not greeted, but arrested and the Helldivers confiscated by the Aeronautica Militare. She carries the insignia of the Italian Navy in blue and white painted by the US Navy in their honour. Giancarlo Garello

carrier *Midway* (CV-41) on 2 December of that year, taking passage aboard her across the Atlantic and into the Mediterranean.

On 19 December the two Helldivers were launched from the American carrier from a position south of Marseilles, in an historic 'first' for the Italian Navy. However, the home country, at the instigation of the Air Force, treated the two pilots more like criminals than heroes upon their landing at Capodichio airfield near Naples.

4699. The former suffered engine failure, crash-landed and was written off on 3 May 1955, but MM. 4699 continued in service in a variety of roles with the Air Force until 19 February 1959.

A further allocation of SB2C-5 Helldivers reached the Air Force early in 1953, and were assigned serial numbers in the range MM. 4682 to MM. 4697. With a total of forty-two machines (less losses), the Air Force was able to expand its anti-

the addition of a towing pod under the starboard wing, to sleeve target-towing duties. In this role they continued to give both Army and Navy units AA practice.

Wear-and-tear and accidents gradually reduced the number of Italian Helldivers flying, 87 Gruppo retiring its final five SB2C-5s as early as 1954, while 86 Gruppo started replacing theirs in the same year, MM. 80021, MM. 80034 and MM. 80038 being casualties that year, reducing the total to thirty-two. This number of Hell-divers had come down to twenty-three by August 1956, when 161a *Squadriglie*, by then based at Capodichino airfield near Naples, began to receive its Grumman S2F Tracker replacements.

The final casualty was MM. 4686, whose engine caught fire while on a liaison flight from the Grottaglie base to Catania.

The pilot, Lieutenant Giannotti, and his observer, Warrant Officer Castro, both managed to bail out safely over Maropati, Reggio Calabria, and the Helldiver crashed and exploded. The last Italian Air Force Helldiver was retired from service on 26 February 1959.

However, these were *not* the only Hell-divers which were supplied to Italy. In its continuing attempts to get its own independent air arm, the Italian Navy had authorized the stockpiling of spare parts, and in 1953 went so far as to obtain, under the Mutual Defence pact, a further forty-six Helldivers which had been declared surplus to US Navy requirements. These were mainly SB2C-4s, but with some SB2C-4Es and SB2C-5s as well, all of which had seen some war service, but only two of which had more than 500 hours' flying time 'on the clock'.

These were duly shipped from the States to Italy aboard the US carrier *Tripoli* (CVE-64) which put into La Spezia on 13 September 1953. The carrier was met by an Italian naval party commanded by Captain Roncallo, who was told by his American counterpart with wry humour, 'This time we are delivering them direct to your door; don't let them get stolen, as in Naples!' They were duly disembarked in record time, eight hours, and, from 26 November onwards, were transhipped via pontoons to Morola where they were stored in a large warehouse. Over the next ten months they were all totally dismantled, and the fuselages, wings, tailplanes, engines, propellers, instruments and electronics, along with landing gear and other equipment, transferred to a Navy warehouse at the La Spezia Navy arsenal where they were carefully stacked and labelled until the Navy would get the word to utilize them.

Italian Helldiver No. 6 of 86 Gruppo in flight. On her tail she carries the unit badge of an Eagle's talon reaching out for a Shark's fin, symbolic of their anti-submarine status in Italian service – a wholly unique and unexpected role for a dive-bomber.
Giancarlo Garello

Italian SB2C-5 Helldiver No. MM. 4696, seen in 1956 at Fiumiciono, Italy.
Giancarlo Garello

A Curtiss SB2C-5 Helldiver of 86 Gruppo Antisom, at d'Istrana airfield on 14 July 1958. They were used in the NATO anti-submarine role by 51 A and B Gruppi over the Adriatic Sea helping patrol the large coastline opposite the communist states of Yugoslavia and Albania. Nicola Malizia

Italian no.	USN no.	Type	Italian no.	BuAer no.	Type
80015	132	SB2C-5	N/A	20306	SB2C-4/4E
80016	160	SB2C-5	N/A	20320	SB2C-4/4E
80017	181	SB2C-5	N/A	20715	SB2C-4/4E
80018	203	SB2C-5	N/A	20718	SB2C-4/4E
80019	235	SB2C-5	N/A	20800	SB2C-4/4E
80020	265	SB2C-5	N/A	20820	SB2C-4/4E
80021	272	SB2C-5	N/A	20851	SB2C-4/4E
80022	275	SB2C-5	N/A	20865	SB2C-4/4E
80023	284	SB2C-5	N/A	20879	SB2C-4/4E
80024	289	SB2C-5	N/A	20927	SB2C-4/4E
80025	294	SB2C-5	N/A	20940	SB2C-4/4E
80026	309	SB2C-5	N/A	20949	SB2C-4/4E
80027	311	SB2C-5	N/A	20961	SB2C-4/4E
80028	313	SB2C-5	N/A	20999	SB2C-4/4E
80029	314	SB2C-5	N/A	21008	SB2C-4/4E
80030	408	SB2C-5	N/A	21010	SB2C-4/4E
80031	430	SB2C-5	N/A	21013	SB2C-4/4E
80032	464	SB2C-5	N/A	21046	SB2C-4/4E
80033	492	SB2C-5	N/A	21054	SB2C-4/4E
80034	362	SB2C-5	N/A	21180	SB2C-4/4E
80035	148	SB2C-5	N/A	65075	SB2C-4/4E
80036	238	SB2C-5	N/A	65102	SB2C-4/4E
80037	231	SB2C-5	N/A	65151	SB2C-4/4E
80038	570	SB2C-5	N/A	65165	SB2C-4/4E
4682	162	SB2C-5	N/A	65183	SB2C-4/4E
4683	319	SB2C-5	N/A	65184	SB2C-4/4E
4684	325	SB2C-5	N/A	65189	SB2C-4/4E
4685	384	SB2C-5	N/A	65193	SB2C-4/4E
4686	350	SB2C-5	N/A	65257	SB2C-4/4E
4687	–	SB2C-5	N/A	65259	SB2C-4/4E
4688	–	SB2C-5	N/A	82870	SB2C-4/4E
4689	–	SB2C-5	N/A	82872	SB2C-4/4E
4690	–	SB2C-5	N/A	82953	SB2C-4/4E
4691	568	SB2C-5	N/A	83007	SB2C-4/4E
4692	623	SB2C-5	N/A	83015	SB2C-4/4E
4693	627	SB2C-5	N/A	83029	SB2C-4/4E
4694	664	SB2C-5	N/A	83087	SB2C-4/4E
4695	373	SB2C-5	N/A	83104	SB2C-4/4E
4696	421	SB2C-5	N/A	83123	SB2C-4/4E
4697	584	SB2C-5	N/A	83127	SB2C-4/4E
4698	212	SB2C-5	N/A	89140	SB2C-5
4699	404	SB2C-5	N/A	89421	SB2C-5
			N/A	89428	SB2C-5
			N/A	89443	SB2C-5

In fact that word *never* came, and gradually the various parts were pirated by the Air Force to keep their own machines operational. By far the bulk of these dismantled Helldivers were subsequently scrapped to clear space when it became clear that the Italian Navy was never going to be allowed its own Fleet Air Arm by the government. Only one machine survived a little longer, an SB2C-4 (Bu. No. 89140) which was sent to the Augusta Navy Training Centre, minus its outwing sections, tailplane and engine, to be utilized for instruction purposes.[184]

The post-war scene. Far from being stricken immediately the war ended, as some historians would have it, the Curtiss SB2C-4s and -5s continued as the mainstay of the new Navy attack squadrons in the post-war fleet for the next four years. Here Helldivers are packed shoulder-to-shoulder aboard the big carrier Franklin D. Roosevelt **(CVB-24) on 24 April 1946.** National Archives, College Park, MD

Grumman Bearcats with Curtiss SB2C-5s of VA-7A aboard the carrier Leyte (CV-37) in the Mediterranean in July 1947. Ron McMasters

The Curtiss SB2C-5s of VA-7A taxiing down the Leyte's flight deck and unfolding their wings prior to take-off in July 1947. Ron McMasters

(Below) The Curtiss SB2C-5 No. L-317, piloted by Ron McMasters, operating from the carrier Leyte in the Mediterranean, July 1947. Ron McMasters

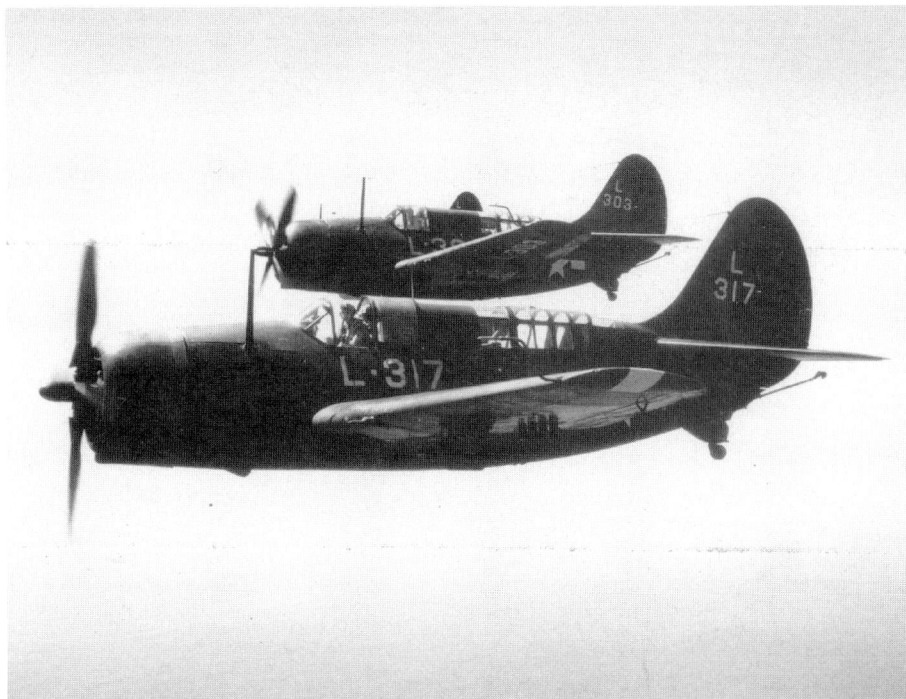

Two SB2C-5s of VA-7A from the Leyte in formation over the Mediterranean in July 1947. Note the carrying angle of the tailhook. Ron McMasters

(Above) The approach by an SB2C-5 of VA-7A landing back aboard the Leyte in the Mediterranean in July 1947. Ron McMasters

(Above right) Hook down, flaps down, wheels down! An SB2C-5 of VA-7A lands back aboard the carrier Leyte in the Mediterranean in July 1947. Ron McMasters

(Right) Safely aboard the carrier Leyte, an SB2C-5 of VA-7A revs forward to make room for the next to land aboard. Mediterranean operations, July 1947. Ron McMasters

Head-on view of a Portuguese Navy SB2C-5 with bomb bay doors open. Lieutenant Duarte Monteiro, Mais Alto

Portugal 1950–1956

Twenty-four Curtiss SB2C-5s, together with the spare parts and equipment, were also supplied to the Portuguese Air Force through the US Military Assistance Programme in 1950.

They were operated by the Portuguese Navy until 1952 and given Navy serial numbers, from AS-1 to AS-24 (AS = Anti-Submarine). They were based at the Centro de Aviação Naval de San Jacinto, then a Navy Base. They arrived in Portugal painted dark blue and continued to be used thus. The Navy painted the rudder with green stripes.

In 1952 these Helldivers were transferred to the Portuguese Air Force who initially operated them from 1953 from Air Base No. 5. In 1954 the fleet was moved to Air Base No. 6, Montijo, near Lisbon, until their retirement. Under Air Force control a white disc under the 'Cruz de Cristo' was added, the National Flag replaced the stripes and the serial number was painted in front of the flag. The usual armament was bombs and rockets. Regrettably none have survived.

Thailand 1951–1957

Six Curtiss SB2C-5s, together with the spare parts and equipment, were also supplied to the Royal Thai Air Force through the U S Military Assistance Programme on 29th June 1951. They were shipped to Thailand aboard the escort carrier *Cape Esperance* (CVE-88) and were commissioned at RTAF Wing 7, under the command of Group Captain Boonchoo Chandrubeksa.

Equivalent Serial Numbers of Portuguese Helldivers

Original US serial no.	Portuguese registration no.
89192	AS-1
?	AS-2
89187	AS-3
89194	AS-4
?	AS-5
?	AS-6
89413	AS-7
?	AS-8
?	AS-9
83542	AS-10
83573	AS-11
89455	AS-12
89389	AS-13
89437	AS-14
83388	AS-15
89196	AS-16
89441	AS-17
89363	AS-18
83481	AS-19
89263	AS-20
?	AS-21
89462	AS-22
89355	AS-23
89297	AS-24

Equivalent Serial Numbers of Royal Thai Air Force Helldivers

Original US serial no.	RTAF registration no.
83206	1/94
83299	2/94
83328	3/94
83410	4/94
83711	5/94
89458	6/94

This unit flew the Helldiver as an attack/bomber aircraft on combat missions against Communist guerrillas, first under Group Captain Han Khampipat (1951–54), then Wing Commander Prakong Pintabutr (1954) and finally Group Captain Usa Chaiyanam (1954–57) after which they were decommissioned. One, serial no. 4/94, still survives to this day.

A Curtiss SB2C-5 of the Royal Thai Air Force in 1953. Royal Thai Air Force

CHAPTER TWENTY

A War of Attrition:
The French Experience

In her post-war campaign against the Communist guerrilla uprising in Indo-China (now Vietnam, Cambodia and Laos) the French Navy had used several squadrons of Douglas SBD Dauntless dive-bombers, with some success.[185] However, by 1948 they were completely worn out,

Curtiss SB2C-5 Helldivers bought from the US Navy in 1950. (Batches of the same mark also served in the adjacent jungle with the Royal Thai Air Force, from 1953 through to 1955.) Forty-eight SB2C-5s were obtained in an initial order from France, and they formed two new Flotilles,

equipped with nine SB2C-5s, and later a further three were added to their strength; they had a total of nine pilots trained to use the Helldiver. They arrived on station on 24 September 1951 and served until 16 May 1952, conducting some 163 combat missions during this period.

French Navy SB2C-5s revving up aboard the carrier Arromanches **off the Indo-China coast November, 1953.**
Musee de l'Air, Paris

and after a further fifty-six battle sorties during which they clocked up 300 hours of combat flying time, the faithful old SBDs finally returned to France aboard the carrier Arromanches on 5 January 1949.

Replacements were later to take their place on the firing line, these being the

3F and 9F, the former seeing the first war service. On 9 April 1951, this unit, which had been equipped with Supermarine Walrus and Dornier 9FT amphibians at Sartrouville, flew to Hyères to be re-equipped with the dive-bombers. Under the command of Lieutenant de Vaisseau Waquet, they were

When Arromanches returned to Indo-Chinese waters for a second commission towards the end of 1952, 9F's Helldivers, commanded initially by Lieutenant de Vaisseau Hervio, were embarked aboard her. They soon found ample employment once they arrived. The siege and fall of

Dien Bien Phu in 1954, while not the actual end of the war, marked the end of the French government's overall intention to go on shouldering, alone and unaided, the burden of protecting the Far East from Communist takeover. Requests for both American (physical) and British (moral) backing had been turned down flat by Eisenhower and Churchill respectively, and France was still too divided to go on without either.

When the first great Viet-Minh offensive had erupted out of the Highlands towards the end of 1950 it had come up

occurred on 22 December 1950, when a Viet-Minh column was observed near to the north coast of Along Bay, near Tien Yen. This enemy concentration could have threatened the vital Red River delta, and a French paratroop regiment was sent to disperse it.

Direct air support was given for the first time in Indo-China, and liberal use was made of napalm. Each aircraft had wing tanks filled with jellied petroleum which they dropped like bombs; on impact the tanks detonated in a sheet of fire and flame, cascading lumps of sticky fire over a

helped the heavily outnumbered French ground forces.

Another significant encounter took place on the 16 and 17 January 1951, when a massed assault by two Communist divisions, the 308th and 312th, was made against French defensive positions on the Red River north-east of Hanoi, at Vinh Yen. Regardless of losses, in the Chinese manner of 'human waves' line after line of infantry flooded forward, was mown down and scattered by automatic fire, and was immediately replaced by others. Such tactics, with their callous disregard for their own casual-

Loading bombs at the French base of Bac Mai, 1954. Castaignos

against French air power, and even in its limited form, the Communist troops had not liked what they had found. The French were then operating close to the sea and their air bases, and they could therefore use what limited aircraft they had to good effect against the massed enemy columns. An early example

wide area and decimating large numbers of the Communists. This initial introduction to close air support was a chastening lesson to the peasant army, and it was soon found that even indoctrinated units were uneasy under threat of air attack, and often broke when it materialized. This slowed down their movements a great deal, and greatly

ties, could easily have overrun any position ultimately, but again it was the introduction of the French dive-bombers that saved the day. Every available aircraft was thrown into the battle with bombs, machine-guns and napalm. At the end of two days' intensive operation it was estimated that some 9,000 Viet-Minh had been killed, 8,000 wounded

(Above) The French Navy used their Helldivers in the Gulf of Tonkin and at Bach Mai airfield near Hanoi, striking at Viet Minh communist troop targets deep inside communist-held territory around the Red River delta. ECP Armees, Paris

Helldivers of 3F lined up at their jungle base; note the Navy tail markings of 3F at this time.
Captain H. de Lestapis

and 600 taken prisoner, out of an attacking force of 22,000 men – and the Communist assault was broken.

Initially the Viet-Minh commander, General Vo Nyguyen Giap, had taken his inspiration and tactics directly from his Communist Chinese backers in the belief that these 'human wave' tactics, as were then currently being practised with some success in the Korean War, would produce the same results in Vietnam. It took a little time for the lesson of the dive-bombers to sink in. These regiments were reconstructed and thrown into a new offensive, together with the 316th Division and the Independent Regiment 1489, in a new assault across the Red River on 14 October 1952. This time they broke through the French defence line after a week of heavy fighting and threatened to flood south and engulf the whole region.

At the isolated French airfield of Na-San, this tidal wave of men met a breakwater in the form of a hastily reinforced defensive position, or 'hedgehog' to use German Eastern Front parlance. It was centred on this airstrip, into which reinforcements, supplies and ammunition were continually flown, and the wave broke against it in vain. The French termed this fortress a 'base aero-terreste' (air-land base), and the defenders were well dug in with mutually supporting defence positions, well laced with automatic weapons. Again, mass infantry attacks resulted in mass slaughter of the attackers. Giap had not studied the Somme or Passchendaele, but he was intelligent enough to realize that this method was not working, and for a time he reverted to guerrilla hit-and-run tactics. As a result the French thought that they had found the answer to jungle fighting – but it was a fateful precedent for both sides.

The Helldivers of 9F served faithfully in this theatre for two years, undertaking a whole variety of missions; these included much true dive-bombing, as well as strafing, and rocket and napalm attacks at low level. In total, 9F carried out no fewer than 824 combat sorties during this commission: this represented some 2,000 hours' flying time, during which period they dropped some 1,442 tons of bombs and expended 100,000 rounds of machine-gun ammunition against Viet-Minh targets of all kinds. In 1953 they returned to France and converted to the Grumman Avenger torpedo-bomber, their places meanwhile being taken by 3F with a new batch of Curtiss SB2C-5s. With this unit the French Navy dive-bombers were to earn undying fame in the last big battle of the war.

French Navy Helldivers of 3F Flotille being refuelled at an inland base during the Indo-China war.
Captain H. de Lestapis

At the decisive battle of Dien Bien Phu, the 3rd Carrier Assault Flotilla's SB2C-5 Helldivers, commanded by Lieutenant Andrieux, operated from the *Arromanches* (Captain Patou) and provided the spearhead of meaningful close air support until 30 April 1954.[186] These dive-bombers were supplemented by the Grumman F6F Hellcat fighters of the 11th Carrier Fighter Flotilla acting in the fighter-bomber role from the same ship. Initially they supplemented the Grumman F8F Bearcat fighter-bombers of the French Air Force's 1/22 Saintonge and 2/22 Languedoc fighter groups, until these were decimated by the Viet-Minh's artillery fire on the besieged airstrip. The dive-bombers also received support from the Navy's 28th Bomber Flotilla equipped with PB4Y2 Privateers and the Air Force's B-26 Marauder light bombers of the 1/25 Tunisie. All air attacks in support of the French garrison were coordinated by Lieutenant Colonel Dossol's bombardment sub-group.[187]

The role of the Air Force in this operation was, of course, crucial, both in basic supply of the fortress and in close air support. This role had been spelt out in a communication sent by General Henri Navarre to Brigadier General René Cogny on 3 December 1953, which read: 'The mission of the Air Force shall be, until further orders, given priority and with the maximum means at its disposal, to the support of our forces in the North-West. The commanding general of the Air Force in the Far East will, to that effect, reinforce the Northern Tactical Air Group.'[188]

What the order ignored was the inability of the severely limited forces at the disposal of the French Air Force and Navy to provide such support in anything like the numbers of sorties required. This was due to the basic miscalculation of the French upper command in Hanoi – one which sealed the fate of the whole campaign – as to the scale and vigour of the Viet-Minh response and their ability to transport by coolie labour, heavy equipment, especially artillery and AA guns, and large numbers of combat divisions, undetected, to the scenes of battle. Dien Bien Phu was a mousetrap set deliberately to entice the Communists into a stand-up battle and then be decimated, like Verdun. Unfortunately for the French, they succeeded only too well, but ended up by luring in not a mouse, but a tiger! This classic underestimation of the enemy, the cardinal sin of warfare, went deep. For example Colonel

Charles Piroth proclaimed loudly before the battle that:

> Firstly, the Viet-Minh won't succeed in getting their artillery through to here. Secondly, if they do get here, we'll smash them. Thirdly, even if they manage to keep on shooting, they will be unable to supply their pieces with enough ammunition to do us any real harm.[189]

Every one of these prophecies proved totally incorrect, and in March, the poor disillusioned colonel, his own guns smashed to atoms or overrun, pulled the pin of a grenade he was holding as acceptance of the blame for his misplaced optimism. So it was upon air support alone that the beleaguered garrison finally depended more and more as the siege progressed. The supporting aircraft were just too few to make up for the four- and five-to-one superiority that the Viet-Minh managed to achieve in heavy artillery, especially as their guns were concealed in firing pits with only their muzzles protruding to fire on a fixed axis. The French heavy guns, while they lasted, were, by contrast, out in the open in a severely restricted space and easily pin-pointed.[190]

Moreover the much-heralded Operation 'Strangle', conducted by the US Air Force to cut similar two-legged supply lines in Korea during 1951–52, had ended in total failure to limit the movement of Communist supplies. In the jungle-clad hills of northern Vietnam these peasant columns were even more difficult to spot, let alone destroy, even if the French could approach, let alone match, the enormous US airborne resources. But of course the French had not a fraction of this strength to deploy against Giap's supply lines, and Giap himself was to boast later, '… our soldiers knew well the art of camouflage, and we succeeded in getting our supplies through.'[191] This art of camouflage even extended to the tying together of the tops of tall trees to form a concealed archway of leaves to hide their movements by day. Special emphasis was always placed by the Communists on concealment from the air. It became the duty of every Viet-Minh soldier when on the march, to carry a kind of framework on his shoulders which was intertwined with leaves and foliage by the man marching behind; thus as the area of the jungle changed vegetation, so the camouflage was altered to suit and blend in perfectly. The biggest asset the Communists had was, of course, their endless

supply of man- (and woman-) power, the impressed peasants of the Dan Cong, who were used totally ruthlessly and with the usual complete indifference for personal loss or injury which so marks Communist-inspired warfare.

Thus in the course of the interdiction missions flown between November 1953 and May 1954 to cut the vital supply links of roads 13B and 41B leading up from the Red River and across the Black River toward Dien Bien Phu, although a heavy toll was taken, the Viet-Minh build-up never slackened. Nor was it conducted without loss to the French aviators, either. Although the air over the fortress itself was a maze of flak, the approach routes leading to it were almost as comprehensively defended, and in December, for example, during the course of 367 combat sorties, forty-nine French aircraft received damaging hits from the ground. This in turn led to dive-bomber attacks being diverted away from either supporting the paratroopers or cutting the enemy supply lines, and into pure anti-flak suppression missions – all of which suited Giap's build-up plans admirably.

Much of the Navy Helldivers' work was interdiction, attempting to sever the vital roads leading into Dien Bien Phu down which so much material and manpower was being carried laboriously for the Communist build-up. On 5 January, for example, a strike was launched against the vital road junction town of Conoi (code-named 'Jezebel') where the main highways of roads 41 and 13B met and became road 41B, the final artery of the Viet-Minh supply route. At 12:30, nine Helldivers were sent in to attack this point, each aircraft being armed with 500lb and 1,000lb bombs. The weather at sea and over the target was perfect and clear, and very accurate attacks were made on Conoi itself, while strafing attacks were also conducted during the return flight, at both Thai Binh and Tutien, where camouflaged vehicles were located and duly burnt.[192]

Similar missions continued; for instance, on 7 January one typical sortie was flown against position 'Melchior', further along road 41B, east of Tuan Giao. The nine Helldivers dropped a variety of bombs on this position, four of them 1,000lb weapons with delay-action fuses of twelve and twenty-four hours so as to cause maximum disruption over a long period. They also dropped six 'Butterfly' M.131 bombs.

Periodically the *Arromanches* had to refuel and replenish, and because the French had virtually no means to replenish whilst under way, she had to leave station every so often and steam north to Hong Kong. During such absences it became the practice for sections of the SB2C-5 unit to fly ashore and continue limited operations from land bases, the principal one being Bach Mai near Hanoi itself. It was from here, on 19 increasingly attacking targets much closer to the besieged garrison. On 29 January, ten SB2C-5s took off between 13:30 and 14:15 hours to bomb and strafe the area just 5 miles (8km) NNE of Dien Bien Phu. However, bad weather over the target area made any degree of accuracy impossible, although twenty 1,000lb and ten 500lb bombs were planted in the target area and strafing was conducted on enemy positions main French defences with twelve out of sixteen 1,000lb bombs, and six out of eight 500lb bombs, and they also strafed the target. Two of the dive-bombers aborted, but delivered their bomb-loads instead on enemy positions at Thai Binh. That January, 3F delivered 436 1,000lb and 385 500lb bombs to the enemy as well as 51 'Butterfly' bombs, 38 loads of napalm and 15,000 rounds of machine-gun ammunition.

The French Navy carrier *Arromanches* (formerly HMS *Colossus*) in Along Bay. A Helldiver is the leading aircraft on the deck, with F6F Hellcats and Grumman TBM Avengers parked aft. Captain H. de Lestapis

January for example, that six Helldivers operated against enemy held positions near Nha Trang in conjunction with Operation 'Arethuse', a sub-operation of the main French offensive, code-named 'Atlante'.

But as the Viet-Minh relentlessly closed in, the dive-bombers found themselves 9 miles (15km) to the south-west of French lines; but the results could not be observed by the pilots.

All the time the enemy was creeping closer. On 3 February a typical mission was sent off involving eight Helldivers: these hit dug-in enemy positions 4 miles (6km) east of the

The same heavy workload continued during February 1954. On the 2nd for example, a Helldiver attack was made on thick plantations some 9 miles NNE of Dien Bien Phu, which was thought to conceal an artillery ammunition dump. Five SB2C-5s made accurate attacks and planted eight bombs on this target, and also on Hill 41 which was being used as an observation post for the Viet-Minh and Chinese regular army gunners. Both targets were hard hit and a large column of white smoke indicated that at least one supply dump had been blown up. The only French Navy air losses up to 11 February were two SB2C-5s damaged by accidents during landing; in one of these, on 8 February, Helldiver 98.367 missed a wire landing on the carrier and went into the crash barrier, which necessitated a new engine and prop. But these were the normal working hazards of intensive carrier operations. Immunity over the target was, however, soon to become of luxury of the past, as the number of AA guns deployed by Giap's forces steadily increased and improved in accuracy.

The *Arromanches* had to return to Hong Kong between 15 and 23 February, and once more a detachment of six Helldivers

worked out of Bach Mai, while some Hellcats were based at Cat Bi Field, near Haiphong. Those aircraft which remained aboard the carrier were rotated to give all flying personnel a break from what was becoming an increasingly wearing and hazardous period of non-stop combat flying. The sortie rate of the dive-bombers ashore in no way lessened either, ranging from between eight and twelve sorties each day, except for the 23rd, when the weather prevented most flying.

The aircraft carrier returned to her flying-off position, and the pounding continued unabated. In one attack on 1 March, five Helldivers dropped eight 260lb bombs on enemy foxholes in the region 19 miles (30km) north-east of Dien Bien Phu itself – the later, the Army observation posts signalled back that the attackers' accuracy had been 'excellent'. The outward leg was flown at 7,000ft (2,130m) and they returned at 11,000ft (3,350m).

It was a different story on 13 March, however. Two strikes were made by the SB2C-5s this day, launchings taking place between 15:30 and 16:00 respectively. The first attack was carried out by four Helldivers, and these hit artillery pieces about 4 miles (6km) east of Dien Bien Phu with four 500lb CR bombs: the resulting explosions were observed to be directly on target, and were photographed, evidence to magnificent shooting. The second mission was carried out by a further five Helldivers against more enemy guns located 4 miles north-east and 4 miles south of the two extremities of the French fortress, now itself under a veritable hail of explosive shells. Ten 1,000lb and fourteen 500lb bombs were delivered against these Communist guns in an effort to bring some relief to the French troops who were the victims of this terrible pounding, but in this second attack detailed observations of the results could not be made. The reason

The deck park of the Arromanches, **with Helldivers** *(left)* **and Hellcats** *(right)*. Captain H. de Lestapis

A Helldiver with wings folded being readied at dawn while a rescue helicopter hovers off the Arromanches's starboard side. Captain H. de Lestapis

given was the same for both: 'Violente Reaction de la D. C. A. sur l'objectif.'[193] Why was the flak so intense this time? The answer was clear – the main Viet Minh assault was now under way.

Just prior to the main Communist assault which went in on 13 March, the commander-in-chief of the French Far-Eastern Air Force, General Henri Charles Lauzin, sent a telegram asking that all the ground-support pilots should take 'exceptional risks' in support of their brothers on the ground. The Navy dive-bomber pilots were to respond to this exhortation with a will and much vigour and bravery, and at considerable cost to themselves. The ensuing battle has often been compared to Verdun of World War I, or to those of Stalingrad, Bataan, Corregidor, Tobruk and L'Orient during World War II; however, as far as the air element involved was concerned, and the nature of the terrain in which it was fought, and the vital part played by dive-bomber aircraft in it, the Dien Bien Phu battle much more resembled the defence of Kohima and Imphal than any of these other examples. Unfortunately for the French, their forces, both on the ground and in the air, were very

much smaller than those involved on the Indo-Burmese border in 1944, although their isolation was as complete, and the RAF Vengeance dive-bombers were as accurate and rigorous in support in this former struggle, as were the French Navy Helldivers in the latter one. Their opponents were similar in that both the Japanese and the Viet-Minh were fanatical and had great tenacity, they took advantage of the natural terrain and cover, and were experts in trench warfare of great sophistication.[194]

However, General Giap had the incalculable advantage of a large supply of modern anti-aircraft weapons, especially the superb Soviet-built, Chinese-supplied and manned, 37mm AA gun, a weapon that could reach up to 8,500ft (2,590m). These they sited in an ever-tighter ring on the hills that overlooked the French base; indeed, before the siege had been under way for many days, some enemy flak gunners were firing *down* on the transport and ambulance aircraft running the gauntlet in and out of the airstrip before it became untenable. The termination of the Korean War had released a huge supply of such weapons, and Red China made no bones about supplying

both them and the skilled gunners to use them, to further Giap's cause.

By March 1954, for instance, a complete and fully equipped Chinese AA regiment with sixty-four 37mm guns had arrived outside Dien Bien Phu and was quickly emplaced. They reinforced the twenty such weapons already in place there, and were manned almost exclusively by the Chinese themselves – it was in fact a Red Chinese unit, although a few Viet-Minh officers and men were seconded to give it 'respectability' and to fool both neutral nations and naïve observers and media commentators in the West. Each of the sixty-four guns was dispersed singly around the perimeter so that the presence of the Chinese was heavily diluted among the other forces present. These AA gunners were ultimately to report to their masters in Peking that they had hit sixty-two French aircraft during the course of the battle, almost one apiece.

Nor did their gun sites remain static, so they could be located and taken out – but were moved constantly from one position to another so that they could *not* be pinned down and eliminated by systematic counter-strike area bombing, always the standard Western answer to any such problem, and always ineffective. This measure effectively counteracted the French policy of 'saturation' air attacks, in which twelve to fourteen aircraft would concentrate their altitude bombing on a particular target area where it was thought the enemy guns were positioned. The Viet-Minh won this game of 'blind-man's bluff' easily, and even the liberal use of napalm had relatively little effect on the sodden and rain-soaked jungle once the rainy season started in earnest.

The actual choice of the ground on which to fight worked against the French from the very start. Located in a valley at the far north-western end of Vietnam, close to the Laotian border, it was at the maximum range of the single-engined dive-bombers and fighter-bombers flying from the Arromanches, and this reduced the time they could spend over the target, even without the heavy flak. Thus they could not linger over the battlefield to the same extent as the American Douglas AD-1 Skyraider dive-bombers of another generation were able to do in the later conflicts in this area. The distance to be travelled and the need to conserve fuel also restricted the choices of approach to the target zone, and thus further assisted the Communist gunners in predicting their approach courses.

Nothing daunted, nine Helldivers carried out an attack on 14 March on artillery positions east of the fortress, dropping twenty-eight 260lb and five 250lb bombs. They also contributed two dozen rockets and ten anti-personnel clusters, to the discomfiture of the Viet Minh, and claimed one major artillery battery damaged by direct hits. Again they reported fierce counter-fire from both 12.7mm and 37mm flak guns. Two Helldivers were hit and damaged at this time: on 13 March, Lieu-

bristling with artillery and mortars, Hill 'Gabrielle'[195] had initially faced the enemy full of confidence. But after a devastating bombardment which pulverized its defences, it was overrun by massed assaults and fell in a very short time indeed, with heavy casualties. The dive-bombers did all they could to assist in its defence, in spite of the fact that the valley was closed down by low cloud which reduced visibility enormously.

Other sorties were made on this crucial day by the SB2C-5s. In the first – the only one to reach the target due to the weather

of the basin, while a third pair bombed Gia-Phu instead, for the same reasons.

In addition to the Navy dive-bombers two of the few surviving Bearcats managed to get airborne, although one was then destroyed. Also, a bold dive-bombing attack was made by 11F's Hellcats, each carrying a single 500lb bomb; though sadly, Lieutenant Lespina's aircraft was destroyed when hit by flak over the battle zone.

The weather conditions eased during 31 March, allowing the Army Bearcats and Navy Helldivers to renew their close support missions in strength. Their role was

Helldivers *(right)* **and Hellcats warm up on the deck of the** Arromanches. Captain H. de Lestapis

tenant Bellone's aircraft was hit by both 12.7mm fire and two 37mm cannon shells while he was directly over Dien Bien Phu village itself. He made it back to base and his aircraft was repaired within three days. The second Helldiver hit was that of Lieutenant Rougevin on the 14th, while some 6km from the target, a single 12.7mm shot being received in the engine; but he, too, got back safely and his Helldiver was operational once more within five days.

On 15 March the first of the strong outlying French positions fell to the enemy. Well garrisoned with seasoned troops, and

conditions – four Helldivers braved the crescendo of flak to bomb artillery pieces 8km, 038 degrees from the hill, hitting these targets with four 500lb bombs and sixteen rockets, but the aircrew could not see the results due to the heavy AA fire. Four 1,000lb bombs were put on target close by 'Gabrielle' itself. Two more dive-bombers had their mission 'weathered out', but placed their 1,000lb bombs on the enemy in the Son-La region instead. The same fate befell another pair of Helldivers, which delivered their eight 500lb bombs and eight rockets on enemy emplacements 25km east

now to help the French offensive to regain three of their defence bunkers, D2, E1 and E2, to the east of the main fortress complex, which had been overrun by the enemy the preceding day. These bloody attacks were the greatest victories achieved by the French during the whole battle. Regardless of the huge flak barrage, the Navy Helldiver pilots did their utmost to assist their comrades below. As in many instances in World War II, it was again found by the troops on the ground that the Navy dive-bomber pilots pressed in much more closely to carry out their attacks than

(Above) **Helldiver of 3F on the catapault of the Arromanches, ready for launching in the Gulf of Tonkin.** Captain H. de Lestapis

Deck crewmen of 3F reading mail from home at Bach Mai airfield. Captain H. de Lestapis

did their Air Force opposite numbers. Indeed, Bernard Fall was to state that the *Arromanches* SB2C-5 pilots '… were much loved by the paratroopers for the risks they took, and for which they paid a higher price in lives than did the Air Force …'.[196]

And on this day of brief French victory, when the Viet-Minh were swept off all three positions and retreated with heavy losses, the cost to 3F was especially high because the Helldiver of Lieutenant de Vaisseau Jean Dominique Andrieux himself, along with his navigator/gunner Petty Officer 2nd Class Jannie, was caught in that deadly

the dive-bomber squadron was taken by Lieutenant de Vaisseau de Lestapis and the missions continued unabated.

Several observers have pointed out that the French themselves contributed in some part to the incredible accuracy of the Viet-Minh flak gunners in that their radio security was '… lax to the point of criminal' according to one eyewitness account. For example, Charles Favrel reported on a supply drop mission he flew when he was amazed to hear – as his article in *Le Monde* subsequently read – '… transmitted quite clearly, messages and orders to aircraft

perimeter ruled out much in the way of diversion over the target, once the dive-bombers were committed to their attack run.

Undeterred by the loss of their gallant leader, the Navy Helldivers continued their operations and the improved weather on 1 April enabled combat sorties to be increased to ninety-nine; this helped to stabilize the newly regained front line and also raised the defenders' morale considerably – and it certainly needed a boost. The continual enemy heavy artillery bombardment was rapidly reducing the French ability to reply in kind, and by 2 April so great a loss

March 1954, Bearcats and Helldivers of 3F being prepared for another mission at Bach Mai airfield near Hanoi.
Captain H. de Lestapis

crossfire pumping from the 37mm guns over strongpoint 'Beatrice'. Their aircraft disintegrated and both men were killed. It was a heavy blow, even though it was the only loss from seventeen sorties flown by the Helldivers that day. His place as commander of

arriving on bombing missions or bringing in reinforcements or material!'[197] Broadcasting such information meant that guns could easily be laid in anticipation of each incoming French flight. But even without such slackness, the limited and ever-shrinking

had been taken by the French gunners that it was directed that almost all the counter-battery work would, in future, have to be conducted by the dive-bombers alone in the area facing 'Beatrice'. Only twelve French heavy guns remained in service at this point, and even this level of strength was not to be maintained for very much longer.

Bombs and napalm therefore rained down on the last remnants of the Viet-Minh regiments holding positions on the flanks of the retaken positions – but French troop losses had also been relatively heavy, and theirs could not be replaced, save in small driblets, whereas fresh Communist formations were arriving on the battle scene every day. Moveover, the enemy

heavy guns which were used to support the main fortress. As the Viet-Minh mined ever closer, the trenches were turned into underground galleries and forts of their own, from which sudden massed assaults could be thrown in, time and again, with hardly any warning. The French Navy Helldivers were continually being called in to try and smash these strong assault-points located on the very edge of the defenders' wire. Only precision attacks would suffice and only dive-bombing gave such precision: it was there-

front. They were taken under heavy and accurate fire by the 37mm gun batteries concentrated around strongpoint 'Beatrice'. The leading Helldiver, piloted by Lieutenant de la Ferrière, managed to bore right through the tracers and shells, unscathed, at almost zero feet, and he escaped the trap; but his wing-man, Ensign Jean Marie Laugier, was not so fortunate. Hit repeatedly, the second Helldiver went into the ground, followed all the way down by further shell-bursts. Even if there had been survivors, the

Helldiver SB2C-5 of 3F Flotille over typical jungle terrain during the fighting around Dien Bien Phu in March–April 1954. ECP Armees, Paris

artillery was benefiting more and more from misplaced French ammunition drops on the shrinking perimeter, which they retrieved, while the few surviving French guns were becoming starved of shells.

To the south of the main bastion another isolated garrison, 'Isabelle', held out alone, although it was being steadily infiltrated by the saps and trenches of the enemy, much as their compatriots to the north had been. Almost completely cut off from their fellow defenders around the now useless airstrip, the garrison of 'Isabelle' fought on in their own private war, although they still had the bulk of the

fore the Helldivers or nothing. The overall area of 'Isabelle' was, of course, far less than even Dien Bien Phu itself, so massed aerial high-level bombing would, once more, have been out of the question, even if the means to provide it had existed.

Attack followed attack and French losses grew. On the 4th, a heavy enemy assault was repulsed, with 2,000 Communist troops killed, and this part of the battle alone was estimated to have cost Giap some ten thousand casualties in all. Then on 9 April, two SB2C-5s made a low-level sortie along road 41B, seeking out and destroying Viet-Minh supply convoys approaching the battle

wreck was deep within enemy-occupied jungle, and they could not have been rescued by the grieving paratroops.

It should not be thought that all the dive-bombers' efforts at cutting the enemy supply routes were total failures. One Viet-Minh deserter from their 209th regiment clearly stated that supplies had been increasingly hard to obtain due to such 'road-cuts'. Losses among the hapless and defenceless 'Dan Cong' human mules must have been very heavy indeed. Nonetheless, by now the Viet-Minh had some 300 guns in position, when the French estimated that they only had sixty!

(Above) **French Navy pilots examine their wing bomb fusing prior to another mission.** ECP Armees, Paris

Another headache for the Viet Minh. Helldivers of 3F taking off from the Arromanches in the Gulf of Tonkin during operations against the communist insurgents in French Indo-China. US Navy via National Archives

Another counter-attack was mounted by the French defenders at dawn on 10 April, against strongpoint 'Elaine 1'. The leading assault troops went 'over the top' in true World War I manner at 06:10 that morning, and were aided by the work of the SB2C-5s which attacked precisely on schedule. Four Helldivers dropped twelve 500lb bombs and two 1,000lb bombs in

The enemy responded in kind with a heavy artillery barrage, and for a while the issue could have gone either way. Finally, further French troops equipped with flame-throwers decided the balance, and the Communists on 'Elaine 1' were wiped out. The Helldivers returned to help finish them off by decimating what was left of an enemy battalion still resisting on nearby

northerly, became cut off from the rest of the command, between them only the shell-swept wastes of the airfield, already itself bisected by further enemy trenches and concealed machine-gun nests. A desperate attempt to break out was planned for dawn on Easter Monday, 19 April. In a preparatory strike the day before, the Helldivers once more earned the accolade of the

Line-up at Bach Lai of French Navy aircraft ready for further operations against Viet-Minh guerrilla forces in Indo-China, 1954. Captain H. de Lestapis

the first attack, hitting enemy positions a bare 400m from 'Elaine 1', close support at its best, while a second attack placed sixteen 500lb bombs around 'Beatrice'. These bombs, and additional napalm deliveries, struck the enemy support units lurking in the gullies and trenches to the west of the battlefield, thus cutting off the Communists from any reinforcements at a critical juncture.

'Phoney Mountain', a dominant hill closeby. Their excellent work in this day's operation earned the Helldiver squadron a special citation from Général de Castries himself. This proved the high-point of the French defence.

The remorseless enemy sapper work had continued, and now it was the turn of the northern defence works to disappear into their cold embrace. 'Huguette 1', the most

troops on the ground for their accuracy, the bombs having to be delivered within a tiny 200m gap separating the isolated post and friendly positions. At heavy cost the garrison broke through – but the ring had been pulled even tighter around the survivors.

As the noose pulled steadily tighter and the fate of the French garrison became ever more inevitable, the anger and frustration of the defenders turned increasingly against the staff officers back in the safety of Hanoi; many felt – and with some justification – that they had no conception of what the garrison was enduring, or how to help them. Bitter comments were also made about the support being received from the French Air Force, although charges of 'excessive prudence' fell short of allegations of cowardice. Despite this, the garrison's high regard for the Navy dive-bombers never diminished, and the Helldiver crews were explicitly exempted from any such talk.[198]

The end was now rapidly approaching. The garrison was exhausted, both mentally and physically; there was no place to hide from the continuous shelling, the troops melted away in attack after attack due to death and wounds, and replacements of men, material and ammunition dropped all the time. The loss of 'Huguette 1' had reduced the already minute area of the supply dropping zone by half, and it was decided to take a desperate gamble to relieve this situation by making another counter-attack. Finding the men to make such an attack by this time proved very difficult, and the openness of the terrain itself was to prove deadly for such an effort. Nonetheless, that effort was made at 14:00 on 23 April.

The Viet-Minh concentrations on the target were dive-bombed by 11F and almost entirely wiped out. Four Helldivers conducted the initial strike, but their eight 1,000lb bombs were not so accurately delivered on this occasion due to the hail of flak that met them. Two further SB2C-5s followed them down, and this time their heavy bombs were delivered precisely into the enemy trenches east of 'Huguette 6', the exactness of their delivery drawing appreciative radio signals from the French troops positioned close by. A further vertical attack was then made in the classic style by two more Helldivers which bombed the enemy occupying 'Huguette 7', their attack dives being made at almost 90 degrees down from 8,000ft (2,440m) and 6,000ft (1,830m), to point-blank release, and the four heavy bombs landed exactly on target, duly obliterating the bulk of the Communist defenders.

The dive-bombing of 'Huguette 1' had been devastating indeed. But when the paratroops emerged to cross the open ground between their positions and the smoking and defenceless enemy strongpoint, they immediately ran into heavy fire from a concealed and unsuspected machine-gun nest hidden in a wrecked aircraft on the edge of the runway. The assault

faltered and then halted. The enemy artillery then joined in and the Chinese flak gunners turned their automatic weapons away from the sky and swept the runway itself. There was nothing left but withdrawal, and the day was a failure.

The Helldivers continued to give support during the final days of the battle, but they were not able to take part in the final débâcle, because the involvement of the dive-bombers ended on 30 April when the Arromanches was relieved off the coast by the carrier Bois Belleau (Captain Mornu). The Helldivers were exchanged between the two ships for the Chance-Vought Corsairs of 11F. The Helldivers of 3F had conducted 186 sorties, or 283 hours of combat flying time during this last period of operations, from 24 May, when they had re-embarked aboard Arromanches, until 2 June; and they had dropped 372,700lbs of bombs and fired 4,350 rounds of 20mm cannon shell.

Four of the nine SB2C-5s embarked by Bois Belleau remained operational for a time, until 2 June, taking part in patrol work before that vessel also left for Hong Kong. This detachment of Helldivers again worked from Bach Mai airfield and on 1 May they flew nine sorties in conjunction with Operation 'Castor' at Dien Bien Phu, and other missions against Hill 13 at Thai Nguyen. Further combat sorties took place in May, led by Lieutenant Bellone, objectives having included positions at Yen Bay and Cat Bi, while all nine were fit by the 6 May to strike at Hill 41, at Tuan Giao. On the 7th the Bois Belleau launched a heavy strike of twelve Helldivers which bombed the Viet-Minh during the final winding-up of the battle.

These had been the last carrier-borne strikes, but the Bach Mai aircraft continued dive-bombing operations on their own after that date. On 9 May four sorties were flown against Son La, on the 10th two against Thai Nguyen and four against villages near

Hai Duong, four more against Yen Bay on the 11th, while on the 12th the Helldivers were more active than ever. Four separate strikes were made on Hill 41 near Tuan Giao, four more against a Viet-Minh battalion caught out in the open 2km southwest of Phuly, and a further four missions were flown against the village of Binh Tru, 20km east of Hanoi, which showed just how close and how bold the enemy had become since their victory, and how they intended to gain as much land as possible prior to the armistice. Three more sorties were mounted against the village of Whiep Xa, and a final three against a village south of Phuly again, for a total of eighteen sorties in one day, three per operational Helldiver.

This continued to be the pattern from 13 May through to 25 May. It was the final period of honourable service for the Curtiss SB2C Helldiver against the forces of tyranny, from that first raid on Rabaul in 1943 when she finally silenced most of the doubters, to the last missions against the Communists nine years later.

Those aircraft that did not return to France aboard their carriers remained in service still longer, although no longer in the front line, their places being taken by the Corsairs until the end of this phase of the take-over of Vietnam came to its end with the partition of 21 July 1954. However, that proved to be the end of the conflict only as far as the trusting Americans and disillusioned French were concerned; for Giap and Ho-Chi-Minh, and their backers in Peking and Moscow, it was merely a breather between achieving the first part of their aims and the final and total take-over of the whole country. When round two commenced some years later it was America and another dive-bomber, the Douglas AD-1 Skyraider, who took up the burden – but that is no part of the Helldiver story.

French Navy SB2C Units in Indo-China 1952–54				
Campaign	Flotille	Strength	Commander	Bases
1951–52	3F	9 SB2C-5s	L de V Waquet	Arromanches
		9 Pilots	L de V Gautriand from 17-10-51	from 24-9-51 to 16-5-52
1952–53	9F		L de V Hervio	Arromanches from 1952–53
1953–54	3F	12 SB2C-5s	L de V Marmier from	Arromanches until 10-7-53
		14 Pilots	29-9-53 to 20-10-53	
			L de V Andrieux to 31-3-54	Bois Belleau to 9-10-54
			L de V de Lestapis to 26-4-54	
			L de V Fatou to 9-10-54	

HELLDIVER MEN: Captain M. H. de Lestapis, French Navy

A veteran pilot who served right through the Dien Bien Phu campaign, Captain de Lestapis flew the Helldiver for the whole of this last period of combat action for the SB2C. He gave me these eyewitness views of those actions and the background to them:

'The 3rd Flotille (3F) took part in the first Indo-China campaign, between 24 September and 16 May 1952 and was commanded by Lieutenant de Vaisseau (L.V.) Waquet. They were equipped with nine SB2C-5 Helldivers, six operational and three spares, and had nine pilots. In the course of this campaign they flew 136 combat missions. In the campaign of 1952–53 the 9th Flotille flew from the *Arromanches*. For the 1953–54 campaign I flew with 3F and we arrived on station aboard the same carrier on 29 September 1953, and left on 9 October 1954, the *Arromanches* being replaced by the *Bois Belleau* as parent carrier on 10 July 1954 at the end of the fighting.'

'3F at this date was equipped with twelve SB2C-5 Helldivers, and we had fourteen pilots. The commanders in this period were L.V. Marmier who was relieved by L. V. Andrieux on 20 October 1953 until 31 March 1954; when he was lost, I took over as temporary CO between that date and 26 April, and then L. V. Fatou took over and led the unit until the end of the campaign. I served as second-in-command of 3F from 20 October 1953 until 31 March 1954, taking over in the interim on the death of L. V. Andrieux when his Helldiver was hit by flak over Dien Bien Phu. I reverted to second-in-command on my replacement by L. V. Fatou. During the whole of this campaign I flew 134 combat missions, totalling 412 hours of flying, an average of 42 combat hours per month, but during the intense period of the siege of Dien Bien Phu itself I flew sixty hours per month and conducted forty-five combat missions over that battlefield.'

'Our aircraft carrier, the *Arromanches* (the former British *Colossus*) was always a very good base for us to work from in the summer months. She had well equipped workshops for efficient carrying out of repairs and the strong English influence of her origins, which meant for us pilots luxury like bathrooms provided with fifteen stalls to keep us fresh in the heat. Our captain, Capitaine de Vaisseau Patou, was a man of very great qualities, and was famous in our Navy for his abhorrence of all types of alcohol, as well as being a strict vegetarian.'

'We were based aboard the ship for most of the main time of the campaign, but later had detachments based ashore at Tou Kin and at North Annam, although our carrier stayed as close as possible offshore to support us. In the period up to 31 January 1954, I note from my diaries that we almost always returned from our missions to the carrier. I note, however, some fourteen landings at Cat-Bi (the airfield of Haiphong) and at Bach Mai (the airfield of Hanoi) for various reasons.'

'From the 1 February, our aircraft were more and more concerned with rescuing the Dien Bien Phu garrison, and because of the distances involved we were often called upon to operate from Cat Bi airfield. The flotilla was also based for a while at Vientiane, the capital of Laos, from 16 to 28 February, where we operated in defence of the garrison of Lonaug Pralang and of Muong Sai which were under attack from the Viet Minh and which we otherwise could not have reached.'

'On 15 March, due to the gravity of the situation at Dien Bien Phu, the two flotilles were disembarked from the carrier and 3F (SB2C-5s) sent to Hanoi and 11F (F6Fs) to Haiphong. This decision was made due to the long flight times from the carrier to the besieged garrison, a journey of three to five hours, whereas from Hanoi the flight times were two to three hours, depending on conditions. In addition, on the 15th March one of my friends, L. V. de Maindreville, was killed on the long return flight from Dien Bien Phu – at the extreme end of the Hellcats range – in trying to reach the carrier in the Bay of Along.'

'During the period of Maximum Activity at Dien Bien Phu, the carrier remained in the anchorage in the Bay of Along while her aircraft operated from the ground fields. During such missions we kept in close radio contact with the *Arromanches* at all times, receiving our liaison orders direct from her. I note from my diary that we returned to the carrier on 30 April after the fortress had fallen and we had completed other support missions. For a few missions we flew from the carrier *Bois Belleau*, which arrived on station from Hong Kong until 14 July.'

'The missions we Helldivers carried out initially were the cutting of supply routes, mainly along R.P. 41, Hoa Buih – Sonla – Dien Bien Phu, down which the Viet Minh were moving their troops and supplies. We hit the roads themselves where they went through gorges and could be blocked. We also bombed specific objectives such as storage areas, workshops and ammunition depots, which were reported by our photo-reconnaissance flights, and from their aerial photographs we were briefed before take-off on our targets. Direct air support

Captain H. de Lestapis on return from yet another mission against the Viet Minh entrenched around the beleaguered French garrison at Dien Bien Phu, March 1954. Captain H. de Lestapis

of our ground troops also featured, of course, and targets of opportunity as directed by our spotter aircraft over the battlefields. Some of our missions were just reconnaissance flights fixed before our departure, though these were rare. Some missions against fixed objectives were directed by the main Air Force control at Tonkin.'

'In the Dien Bien Phu battle the majority of our missions were carried out against the enemy saps and entrenchments that were encroaching upon the defensive perimeter. However, initially our aircraft continued to carry out other missions, again, principally road-cutting along route 41. For example, my combat diary between 28 January and 7 May 1954, shows sixty-four missions, of which just forty-five were over Dien Bien Phu. For these latter our attacks were mainly made against the enemy artillery batteries and AA batteries, but also included bombing pre-selected objectives from photos, protecting our parachutists on drops into the combat zone; observer missions in support of our own artillery; low strafing of the Viet Minh trenches; and – once – dropping packages of whisky and revictualling the survivors of the Legionaries towards the end of the battle. After the fall of Dien Bien Phu we reverted to missions cutting RP. 41 and conducted actions in the Tonkin delta.'

'The SB2C Helldiver was a good aircraft for these types of mission, and had an excellent serviceability as well as good landing qualities (there were hardly any accidents despite the intensity of our operations). They were also tough aeroplanes, and out of twenty of our machines which were hit by enemy AA shells, only two actually crashed. The Helldiver was very stable in the dive, also, despite somewhat primitive sighting facilities – just a simple grill on the cockpit, which did not indicate the angle of dive – our Helldivers were very precise and accurate dive-bombers, though this in part was due to our intensive training. They were also manoeuvrable with a heavy bomb load. They lacked the rapid movement necessary for the escape out after the attacks against targets ringed with anti-aircraft guns, however, for they were slow aircraft, and they were also unsuitable for night intervention missions in among the mountains, which would have been very useful in blocking enemy supply lines.'

'Our normal bomb-load was usually three bombs of 1,000lb, or four bombs of 500lb as well as rockets on rails under our wings. Also, on rare occasions, we carried 250lb anti-personnel bombs and, also rarely, napalm in two tanks under the wings. Our 20mm cannon were always loaded of course. The bombs were mainly fitted with delayed-action fuses (1–2 seconds) to facilitate our safe exit from low pullouts. For road-cutting missions our aircraft carried a mixture of bombs with delayed action fuses (4–5 second) permitting the dropping at low altitudes, and also of long delayed-action fuses (4 to 12 hours), to inhibit the repair work on the roads, we also dropped anti-personnel grenades (cluster bombs) to constrain the workforce of coolies.'

'The dive attacks were our main role, and our normal angle of attack was between 50 and 60 degrees which we found gave us excellent precision of aim. We generally commenced our dives from 6,000–8,000ft, through over Dien Bien Phu we had to attack from much lower due to the low cloud base and the need to take advantage of it. The height of bomb release varied depending on the type of target being attacked, and the conditions, and the precision required. We rarely descended below 1,000ft due to the risk of malfunction of the bombs. Against badly defined objectives, which were the most frequent type, the base of our pullouts was about 1,500ft; against AA sites we often released at 3,000ft. During the course of the Dien Bien Phu battle when some of our aircraft were shot down by the intense flak fire we were instructed by the Air Force high command not to descend below 5,000ft, and this the Air Force aircraft obeyed. But we Navy pilots thought this order was stupid (stupide) and our Helldivers just ignored it.'

'For road-cutting missions a much shallower angle was adopted; we constrained ourselves to a 20-degree angle of attack, and our let-down approach was longer as we followed the road, releasing at low altitudes, between 200 and 300ft only. On patrol missions we flew with our gunners and in formation of two or four aircraft. When four aircraft were present against an objective defended by AA batteries, we split into two groups of two to divide their fire, and each group attacked from a different direction. At Dien Bien Phu there was no space for this type of luxury and we often kept two aircraft over the area on patrol ready to be called upon. It was important for morale of the garrison that we maintained an aerial presence overhead as long as possible on each mission. It was risky, but we felt it was an important duty.'

'The journey from Bach Mai (Hanoi) to the Dien Bien Phu battlefield was the main problem, and made the enemy's task easier and ours harder. The weather could be good when we left, but by the time we reached the war zone the clouds might have closed down, giving us very bad visibility. Also the bad weather in the valleys on our approach path gave overlapping cloud layers, so that often our route was barred by cumulus and nimbus through which we had to find a gap – we had to turn aside to fly over the 7,000ft mountains en route, and for the latter part of the flight we went up to 10,000–13,000ft to pass the barrier of clouds'

'Dien Bien Phu sits in a basin, and we often had to fly high to reach the area and then descend steeply through the clouds to maintain our patrols. Thanks to our excellent training we made contact on almost every one of our flights despite these conditions. I note in my diary three occasions when long delays were caused by these conditions, but due to our increasing familiarity of the terrain over which we flew almost daily, we always got through to the battle zone.'

'Locating our targets was very difficult. The camouflage of the Viet Minh was absolutely remarkable. Often our aerial photographs designating target areas showed absolutely no trace of the enemy at all, although we knew he was there. We hardly ever saw any traffic activity on our patrols (most of the movement of troops, ammunition and supplies was done by the enemy at night). The triumph of the Viet Minh was that their build-up was practically invisible and the scale of it surprised the French garrison. All their gun sites and batteries were also intensely camouflaged, and they just fired through slits on fixed bearings. Their AA positions were moved almost daily to avoid pinpointing.'

'Thus our missions, although accurate, showed paltry returns along RP. 41. After we had passed over the road and blocked it in the ravines, like ghosts overnight the damage was repaired and refurbished by the army of men and women from the army of coolies they forced to work for them and which were stationed at all the sensitive points of the route. Our blows against bridges along the road were most successful to the bombardment plan, but at the end of the day, were insufficient to stop the flow of supplies.'

'As I have said, our Helldivers were good and accurate precision bombers for attacking visible objectives, but these were few and far between, and often we were ordered to drop our ordnance on a fixed square marked on the photographs. We could not see if we were successful in these attacks except by the resulting flames and explosions if we hit something important. Our precision of aim was much appreciated by the garrison at Dien Bien Phu. At the height of the battle we mostly bombed from 1,000ft Viet Minh positions only 50ft from the defending legionaries. The ground around the fortress was very ploughed up making identification most difficult. One outstanding attack was made by L. V. Duvillier (who was later killed in his aircraft) who sent one of his rockets exactly down the muzzle slit of an enemy artillery piece and totally destroyed it.'

'As far as I was concerned I personally attacked an AA gun position which surrounded me with intense return fire, with cones of tracer all about me. My team member who followed me down in my dive had the satisfaction of seeing my bombs explode exactly on this gun battery and wipe it out. I have some other memories of equally precise attacks. The enemy AA was intense, and in all, at Dien Bien Phu we lost three Hellcats and three Helldivers to their fire. Eventually the Air Force tried high-level bombing by a formation of B-26s, but after this bombing, during which the enemy guns stayed silent, they soon regained their former strength and continued as before.'

'Our own Helldiver losses included L. V. Andrieux, the commander of 3F, on 31 March 1954. He led a group of four SB2Cs over Dien Bien Phu and found the target area invisible due to cloud cover. The garrison desperately needed support, and Andrieux led two aircraft down through the murk in the hope of piercing it. They found a gap in the clouds close by RP. 41 to the east of the fortress and decided to follow the road to the target area below the ceiling. They had just got a poor glimpse of Dien Bien Phu at a distance when the enemy AA opened up on them and Andrieux's aircraft blew up in the air from direct hits. His team-mates who were following managed to pull up out of the fire cone without mishap and lose themselves in the clouds over the mountains.'

'On 9 April 1954, Enseigne de Vaisseau Laugier was part of a four-plane Helldiver patrol. They divided into two groups of two to make diverging attacks from two different angles but he ran right into the middle of a wall of AA shells. His aircraft was hit along her fuselage, and although he bailed out promptly, he was too low for his parachute to have time to open.'

CHAPTER TWENTY-ONE

Survivors

Pride of place must go to the last flying Helldiver, which belongs to the Confederate Air Force. She is an SB2C-5 (Bu. No. 83725; Mfg serial number 83589, under contract No. 1609, and Registered as N. 92879). She was originally accepted into service on 5 July 1945, being delivered to NAS Alameda in the pool from August

Corpus Christi, Texas in April 1948. She was stricken from active service on 31 August 1948, and was declared surplus to requirements on 14 October 1948.

She first went to the Montana School of Aeronautics at Helena; her second civilian home was with the Air Museum, Ontario, California from 29 May 1963,

severe that many did not expect her to fly again – but as ever the 'Beast' proved a tough old bird.

Under the guidance of Colonel Bob Richeson, all the many pieces were carefully collected together and taken from Harlingen to Breckenridge, Texas. Here at Ezell Aviation painstaking work over the

The SB2C-5 Helldiver restored by the West Texas Wing of the Confederate Air Force. A fine view dedicated to Colonel Jake Miller. Philip Makanna via Confederate Air Force

1945 until February 1946. She was then assigned to the pool of CASU-5 at San Diego until October 1946; to the pool FASRON from 7 October 1946 until February 1947. She joined VA-1B in February and served with that squadron until September 1947, before being returned to the pool FASRON-5 from 5 October 1947 until March 1948, then to the Pool NAS

still carrying the markings of the Reserve Squadron. She was subsequently purchased by Colonel Robert L. Griffen on 19 October 1971 and donated by him to the Confederate Air Force Ghost Squadron on 20 December 1971. Working up for the CAD Air Show in 1982, she had to make a 'hard' emergency landing at Harlingen, Texas, and the damage she suffered was so

next six years, totalling thousands of volunteer man hours, was devoted to her restoration and she was completely rebuilt by the West Texas Wing, CAF, at a cost of over $200,000, a task co-ordinated by Colonel Nelson Ezell. She was given the paint scheme of the carrier *Franklin* (CV-13) from which the late Colonel J. E. 'Jake' Miller, West Texas wing leader in 1984,

(Above) A well-known survivor is this Curtiss SB2C-5 (Bu. No. 83725, Mfg SN 83589, N-92879), belonging to the West Texas Wing of the Confederate Air Force. The -5 was the ultimate Helldiver variant and although production began in February 1945, only a few active carrier combat squadrons had re-equipped with this model to attack targets in mainland Japan before the war's end. However, they soldiered on as the principal strike aircraft in the post-war US Navy and were not finally phased out of service until 1948–49. This particular aircraft was accepted into the Navy on 5 July 1945, and wore the markings of VB-6 which served aboard the USS Hancock (CV-19). It is known she was with VA-1B at Naval Air Station Alameda, California, from 7 July and was assigned to various pools between then and April 1948, when her final assignment was at NAS Corpus Christi, Texas, before being stricken on 31 August 1948 and pronounced surplus on 14 October 1948.

She then went to the Montana School of Aeronautics, Helena, for many years, being transferred on 29 May 1963, to the Air Museum in California. On 19 October 1971, she was purchased by Colonel Robert Griffin, who duly donated her to the CAF on 20 December 1971. She was a popular exhibit but crashed at the Texas Air Show at Harlingen in 1982.

Since then 83725 has been lovingly restored to flying condition once more, by a team of enthusiasts under the co-ordination of Colonel Nelson Ezell, and at a cost of over $200,000. This photograph was taken at the Dayton Air Show, Dayton, Ohio, in 1992.
Peter C. Smith via Stanley Vaughn III

Head-on view of the SB2C-5 Helldiver, Bu. No. 83725, restored by the West Texas Wing of the Confederate Air Force. Its first flight was on 27 September 1988. Re-built to airworthiness again after the accident at Harlingen. Colonel Ted Short, CAF had flown Helldivers. She took her first flight with Ezell at the controls, on 27 September 1987. She is usually based at Midland International Airport, Midland, Texas, but under the care of operations officer Colonel Ted W. Short at Richland Hill, Texas.

The next-best preserved Helldiver is the static display at the Naval Air Museum Pensacola. This is an SB2C-5 (Bu. No. 83479) belonging to the National Air and Space Museum, and for years it was stored at their site at Silver Hill, Maryland. In 1976 it was passed to the Naval Air Museum at NAS Pensacola where it was carefully restored, and at the time of writing (1997) it is exhibited (marked H-212) as part of the carrier flight deck display.

The Greek Helldiver of the Royal Hellenic Air Force, an SB2C-5 (Bu. No. 83321) was exhibited as a static open-air display from 1967 to 1996. In spite of the favourable climate its condition has deteriorated, and plans were in train to transfer it to Tatoi air base where it is hoped that it will be restored properly and put back on display once more.

The Royal Thai Air Force Museum is on the Phanonyothin road, just to the south of the Don Muang airport and their SB2C-5 Helldiver (No. 366, Bu. No. 83410) was a static display from 1967 to 1998. The bomb-bay doors were displayed open, and rails for four rockets could be seen under each wing. Two 20mm cannon and two 0.3in machine guns were its armament. The aircraft is painted black overall, although the rudder might be dark blue. It has six Thai roundels and the national flag on the rudder. The propeller is black, rear tips yellow, front sides and outer tips yellow, then striped black/yellow/black/yellow as far as on the rear. There is a white band and a white numeral '4' on the rear fuselage. It was not in the best of condition when seen and presumably continues to deteriorate. There were plans to transfer it back again to the Royal Thai Air Force where it is hoped it will be properly restored and again displayed.

The SB2C-3 Helldiver on display at the USS *Intrepid* Sea-Air-Space Museum at W. 46th Street and 12th Avenue, New York, is a *total* fake. Displayed as USN 44, it is in fact a full-scale replica manufactured by Starr Aircraft of California City, a southern California company. It is built with an aluminium frame and skin mounted on a tubular structure, with a metal or fibreglass engine and prop replicas, and with an

The Greek Navy SB2C-5 Helldiver on display at the Hellenic Air Force Museum, Athens, Greece, in April 1984. Commander J. Alton Chinn, USN, Rtd

authentic paint scheme – but in places you can see light through it, which gives the game away to the discerning! The company does a huge amount of work for the movie studios in aviation and space subjects. At the time of writing (1997) the cost of such replicas was around $50,000, but they do have corrosion protection; they are guaranteed to withstand winds up to 100mph (160km), although this is not a problem in New York City.

There are numerous crash sites for Helldivers across the States. Most significant must be Lake Washington, near Seattle, which might yet be to the Helldiver what Lake Michigan is to the SBD Dauntless. Certainly three wrecks have been recovered from there – all in 1990 – by Mike Rawson of Minneapolis: an SB2C-1 and two former ex-USMC SB2C-1Cs formerly A-25As, Nos. 42-80449, Bu. No. 76805, and Bu. No. 75552. Other include the two

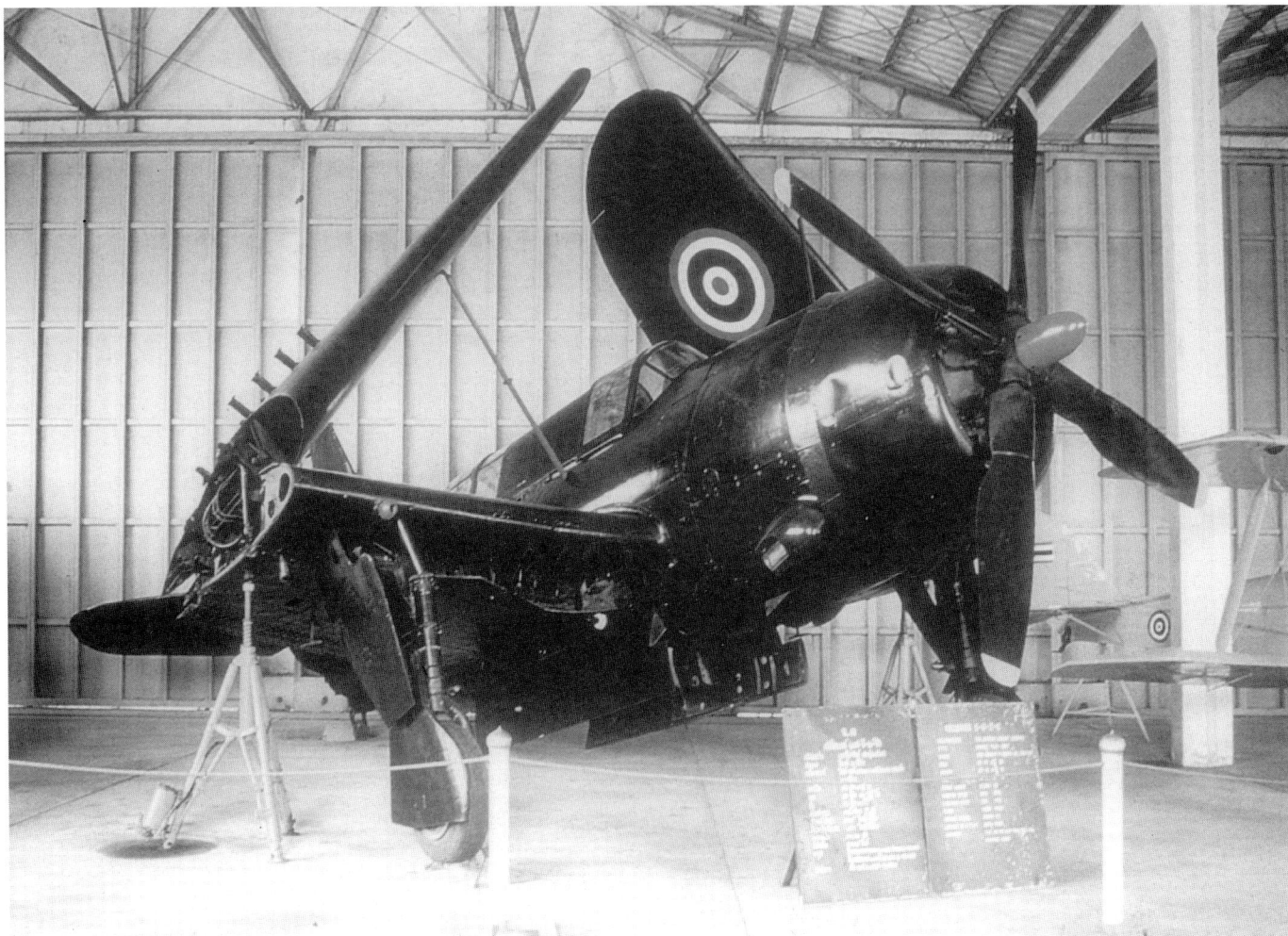

(Above) **The SB2C-5 of the Royal Thai Air Force preserved at the Royal Thai Air Force Museum, Bangkok, October, 1997.** Simon Watson

recovered by Ted Darcy from a dump in Hawaii in 1987; these have not been restored. Also, there is the French Navy SB2C-5 Bu. No. 89255, which was for disposal at Lann-Bihone Airfield, L'Orient between 1965 and 1970; this plane was not restored, either. Of course while most of the wreck sites in the USA yield 'spare parts' to a greater or lesser extent, they cannot be counted among the 'survivors', but parts from some more obscure wrecks may yet feature in future rebuilds.

A much better hope for future restoration is SB2C-3 Bu. No. 19075, registered as N4250Y, which is currently stored in the open at Charles F. Nichols Yankee Air Corps at Chino. Originally one of the exhibits at the Ed Maloney Air Museum at Ontario, California between 1967 and 1978, this machine passed to David C. Tallichet and Yesterday's Air Force at Chino, where it was displayed between 1979 and 1981. It was registered by MARC from July of that year until 1987 when it went to its present location.

The replica Curtiss SB2C Helldiver built for the USS Intrepid **Air/Sea Museum, New York, by Starr Aircraft of California City, California.** Lawrence Webster

US Navy and Marine Corps SB2C Units at the End of World War II, 7 September 1945

UNIT	CARRIER/BASE	TYPE	NO	NOTES
VB-1	*Bennington* (CV-20)	SB2C-4	1	Pacific fleet
		SB2C-4E	13	
VB-5	Klamath Falls	SB2C-4E	6	Ready 10 October CVG-5
		SBW-4E	9	
VB-6	*Hancock* (CV-19)	SB2C-4E	15	Pacific fleet
VB-7	Astoria	SB2C-4E	15	Ready 15 September CVG-7
VB-10	*Intrepid* (CV-11)	SB2C-5	11	Pacific fleet
VB-11	Santa Rosa	SB2C-4E	15	Ready 30 August CVG-11
VB-13	*Bunker Hill* (CV-17)	SB2C-4	7	Carrier in NYD Seattle;
		SB2C-4E	6	aircraft at Puunene
		SB2C-5	3	
VB-14	Kahului	SB2C-4	2	
		SB2C-5	13	
VB-15	Los Alamitos	SB2C-5	8	Ready
VB-16	*Randolph* (CV-15)	SB2C-4E	14	Pacific fleet
VB-17	Alameda	–	–	Ready 19 December
VB-18	San Diego	SB2C-4E	12	Ready
		SB2C-5	6	
VB-19	*Hornet* (CV-12)	SB2C-4	5	Carrier NYD San Francisco
		SB2C-4E	5	Aircraft Kahului
VB-80	Ream Field	SB2C-4	7	Ready 12 September
		SBF-4E	10	
VB-82	Alameda	–	–	Ready 20 December
VB-83	*Essex* (CV-9)	SB2C-4	3	Pacific fleet
		SB2C-4E	12	
VB-84	Los Alamitos	SB2C-4E	6	Ready 12 November
		SB2C-5	8	
		SBW-4E	10	
VB-85	*Shangri-La* (CV-38)	SB2C-4E	15	Pacific fleet
VB-86	*Wasp* (CV-18)	SB2C-5	9	Aircraft Eniwetok
VB-87	*Ticonderoga* (CV-14)	SB2C-4E	15	Pacific fleet
VB-88	*Yorktown* (CV-10)	SB2C-4E	12	Pacific fleet
VB-89	*Antietam* (CV-36)	SB2C-5	15	Pacific fleet
VB-92	Saipan	SB2C-4	13	
		SB2C-4E	2	
VB-93	*Boxer* (CV-21)	SB2C-5	14	Hawaii
VB-94	*Lexington* (CV-16)	SB2C-4E	14	Pacific fleet
VB-95	Hilo	SB2C-4	1	
		SB2C-4E	11	
		SB2C-5	4	
VB-96	Ventura	SB2C-4E	3	Replacement training
		SBF-4E	4	

UNIT	CARRIER/BASE	TYPE	NO	NOTES
	Ventura	SBW-4E	3	
VB-98	Los Alamitos	SB2C-4	4	
		SB2C-4E	23	
		SB2C-5	3	
VB-99	Saipan	SB2C-4	17	
		SB2C-5	1	
VB-100	Barbers Point	SB2C-4	9	Replacement training
		SB2C-5	8	
VB-151	Corvallis	SB2C-4E	8	
VB-152	Brown Field	SB2C-4E	7	
NACTU	Barbers Point	SB2C-4E	1	Pacific
SERVRON 42	El Centro	SB2C-4E	29	Assigned marine carriers
VMF-214	El Centro	SBW-4E	2	Ready February 1945
VMF-452	El Centro	SBW-4E	2	Ready February 1945
SERVRON 44	Ewa	SB2C-4	3	Marine Air Support Group 44
VMSB-333	Ewa	SB2C-4	13	
		SB2C-4E	1	
SERVRON 46	El Toro	SB2C-4	1	Marine Air Support Group 46
		SBF-4E	1	
		SBW-4E	1	
VS-46	Barbers Point	SB2C-4E	12	Replacement training
VS-53	Barbers Point	SB2C-4E	18	Replacement training
VS-66	Majuro	SB2C-4	10	
	Roi	SB2C-4	5	
VS-69	Barbers Point	SB2C-4E	18	
HEADRON 2	Camp Kearney	SB2C-4E	12	Fleet Air Wing 14
		SBF-4E	6	
		SBW-4E	1	
VMSB-244	Cotobato	SB2C-4	31	
VMJ-3	Ewa	SB2C-4	1	Third Marine Air Wing
VMSB-343	Ewa	SB2C-5	22	
HEDRON 4	Kwajalein	SB2C-4	1	Fourth Marine Air Wing
VMSB-245	Ulithi	SB2C-4E	24	
HEDRON 94	Engebi	SB2C-4	2	MAG-94
VMSB-331	Majuro	SB2C-4	2	
		SB2C-4E	10	
	Eniwetok	SB2C-4E	12	
–	San Diego	SBW-4E	1	Marine Fleet Air West Coast
HEDRON 41	El Toro	SBF-4E	3	
		SBW-4E	3	
SERVRON 41	El Toro	SBW-4E	1	MAG- 41
VJ-19	Engebi	SB2C-3	3	
CASU-1	Pearl Harbor	SB2C-4E	1	Service unit
CASU-2	Barbers Point	SB2C-4	1	Service unit
		SB2C-5	1	
CASU-5	San Diego	SB2C-5	3	Service unit
		SBF-4E	3	
CASU-6	Alameda	SB2C-4E	5	Service unit
		SB2C-5	1	
CASU-7	Seattle	SB2C-4E	2	Service unit
CASU-13/14	Saipan	SB2C-4	16	Service Unit
		SB2C-4E	1	
		SB2C-5	1	
CASU-50	Pasco	SB2C-4E	1	Service unit
CASU-53	Holtville	SBW-4	1	Service unit
CASU-54	Fallon	SB2C-4	1	Service unit
		SB2C-4E	3	
CASU-64	Watsonville	SB2C-4E	8	Service unit
CASUF 69	Twenty-Nine Palms	SBW-4E	1	Service unit

UNIT	CARRIER/BASE	TYPE	NO	NOTES
SOSU-3	Alameda	SB2C-4E	2	Service unit
COMAIRPAC	Pearl Harbour	SB2C-5	1	Flag unit
COMUTWINGSERVPAC	Pearl Harbour	SB2C-4	6	Flag unit
VB-3	Oceana	SB2C-5	15	Ready 7 October
VB-4	Wildwood	SB2C-4E	6	Ready 9 November
		SB2C-5	18	
VB-74	Midway (CVB-41)	SBW-4E	26	Carrier Newport News
				Aircraft Ayer and Otis Fields
VB-75	Franklin D. Roosevelt (CVB-42)	SBF-4E	13	Carrier NYD New York
		SBF-4E	22	Aircraft Chincoteague
VB-81	Wildwood	SB2C-4E	24	Ready 30 November
VB-97	Wildwood	SB2C-4E	7	
	Grosse Isle	SB2C-4E	39	
	Wildwood	SB2C-5	1	
VT-97	Hyannis	SB2C-4E	1	
		SB2C-5	2	
VB-150	Lake Champlain (CV-39)	SB2C-5	15	Atlantic
VB-153	Oceana	SB2C-5	15	Ready 26 August
HEDRON 34	Oak Grove	SB2C-4E	9	MAG-34
VMSB-931	Oak Grove	SB2C-4E	24	Replacement training
VMSB-932	Oak Grove	SB2C-4E	24	Replacement training
HEDRON 53	Eagle Mountain Lake	SB2C-4E		
VMFN-531	Eagle Mountain Lake	SB2C-4E	12	
HEDRON 93	Bogue Field	SB2C-4E	3	MAG-93
VMSB-933	Bogue Field	SB2C-4E	24	Replacement training
VMSB-934	Bogue Field	SB2C-4E	24	Replacement training
VJ-5	Guantanamo	SB2C-4E	3	Utility Wing Atlantic
	Cape May	SB2C-4E	4	
CASU-21	Norfolk	SB2C-5	5	Service unit
CASU-22	Quonset	SB2C-4E	1	Service unit
		SB2C-5	4	
CASU-26	Otis Field	SBW-4E	1	Service unit
CASU-27	Chincoteague	SB2C-4E	2	Service unit
FAETULANT	Cape May	SB2C-4E	3	Training
	Grosse Isle	SB2C-4E	2	
	Quonset	SB2C-4E	2	
ASDEVLANT	Quonset	SB2C-4E	2	
XVJ-25	Rockland	SB2C-4E	3	Experimental
	Brunswick	SB2C-4E	13	

SB2C Units on 1 February 1947

UNIT	CARRIER/BASE	TAIL CODE	TYPE	NO.
VA-1B	CVBG-1	M	SB2C-5	32
VA-2B	CVBG-1	M	SB2C-5	28
VA-5B	CVBG-5	C	SB2C-5	32
VA-6B	CVBG-5	C	SB2C-5	32
VA-3A	CVG-3	K	SB2C-5	24
VA-7A	CVG-7	L	SB2C-5	24
VA-9A	CVG-9	PS	SB2C-5	24
VA-17A	CVG-17	R	SB2C-5	24
VA-1A	CVG-1	T	SB2C-5	24
VA-5A	CVG-5	S	SB2C-5	24
VA-11A	CVG-11	V	SB2C-5	24
VA-13A	CVG-13	P	SB2C-5	24
VA-15A	CVG-15	B	SB2C-5	24
VA-21A	CVG-21	RI	SB2C-5	24
Akron-Cleveland	Air Training Command	–	SB2C-4E	4
Buffalo-Syracuse	Air Training Command	–	SB2C-4E	4
Portland	Air Training Command	–	SB2C-4E	3
Various	BuAer Research	–	SB2C-4E	3
			SB2C-5	22
Anacostia	Naval Air Reserve	–	SB2C-4E	8
Atlanta	Naval Air Reserve	–	SB2C-4E	4
Columbus	Naval Air Reserve	–	SB2C-4E	8
Dallas	Naval Air Reserve	–	SB2C-4E	8
Denver	Naval Air Reserve	–	SB2C-4E	3
Glenview	Naval Air Reserve	–	SB2C-4E	8
Grosse Isle	Naval Air Reserve	–	SB2C-4E	8
Jacksonville	Naval Air Reserve	–	SB2C-4E	3
Los Alamitos	Naval Air Reserve	–	SB2C-4E	11
Memphis	Naval Air Reserve	–	SB2C-4E	3
Miami	Naval Air Reserve	–	SB2C-4E	3
Minneapolis	Naval Air Reserve	–	SB2C-4E	8
New Orleans	Naval Air Reserve	–	SB2C-4E	3
New York	Naval Air Reserve	–	SB2C-4E	8
Norfolk	Naval Air Reserve	–	SB2C-4E	3
Oakland	Naval Air Reserve	–	SB2C-4E	8
Olanthe	Naval Air Reserve	–	SB2C-4E	8

SB2C Units on 1 February 1948

UNIT	CARRIER/BASE	TYPE	NO	NOTES
VA-3A	*Kearsarge* (CV-33)	SB2C-5	20	
VA-7A	*Leyte* (CV-32)	SB2C-5	20	To re-equip with
VA-9A	*Leyte* (CV-32)	SB2C-5	20	F4U-4s prior
			to ADs.	
VA-ATU-4	Naval Air Advanced Training	SB2C-5	20	
	BuAer	SB2C-5	16	(RD & DE)
		SB2C-4E	5	

US Navy SB2C Carrier VB Squadron Deployment

(Key: CP = Central Pacific; MI = Marianas campaign; EP = Eastern Pacific; WP = Western Pacific; JPN = Action over Japan; SP = South Pacific; ATL = Atlantic; PI = Philippines campaign; SEA = South East Asia; OKI = Okinawa campaign; CARB = Caribbean; GULF = Gulf of Mexico; SATL = South Atlantic; MED = Mediterranean; WI = West Indies; N/A = No aircraft on establishment.)

UNIT	DEPLOYMENT DATES	TYPE	WING	OPS. AREA	CARRIER
VB-1	29-5-44 – 3-6-44	SB2C-1C	CVG-1	CP	Yorktown (CV-10)
	6-6-44 – 27-6-44	SB2C-1C	CVG-1	CP-MI	Yorktown (CV-10)
	30-6-44 – 1-8-44	SB2C-3	CVG-1	CP-MI	Yorktown (CV-10)
	4-8-44 – 10-8-44	SB2C-1C	CVG-1	CP	Yorktown (CV-10)
	11-8-44 – 17-8-44	SB2C-1C	CVG-1	EP	Yorktown (CV-10)
	1-7-45 – 7-11-45	SB2C-4	CVG-1	WP-JPN	Bennington (CV-20)
VB-2	8-3-44 – 20-3-44	SB2C-1C	CVG-2	CP	Hornet (CV-12)
	22-3-44 – 6-4-44	SB2C-1C	CVG-2	SP-CP	Hornet (CV-12)
	13-4-44 – 4-5-44	SB2C-1C	CVG-2	CP-WP	Hornet (CV-12)
	6-6-44 – 27-6-44	SB2C-1C	CVG-2	CP-MI	Hornet (CV-12)
	30-6-44 – 9-8-44	SB2C-1C	CVG-2	CP-MI	Hornet (CV-12)
	28-8-44 – 29-9-44	SB2C-3	CVG-2	WP	Hornet (CV-12)
	1-10-45 – 22-10-45	SB2C-4	CVG-2	EP-ATL	Shangri-La (CV38)
VB-3	16-4-44 – 22-4-44	SB2C-1C	CVG-3	EP	Essex (CV-9)
	24-10-44 – 3-11-44	SB2C-3	CVG-3	CP-WP	Yorktown (CV-10)
	5-11-44 – 24-11-44	SB2C-3	CVG-3	WP-PI	Yorktown (CV-10)
	11-12-44 – 24-12-44	SB2C-3	CVG-3	WP-PI	Yorktown (CV-10)
	30-12-44 – 26-1-45	SB2C-3	CVG-3	WP-SEA	Yorktown (CV-10)
	10-2-45 – 1-3-45	SB2C-4	CVG-3	WP-IWO-JPN	Yorktown (CV-10)
	7-3-45 – 31-3-45	SB2C-3	CVG-3	WP-CP-EP	Lexington (CV16)
VB-4	21-5-43 – 16-6-43	SB2C-1	CVG-5	CARB	Yorktown (CV-10)
VB-5	12-2-45 – 7-3-45	SB2C-4	CVG-5	EP	Franklin (CV-13)
	3-3-45 – 13-3-45	SB2C-4	CVG-5	CP-WP	Franklin (CV-13)
	14-3-45 – 25-3-45	SB2C-4	CVG-5	WP-OKI-JPN	Franklin (CV-13)
	18-3-46 – 20-5-46	SB2C-5	CVG-5	EP	Lexington (CV16)
VB-6	21-5-43 – 16-6-43	SB2C-1	CVG-5	CARB	Yorktown (CV-10)
	6-4-45 – 14-4-45	SB2C-4	CVG-6	WP-OKI-JPN	Hancock (CV-19)
	13-4-45 – 21-4-45	SB2C-4	CVG-6	WP-CP	Hancock (CV-19)
	13-6-45 – 26-6-45	SB2C-4	CVG-6	CP-Wake-WP	Hancock (CV-19)
	1-7-45 – 21-10-45	SB2C-4	CVG-6	WP-JAPAN	Hancock (CV-19)
VB-7	21-6-44 – 8-7-44	SB2C-3	CVG-7	CARB	Hancock (CV-19)
	31-7-44 – 30-8-44	SB2C-3	CVG-7	ATL-EP	Hancock (CV-19)
	24-9-44 – 5-10-44	SB2C-3	CVG-7	CP-WP	Hancock (CV-19)
	6-10-44 – 9-11-44	SB2C-3	CVG-7	WP-Leyte	Hancock (CV-19)
	14-11-44 – 27-11-44	SB2C-3	CVG-7	WP-PI	Hancock (CV-19)
	11-12-44 – 24-12-44	SB2C-3	CVG-7	WP-PI	Hancock (CV-19)
	30-12-44 – 26-1-45	SB2C-3	CVG-7	WP-SEA	Hancock (CV-19)
	28-1-45 – 15-2-45	SB2C-3	CVG-7	WP-CP-EP	Ticonderoga (CV-14)
	19-2-46 – 6-3-46	SB2C-5	CVG-7	EP	Hancock (CV-19)
VB-8	7-10-43 – 1-11-43	SB2C-1	CVG-8	CARB	Intrepid (CV-11)
	3-12-43 – 10-1-44	SB2C-1	CVG-8	ATL-EP	Intrepid (CV-11)

UNIT	DEPLOYMENT DATES	TYPE	WING	OPS. AREA	CARRIER
	22-3-44 – 6-4-44	SB2C-1C	CVG-8	SP-CP	*Bunker Hill* (CV-17)
	13-4-44 – 4-5-44	SB2C-1C	CVG-8	CP-WP	*Bunker Hill* (CV-17)
	6-6-44 – 25-6-44	SB2C-1C	CVG-8	CP-MI	*Bunker Hill* (CV-17)
	11-7-44 – 9-8-44	SB2C-1C	CVG-8	CP-MI	*Bunker Hill* (CV-17)
	28-8-44 – 1-10-44	SB2C-1C	CVG-8	WP	*Bunker Hill* (CV-17)
	6-10-44 – 27-10-44	SB2C-1C	CVG-8	WP	*Bunker Hill* (CV-17)
VB-9	13-10-44 – 18-10-44	SB2C-3	CVG-9	EP	*Yorktown* (CV-10)
	10-2-45 – 4-3-45	SB2C-3	CVG-9	WP-IWO-JPN	*Lexington* (CV-16)
	14-3-45 – 15-5-45	SB2C-4	CVG-9	WP-OKI-JPN	*Yorktown* (CV-10)
	25-5-45 – 14-6-45	SB2C-4	CVG-9	WP-OKI	*Yorktown* (CV-10)
	19-6-45 – 7-7-45	N/A	CVG-9	WP-CP-EP	*Hornet* (CV-12)
VB-10	10-2-45 – 16-2-45	SB2C-4	CVG-10	EP	*Intrepid* (CV-11)
	3-3-45 – 13-3-45	SB2C-4	CVG-10	CP-WP	*Intrepid* (CV-11)
	14-3-45 – 19-4-45	SB2C-4	CVG-10	WP-OKI-JPN	*Intrepid* (CV-11)
	4-5-45 – 19-5-45	SB2C-4	CVG-10	WP-CP-EP	*Intrepid* (CV-11)
	28-6-45 – 5-7-45	SB2C-4	CVG-10	EP	*Intrepid* (CV-11)
	30-7-45 – 2-11-45	SB2C-4	CVG-10	CP-WP	*Intrepid* (CV-11)
VB-11	30-3-44 – 3-4-44	SB2C-1C	CVG-11	EP	*Wasp* (CV-18)
	4-10-44 – 30-10-44	SB2C-3	CVG-11	WP-Leyte	*Hornet* (CV-12)
	2-11-44 – 22-11-44	SB2C-3	CVG-11	WP-PI	*Hornet* (CV-12)
	11-12-44 – 24-12-44	SB2C-3	CVG-11	WP-WI	*Hornet* (CV-12)
	30-12-44 – 26-1-45	SB2C-3	CVG-11	WP-SEA	*Hornet* (CV-12)
VB-12	20-1-45 – 26-1-45	SB2C-4	CVG-12	EP	*Randolph* (CV-15)
	29-1-45 – 6-2-45	SB2C-4	CVG-12	CP-WP	*Randolph* (CV-15)
	10-2-45 – 1-3-45	SB2C-4	CVG-12	WP-IWO-JPN	*Randolph* (CV-15)
	30-3-45 – 4-6-45	SB2C-4	CVG-12	WP-OKI-JPN	*Randolph* (CV-15)
VB-13	20-3-44 – 19-4-44	SB2C-1C	CVG-13	CARB	*Franklin* (CVB-13)
	5-5-44 – 19-5-44	SB2C-1C	CVG-13	ATL-EP	*Franklin* (CVB-13)
	31-5-44 – 6-6-44	SB2C-3	CVG-13	EP	*Franklin* (CVB-13)
	16-6-44 – 22-6-44	SB2C-3	CVG-13	CP	*Franklin* (CVB-13)
	30-6-44 – 9-8-44	SB2C-3	CVG-13	CP-MI	*Franklin* (CVB-13)
	28-8-44 – 21-9-44	SB2C-3	CVG-13	WP	*Franklin* (CVB-13)
	24-9-44 – 1-11-44	SB2C-3	CVG-13	WP-Leyte	*Franklin* (CVB-13)
	11-11-44 – 21-11-44	SB2C-3	CVG-13	WP-CP	*Franklin* (CVB-13)
	23-11-43 – 29-11-43	SB2C-3	CVG-13	EP	*Franklin* (CVB-13)
VB-14	31-1-44 – 27-2-44	SB2C-3	CVG-14	CARB	*Wasp* (CV-18)
	15-3-44 – 28-3-44	SB2C-1C	CVG-14	ATL-EP	*Wasp* (CV-18)
	30-3-44 – 3-4-44	SB2C-3	CVG-14	EP	*Wasp* (CV-18)
	3-5-44 – 8-5-44	SB2C-1C	CVG-14	CP	*Wasp* (CV-18)
	15-5-44 – 26-5-44	SB2C-1C	CVG-14	CP-Marcus	*Wasp* (CV-18)
	6-6-44 – 25-6-44	SB2C-1C	CVG-14	CP-MI	*Wasp* (CV-18)
	30-6-44 – 1-8-44	SB2C-1C	CVG-14	CP-MI	*Wasp* (CV-18)
	28-8-44 – 29-9-44	SB2C-3	CVG-14	WP	*Wasp* (CV-18)
	4-10-44 – 30-10-44	SB2C-3	CVG-14	WP-Leyte	*Wasp* (CV-18)
	2-11-44 – 10-11-44	SB2C-3	CVG-14	WP-PI	*Wasp* (CV-18)
VB-15	13-1-44 – 2-2-44	SB2C-1C	CVG-15	CARB	*Hornet* (CV-12)
	14-2-44 – 27-2-44	SB2C-1C	CVG-15	ATL-EP	*Hornet* (CV-12)
	29-2-44 – 4-3-44	SB2C-1C	CVG-15	EP	*Hornet* (CV-12)
	3-5-44 – 8-5-44	SB2C-!C	CVG-15	CP	*Essex* (CV-9)
	15-5-44 – 26-5-44	SB2C-1C	CVG-15	CP-Marcus	*Essex* (CV-9)
	6-6-44 – 6-7-44	SB2C-1C	CVG-15	CP-MI	*Essex* (CV-9)
	14-7-44 – 9-8-44	SB2C-1C	CVG-15	CP-MI	*Essex* (CV-9)
	28-8-44 – 1-10-44	SB2C-3	CVG-15	WP	*Essex* (CV-9)
	6-10-44 – 30-10-44	SB2C-3	CVG-15	WP-Leyte	*Essex* (CV-9)
	1-11-44 – 17-11-44	SB2C-3	CVG-15	WP-PI	*Essex* (CV-9)
	20-11-44 – 6-12-44	SB2C1C	CVG-15	WP-CP-EP	*Bunker Hill* (CV-17)
VB-16	23-1-45 – 3-3-45	SB2C-4E	CVG-16	CARB	*Bon Homme Richard* (CV31)
	19-3-45 – 30-3-45	SB2C-4E	CVG-16	ATL-EP	*Bon Homme Richard* (CV31)
	1-4-45 – 5-4-45	SB2C-4E	CVG-16	EP	*Bon Homme Richard* (CV31)

UNIT	DEPLOYMENT DATES	TYPE	WING	OPS. AREA	CARRIER
	1-8-45 – 11-9-45	SB2C-4E	CVG-16	WP-JPN	Randolph (CV-15)
	1-10-45 – 15-10-45	SB2C-4E	CVG-16	EP-ATL	Randolph (CV-15)
VB-17	13-7-43 – 10-8-43	SB2C-1	CVG-17	CARB	Bunker Hill (CV-17)
	7-9-43 – 25-9-43	SB2C-1	CVG-17	ATL-EP	Bunker Hill (CV-17)
	28-9-43 – 2-10-43	SB2C-1	CVG-17	EP	Bunker Hill (CV-17)
	21-10-43 – 5-11-43	SB2C-1	CVG-17	SP	Bunker Hill (CV-17)
	8-11-43 – 13-11-43	SB2C-1	CVG-17	SP-Rabaul	Bunker Hill (CV-17)
	15-11-43 – 10-12-43	SB2C-1	CVG-17	CP-Tarawa	Bunker Hill (CV-17)
	23-12-43 – 6-1-44	SB2C-1C	CVG-17	SP	Bunker Hill (CV-17)
	19-1-44 – 4-2-44	SB2C-1C	CVG-17	CP-Marshall	Bunker Hill (CV-17)
	12-2-44 – 26-2-44	SB2C-1C	CVG-17	CP-Truk	Bunker Hill (CV-17)
	28-2-44 – 6-3-44	SB2C-1C	CVG-17	CP	Bunker Hill (CV-17)
	6-3-44 – 10-3-44	N/A	CVG-17	EP	Essex (CV-9)
	10-2-45 – 4-3-45	SB2C-3	CVG-17	WP-IWO-JPN	Hornet (CV-12)
	14-3-45 – 30-4-45	SB2C-3	CVG-17	WP-OKI-JPN	Hornet (CV-12)
	9-5-45 – 14-6-45	SB2C-3	CVG-17	WP-OKI	Hornet (CV-12)
	19-6-45 – 7-7-45	SB2C-3	CVG-17	WP-CP-EP	Hornet (CV-12)
VB-18	24-2-44 – 28-2-44	SB2C-1C	CVG-18	EP	Lexington (CV-16)
	16-8-44 – 24-8-44	SB2C-3	CVG-18	CP	Intrepid (CV-11)
	28-8-44 – 1-10-44	SB2C-3	CVG-18	WP	Intrepid (CV-11)
	6-10-44 – 9-11-44	SB2C-3	CVG-18	WP-Leyte	Intrepid (CV-11)
	14-11-44 – 29-11-44	SB2C-3	CVG-18	WP-PI	Intrepid (CV-11)
	3-12-44 – 20-12-44	SB2C-3	CVG-18	WP-CP-EP	Intrepid (CV-11)
	30-9-45 – 18-10-45	N/A	CVG-18	EP-GULF	Ranger (CV-4)
	30-10-45 – 18-11-45	N/A	CVG-18	GULF-ATL	Ranger (CV-4)
	30-8-46 – 11-12-46	SB2C-5	CVG18	CARB-SATL	Leyte (CV-32)
VB-19	24-2-44 – 28-2-44	SB2C-1C	CVG-19	EP	Lexington (CV-16)
	23-6-44 – 1-7-44	SB2C-3	CVG-19	CP	Intrepid (CV-11)
	11-7-44 – 9-8-44	SB2C-3	CVG-19	CP-MI	Lexington (CV-16)
	28-8-44 – 1-10-44	SB2C-3	CVG-19	WP	Lexington (CV-16)
	6-10-44 – 30-10-44	SB2C-3	CVG-19	WP-Leyte	Lexington (CV-16)
	1-11-44 – 9-11-44	SB2C-3	CVG-19	WP-PI	Lexington (CV-16)
	23-11-44 – 6-12-44	SB2C-3	CVG-19	WP-CP	Enterprise (CV-6)
	17-3-46 – 1-4-46	SB2C-5	CVG-19	EP-CP	Hancock (CV-19)
	2-4-46 – 20-8-46	SB2C-5	CVG-19	WP	Antietam (CV-36)
VB-20	16-8-44 – 24-8-44	SB2C-3	CVG-20	CP	Enterprise (CV-6)
	28-8-44 – 21-9-44	SB2C-3	CVG-20	WP	Enterprise (CV-6)
	24-9-44 – 1-11-44	SB2C-3	CVG-20	WP-Leyte	Enterprise (CV-6)
	5-11-44 – 22-11-44	SB2C-3	CVG-20	WP-PI	Enterprise (CV-6)
	11-12-44 – 24-12-44	SB2C-3	CVG-20	WP-PI	Lexington (CV-16)
	30-12-44 – 26-1-45	SB2C-3	CVG-20	WP-SEA	Lexington (CV-16)
	30-9-46 – 23-11-46	SB2C-5	CVG-20	CARB	Philippine Sea (CV-47)
VB-74	1946	SBW-4E	CVBG-74	ATL	Midway (CV-41)
VB-75	8-1-46 – 20-3-46	SB2C-5	CVBG-75	CARB	Franklin D. Roosevelt (CV-42)
	19-4-46 – 10-6-46	SB2C-5	CVBG-75	CARB-ATL	Franklin D. Roosevelt (CV-42)
	8-8-46 – 4-10-46	SB2C-5	CVBG-75	MED	Franklin D. Roosevelt (CV-42)
VB-80	26-6-44 – 23-7-44	SB2C-3	CVG-80	CARB	Ticonderoga (CV-14)
	30-8-44 – 24-9-44	SB2C-3	CVG-80	ATL-EP	Ticonderoga (CV-14)
	18-10-44 – 30-10-44	SB2C-3	CVG-80	CP-WP	Ticonderoga (CV-14)
	1-11-44 – 17-11-44	SB2C-3	CVG-80	WP-PI	Ticonderoga (CV-14)
	23-11-44 – 1-12-44	SB2C-3	CVG-80	WP-PI	Ticonderoga (CV-14)
	11-12-44 – 24-12-44	SB2C-3	CVG-80	WP-PI	Ticonderoga (CV-14)
	30-12-44 – 24-1-45	SB2C-3	CVG-80	WP-SEA	Ticonderoga (CV-14)
	10-2-45 – 4-3-45	SB2C-3	CVG-80	WP-IWO-JPN	Hancock (CV-19)
	17-3-45 – 1-4-46	SB2C-4	CVG-80	EP-CP	Hancock (CV-19)
	2-4-46 – 10-9-46	SB2C-5	CVG-80	WP	Boxer (CV-21)
VB-81	11-11-44 – 24-11-44	SB2C-3	CVG-81	WP-PI	Wasp (CV-18)
	11-12-44 – 24-12-44	SB2C-3	CVG-81	WP-PI	Wasp (CV-18)
	30-12-44 – 26-1-45	SB2C-3	CVG-81	WP-SEA	Wasp (CV-18)

UNIT	DEPLOYMENT DATES	TYPE	WING	OPS. AREA	CARRIER
	10-2-45 – 4-3-45	SB2C-3	CVG-81	WP-IWO-JPN	Wasp (CV-18)
	13-12-45 – 31-5-46	SB2C-5E	CVG-81	CARB-ATL	Princeton (CV-37)
	1-6-46 – 30-6-46	SB2C-5E	CVG-81	ATL-EP	Princeton (CV-37)
	3-7-46 – 15-4-47	SB2C-5E	CVG-81	WP	Princeton (CV-37)
VB-82	16-10-44 – 14-11-44	SB2C-3	CVG-82	CARB	Bennington (CV-20)
	15-12-44 – 29-12-44	SB2C-4	CVG-82	ATL-EP	Bennington (CV-20)
	1-1-45 – 8-1-45	SB2C-4	CVG-82	EP	Bennington (CV-20)
	29-1-45 – 8-2-45	SB2C-4	CVG-82	CP-WP	Bennington (CV-20)
	10-2-45 – 1-3-45	SB2C-4	CVG-82	WP-IWO-JPN	Bennington (CV-20)
	14-3-45 – 30-4-45	SB2C-4	CVG-82	WP-OKI-JPN	Bennington (CV-20)
	9-5-45 – 11-6-45	SB2C-4	CVG-82	WP-OKI	Bennington (CV-20)
	19-10-46 – 20-12-46	SB2C-5	CVG-82	ATL-MED	Randolph (CV-15)
VB-83	14-3-45 – 1-6-45	SB2C-3/4	CVG-83	WP-OKI-JPN	Essex (CV-9)
	1-7-45 – 13-9-45	SB2C-4	CVG-83	WP-JPN	Essex (CV-9)
VB-84	24-1-45 – 28-1-45	SB2C-4E	CVG-84	EP	Bunker Hill (CV-17)
	29-1-45 – 7-2-45	SB2C-4E	CVG-84	CP-WP	Bunker Hill (CV-17)
	10-2-45 – 4-3-45	SB2C-4E	CVG-84	WP-IWO-JPN	Bunker Hill (CV-17)
	14-3-45 – 14-5-45	SB2C-4E	CVG-84	WP-OKI-JPN	Bunker Hill (CV-17)
	17-5-45 – 25-5-45	SB2C-4E	CVG-84	WP-CP	Bunker Hill (CV-17)
	28-5-45 – 3-6-45	SB2C-4E	CVG-84	EP	Bunker Hill (CV-17)
VB-85	21-11-44 – 23-12-44	SB2C-3	CVG-85	CARB	Shangri-La (CV-38)
	17-1-45 – 2-2-45	SB2C-3	CVG-3	ATL-EP	Shangri-La (CV-38)
	7-2-45 – 13-2-45	SB2C-3	CVG-85	EP	Shangri-La (CV-38)
	10-4-45 – 20-4-45	SB2C-4	CVG-85	CP-WP	Shangri-La (CV-38)
	21-4-45 – 15-5-45	SB2C-4	CVG-85	WP-OKI	Shangri-La (CV-38)
	25-5-45 – 14-6-45	SB2C-4	CVG-85	WP-OKI	Shangri-La (CV-38)
	1-7-45 – 16-9-45	SB2C-4	CVG-85	WP-JPN	Shangri-La (CV-38)
VB-86	14-3-45 – 25-3-45	SB2C-4	CVG-86	WP-OKI-JPN	Wasp (CV-18)
	28-3-45 – 2-4-45	SB2C-4	CVG-86	WP-CP	Wasp (CV-18)
	5-4-45 – 13-4-45	SB2C-4	CVG-86	EP	Wasp (CV-18)
	13-6-45 – 19-6-45	SB2C-4	CVG-86	EP	Wasp (CV-18)
	11-7-45 – 19-7-45	SB2C-4	CVG-86	CP-WP	Wasp (CV-18)
	21-7-45 – 11-9-45	SB2C-4	CVG-86	WP-JPN	Wasp (CV-18)
	1-10-45 – 15-10-45	SB2C-4	CVG-86	EP-ATL	Wasp (CV-18)
VB-87	22-11-44 – 16-12-44	SB2C-3	CVG-8	CARB	Randolph (CV-15)
	17-12-44 – 31-12-44	SB2C-3	CVG-87	ATL-EP	Randolph (CV-15)
	11-5-45 – 13-6-45	SB2C-5	CVG-87	CP-WP	Ticonderoga (CV-14)
	1-7-45 – 5-10-45	SB2C-5	CVG-87	WP-JPN	Ticonderoga (CV-14)
VB-88	1-7-45 – 20-10-45	SB2C-4	CVG-88	WP-JPN	Yorktown (CV-10)
VB-89	12-8-45 – 1-4-46	SB2C-4	CVG-89	WP	Antietam (CV-36)
	2-4-46 – 13-4-46	SB2C-4	CVG-89	CP-EP	Hancock (CV-19)
VB-93	7-5-45 – 25-6-45	SB2C-5	CVG-93	CARB	Boxer (CV-21)
	23-7-45 – 7-8-45	SB2C-5	CVG-93	ATL-EP	Boxer (CV-21)
	1-9-45 – 4-12-45	SB2C-5	CVG-93	WP	Boxer (CV-21)
	2-4-46 – 13-4-46	SB2C-5	CVG-93	CP-EP	Hancock (CV-19)
VB-94	22-5-45 – 26-5-45	SB2C-5	CVG-94	EP	Lexington (CV-16)
	13-6-45 – 26-6-45	SB2C-5	CVG-94	CP-Wake-WP	Lexington (CV-16)
	1-7-45 – 15-12-45	SB2C-5	CVG-94	WP-JPN	Lexington (CV-16)
VB-150	5-7-45 – 9-8-45	SB2C-5	CVG-150	CARB	Lake Champlain (CV-39)
VB-153	6-5-46 – 2-7-46	SB2C-5	CVG-153	CARB	Kearsarge (CV-33)
	30-10-46 – 7-11-46	SB2C-5	CVG-153	EP	Antietam (CV-36)

US Navy SB2C Carrier VA and VT Squadron Deployment

UNIT	DEPLOYMENT DATES	TYPE	WING	OPS. AREA	CARRIER
VA-5A	31-3-47 – 15-6-47	SB2C-5	CVAG-5	WP	*Shangri-La* (CV38)
VA-5B	22-1-47 – 27-3-47	SB2C-5	CVBG-5	CARB	*Valley Forge* (CV-45)
VA-7A	1-2-47 – 18-3-47	SB2C-5	CVAG-7	ATL	*Leyte* (CV-32)
	19-3-47 – 9-6-47	SB2C-5	CVAG-7	MED	*Leyte* (CV-32)
	30-7-47 – 19-11-47	SB2C-5	CVAG-7	MED	*Leyte* (CV-32)
	9-2-48 – 19-3-48	SB2C-5	CVAG-7	ATL	*Leyte* (CV-32)
VA-9A	31-3-47 – 5-5-47	SB2C-5	CVAG-9	CARB	*Philippine Sea* (CV-47)
	9-2-48 – 26-6-48	SB2C-5	CVAG-9	CARB-MED	*Philippine Sea* (CV-47)
VA-11A	19-11-46 – 25-11-46	SB2C-5	CVAG-11	EP	*Boxer* (CV-21)
	9-10-47 – 11-6-48	SB2C-5	CVAG-1	WP-WORLD	*Valley Forge* (CV-45)
VA-15A	15-11-46 – 25-11-46	SB2C-5	CVAG-15	EP	*Antietam* (CV-36)
	26-11-46 – 15-12-46	SB2C-5	CVAG-15	EP	*Boxer* (CV-21)
	22-2-47 – 24-3-47	SB2C-5	CVAG-15	CP	*Boxer* (CV-21)
	31-3-47 – 11-10-47	SB2C-5	CVAG-15	WP	*Antietam* (CV-36)
VA-75	27-8-48 – 9-10-48	SB2C-5	CVAG-7	CARB	*Leyte* (CV-32)
VA-94	22-10-48 – 23-11-48	SB2C-5	CVG-9	NATL	*Philippine Sea* (CV-47)
VT-74	7-11-45.– 3-1-46	SB2C-4E	CVBG-74	CARB	*Midway* (CV-41)
	1-3-46 – 9-4-46	SB2C-5	CVBG-74	ARCTIC	*Midway* (CV-41)
	18-4-46 – 27-5-46	SB2C-5	CVBG-74	CARB-ATL	*Midway* (CV-41)
VT-75	8-1-46 – 20-3-46	SB2C-5	CVBG-75	CARB	*Franklin D. Roosevelt* (CV-42)
	19-4-46 – 10-6-46	SB2C-5	CVBG-75	CARB-ATL	*Franklin D. Roosevelt* (CV-42)
	8-8-46 – 4-10-46	SB2C-5	CVBG-75	MED	*Franklin D. Roosevelt* (CV-42)

Notes

CHAPTER ONE

1 *See Dive Bomber!, an Illustrated History*, Peter C. Smith, Ashbourne, UK and Annapolis, USA, 1982.

2 *See The Douglas SBD Dauntless*, Peter C. Smith, Ramsbury, UK, 1997.

3 *See Development of the Attack Concept, and The Attack Bomber – the War Years* by Lee M. Parsons, Naval Aviation Confidential Bulletins, Washington DC, 1951.

4 *See* Curtiss Aeroplane Division, Buffalo, VSB (Proposal 'A'), Report No. 7378, dated 28 November 1938.

5 *See* Contract No(s). 1609 (formerly No. 79082) Model SB2C-1 Airplane, Production Inspection Trials, TED No. BIS 2130: Final Report of Navy Department Board of Inspection and Survey, Washington DC (NA83 (InSurv) VSB2C-1/F8-4 (4037-S), Serial 013821), dated 30 October 1944. Copy in author's collection.

6 Ibid.

7 *See* Curtiss-Wright, Buffalo, VSB (Proposal 'A', *op cit*).

8 Ibid.

CHAPTER TWO

9 Which later became Contract No. 1609.

10 *See* memo from M. C. Langslow, secretary, to Department of Defence, Melbourne, (RAAF 9/63/16) MXY550, dated 1944, and Secret Report, Commonwealth of Australia, John Curtin, Minister for Defence to General Douglas MacArthur, C-in-C, South-West Pacific Area, dated 23 August 1944.

11 *See* Curtiss Factory, Buffalo, New York, XSB2C-1, Report Number 8210-A., dated 13 May 1941.

12 Stanley I. Vaughn II to the author, 10 April 1997.

CHAPTER THREE

13 Stanley I. Vaughn II to the author, 10 April 1997.

14 *See* Lee M. Pearson *The Attack Bomber – The War Years* op cit.

15 *See* Curtiss report #20462, Navy Preliminary Demonstration – SB2C-1 #00001, dated 8 January 1943.

16 *See* Chief BuAer to President, Board of Inspection and Survey, dated 24 June 1942, (Aer-E-211-EM; C-79082/C8647- Confidential Memo. Contract 79082 – Model SB2C-1 Airplanes – Production Inspection Trials.

17 A subsequent memo from Rear Admiral Ralph Davison, the acting chief, BuAer to the President Board of Inspection and Survey, dated 2 November 1942 (Aer-E-211-EM/C-79082/C16259) had to amend these figures sharply upwards: 'As a result of subsequent information received from the contractor, this weight may be increased to 13,880 pounds.'

18 Formal acceptance of the aircraft by the Navy was made, for record purposes, on 12 July 1943, subsequent to its crash to a complete loss. However, two other model SB2C-1 aircraft were accepted by the Navy in September 1942.

19 *See* George F. P. Kernahan, *The Trials of the Helldiver*, Journal American Aviation Historical Society, winter, 1991.

20 *See* Confidential Report *Carrier Acceptability Tests on Model SB2V-1 Airplane* no. 00002., contained within Production Inspection Trials, *op cit*. Folder.

21 *See* Project No. 4350 – *Model SB2C-1 Airplane – Pneumatic Tail Wheel*, test of, Confidential Report F1/SEU 42-13, (E-4) (9615) (RWD:Hw), dated 31 March 1943.

CHAPTER FOUR

22 *See* Commander J. E. Clarke, OC Aircraft Armament Unit, Patuxent River, MD, final report on Project No. 63-42 – SB2C-1 Airplane Armament Tests – Production Inspection Trials, Confidential Report NA83/F41-1/F41-5/AA, dated 16 October 1943, *et seq*.

23 *See* J. E. Clark, OC Patuxent River to President of the Board of Inspection and Survey, Armament Test (formerly AAU) Project No. 63-42 – Model SB2C-1 Airplane – Production Inspection Trials – Supplementary Report on Wing Fixed Gun Camera Installation, confidential memo NA83/F41-4/F41-5/AT. PAH/ams, dated 2 May 1944.

24 *See* J. E. Clark, OC AAU, Patuxent River to chief, BuAer – Project TED No. PTR 3277 – SB2C-1 1000# displacement gear – Final Report on, NA83/F41-7/AAU, PAH/ams, confidential memo C-279, dated 8 January 1944.

25 Lee M. Person, *The Attack Bomber – the War Years*, op cit.

26 *See* SD-268-1A, Detail Specification for model SB2C-1 Airplane, revised 1 April 1942.

27 *See* Model SB2C-1 Airplane, Production Inspection Trials, op cit.

28 Amendment #65 to contract no. 1609.

29 AEDS conference report VSB(1) C 0019, dated 21 July, 1943.

30 *See* Vibration Survey of SB2C-1 (#00002) Airplane, Project No. 3446, by aeronautical materials section of the Naval Aircraft Factory Philadelphia, Pennsylvania, report no. AMS(S)-656; dated 25 January 1943.

CHAPTER FIVE

31 NAS Patuxent River letter NA83 VSB2C-1 (FT-AST), dated 18 November 1943.

32 *See* Model SB2C-1 Airplane: Production Inspection Trials, *op cit*.

33 This report was signed by the President of the Board, Rear Admiral Leigh Noyes; Captain E. W. Rounds; Captain L. M. Markham Jr, Recorder of the Board, and Commander W. B. Mechling.

34 *See* Status as of 1 January 1943 of NACA SB2C-1 Flight Investigation AVIA15/2656.

35 Bertram F. Rogers to Commander J. Alton Chinn, 7 March 1985.

CHAPTER SIX

36 *See* memorandum Aer-E-12-LSP C-79052, dated 30 January 1943, from VSB design to head of engineering branch, BuAer- Contact 79082 – Model SB2C-1 Airplane – Serial 00001 Crash of. Commander J. N. Murphy US Navy.

37 *See* H. R. Test No. 42-8. Accelerated Service Test of SB2C-1 Airplane, Final Report of. Commanding officer, NAS Norfolk, Virginia to chief, Bu Aer, dated 26 February 1943, NA8/HR42-8 (96-M1).

38 Memo dated 2 March 1943, M.K.B. to D.G. (AVIA 38/871).

39 Memo dated 3 March 1943, Air Vice Marshal Jones to director general British Air Commission (ABVIA38/871).

40 Memo dated 3 March 1943, E. P Taylor, Washington DC, to D.G. BAC (AVIA 38/871).

41 *See* Richard A. Morley *The Ohio-Curtiss Airplane Connection, Curtiss-Wright, Airplane Division in Columbus 1940–1950*, Journal American Aviation Historical Society, Fall 1991.

CHAPTER SEVEN

42 *See* Wing Commander H. J. Wilson, RAF, Report – Navy SB-2C1 or Army A-25 Dive Bomber, dated 23 March 1943 (AVIA15/2657).

43 Author's italics.

44 *See* Lee M. Pearson, *Development of the Attack Concept*, op cit.

45 *See* confidential memo from F. H. Roberts, inspector of ordnance in charge, Naval Torpedo Station, Newport, Rhode Island, to president Board of Inspection and Survey, Model SB2C-1 Airplane – Torpedo Installation Tests – preliminary report on, dated 22 December 1942.

46 *See* F. H. Roberts, inspector of ordnance in charge, to chief, BuAer NP1/Y-1/F41-9/Br, confidential memo, Model SB2C-1 Airplane – Torpedo Installation Tests – report on, dated 20 January 1943.

47 *See* F. H. Roberts, inspector of ordnance in charge, Newport, Rhode Island, to chief, BuAer NP1/Y-1/SB2C-1/Br, Model SB2C-1 Airplane – Torpedo Installation Tests – Final Report on, dated 6 February 1943.

48 *See* F. G. Fahrion, inspector of ordnance in charge, Newport, Rhode Island, to chief BuAer NP1/Y-1/SB2C-1/Br, Model SB2C-1 Airplane – Recommendations for Use as a Torpedo Airplane, dated 26 February 1943.

49 *See* confidential memo from David I. Hedrick, C.O. US Naval Proving Grounds, Dahlgren, Virginia, to chief, BuAer, dated 23 June 1943, F41-6/(NA7). Serial No. 6152-43 Bomb and Torpedo Handling Equipment, Partial Report of Mock-ups on an SB2C-1 #00149 airplane.

50 Lee M. Pearson, *Development of the Attack Concept*, op cit.

CHAPTER EIGHT

51 *See* Robert Olds, *Helldiver Squadron*, Dodd Mead, New York 1945.

52 Ibid.

53 *See* George F. P. Kernahan, *The Trials of the Helldiver*, op cit.

54 Edward J. McCarten to Commander J. Alton Chinn, 23 November 1982.

55 *See* Captain J. J. Clark to chief, BuAer, Report on Operational Record of SB2C-1 airplane, CV10/F48/C-1021, dated 10 June 1943.

56 *See* Curtiss Wright Corporation, Airplane Division – Columbus Plant, Columbus, Ohio, 9th Revised List of Modification -3, dated 14 August 1943 (AVIA 15/2656).

57 Captain Robert B. Woods, US Navy, rtd, to the author, 27 March 1997.

58 Lee M . Pearson, *The Attack Bomber – The War Years,* op cit.

59 Richard A. Morley, *The Ohio-Curtiss Airplane Connection,* op cit.

60 Ibid.

CHAPTER NINE

61 Commander William S. Palmer, US. Navy rtd., to the author, 29 April 1997.

62 *See* Peter C. Smith, *Jungle Dive-Bombers at War,* London & Washington, 1987, Taipei, 1997.

63 *See* Lieutenant Oliver Jensen, USNR, *Carrier War,* New York, 1945.

64 For a graphic account of this attack, *see* Chapter 3 of Robert Olds' *Helldiver Squadron*, Washington, 1945.

65 *See* Robert Olds, *Helldiver Squadron,* op cit.

66 Commander Geoffrey P. Norman, USN, to Robert Olds, 1945, by permission of Mrs Peggy Olds.

67 Commander William S. Palmer to the author, *op cit.*

68 *See* for example Robert Olds, *Helldiver*; also more recent claims like Geoffrey Norman, *Ships of my Father*, Forbes FYI, and *Mission at Truk Lagoon*, Outside, November 1987 for the author's accounts of dives to find the sunken wreck of this totally mythical Japanese carrier.

CHAPTER TEN

69 *See*, for just one typical example, Robert Stern *SB2C Helldiver in Action*, Carrollton, Texas, 1982, in which it is stated that the RAAF found the Helldiver '…totally unsatisfactory', and that the twenty-six SBWs delivered to the Fleet Air Arm for evaluation '…were found to be unsatisfactory'.

70 *See* Peter C. Smith, *Vengeance!* Shrewsbury 1986.

71 *See* Peter C. Smith, *The Douglas SBD Dauntless*, Ramsbury 1997.

72 *See* George Odgers, *Air War Against Japan 1943–1945*, Canberra 1957.

73 *See* John Curtin, Minister for Defence to General Douglas MacArthur, C-in-C Southwest Pacific Area, dated 23 August 1944, contained in Australian Archives Series AA1969/100; item 452/A69, Shrike Aircraft A69. Dickson A.C.T. Australia.

74 Brig-Gen B. M. Fitch to commander, Allied Air Forces, Shrike dive-bombers from U.S. in possession of R.A.A.F., AG 091.712 (26 Aug 44) DCS, APO 500, dated 26 August 1944.

75 Colonel Perry C. Ragan, Air Adjutant General, to C-in-C, SWPA, AG 091.712 (26 Aug), 1st Ind, dated 8 September 1944.

76 General Douglas MacArthur to Prime Minister John Curtin, dated 16 September 1944.

77 Brig-Gen B. M. Fitch, adjutant general, US Army, to commander, Allied Air Forces, AG 091.721 (16 Sep44) D, 2nd Ind,.

78 Colonel Perry C. Ragan, to commander, Allied Naval Forces, AG 400.318 3rd Inst, dated 29th September 1944.

79 D. S. Crawford, Allied Naval Forces, Southwest Pacific Area to commander Aircraft Seventh Fleet (Logistics), A21 (F-1-5/Bk) serial 00-28-A, dated 5 October 1944.

80 D. S. Fahrney, commander Aircraft Seventh Fleet (Logistics section) to commander Allied Naval Forces, Southwest Pacific Area, dated 12 October 1944.

81 Colonel Perry C. Ragan, to commanding general Far East Air Service Command, APO 923, AG.452.1, 7th Ind, dated 23 October 1944.

82 *See* Squadron Leader D. R. Cuming and Mr G. J. Dailey, directorate of technical services; special duties and performance flight, report on Brief Performance Trials of a Curtiss Shrike (A-25A), Detail No: 192/69/1, dated September 1944; RAAAF HQ File No: 9/41/208; APU File No: A/6/69.

83 Author's italics.

84 *See* Wg Cdr A. N. Hocking to Lt Colonel G. B. Brophy, Re-transfer in SWPA of Ten A-25 Airplanes and Spares,78465 233/209/EQ, dated 2 November 1944.

85 Major E. C. Ackerman to C.O. RAAF HQ, Brisbane, AG 452.01, dated 11 November 1944.

86 Department of Air, Disposal of Shrike Aircraft and Spares, dated 11 November 1944, attached to File 9/63/16.

CHAPTER ELEVEN

87 Italics in original.

88 *See* Captain Eric Brown, *Wings of the Navy*, Shrewsbury, 1987.

89 *See* Chapter 20.

90 *See* BAC to MAP, dated 28 January 1941, MAP 3148 (AVIA38/580).

91 *See* BAC to MAP, Briny 6611, dated 28 June 1941, AVIA38/580.

92 *See* Group Captain H. W. Heslop to Air Marshall Roderic Hill, and Mr C. R. Fairey, Report on Examination of Aircraft, Form 555, Ref: S10C-13, dated 11 November 1941. AVIA38/871.

93 *See* Colonel H. Burchall, BAC, to MAP, dated 28 November 1941; Ref: MAP 10700, AVIA 38/871.

94 *See* Peter C. Smith, *Vengeance*, op cit.

95 *See* Supply of Aircraft to the Fleet Air Arm, ADM116.5534.

96 *See* Naval Air Requirements, ADM1/12126.

97 *See* Supply of Aircraft to the Fleet Air Arm, op cit.

98 Ibid.

99 *See* Lt Cdr Smeeton RN, BAC, Washington DC, to C.N.R. MAP, London, Helldiver Aircraft – Contract Changes to, dated 22 April 1943, S10C-13-SHM 8893, AVIA 15/2656.

100 *See* Smeeton, BAC to CNR, MAP, Helldiver, dated 27 April 1943, Ref: S10C-13-BLBW, 9283, AVIA 15/2656.

101 *See* S.N.R., BAC to C.N.R., MAP, dated 26 July 1943, Briny 8486, AVIA 15/2656.

102 *See* H. Luby, B.A.C., to A.C.N.R. (E), dated 31 August 1943, Briny 8145, AVIA 15/2656.

103 *See* Lieutenant (A) S. J. Miller, BAC, to Mr N. E. Wheeler, MAP, dated 11 September 1943, S10C-13-SJM 2163. AVIA 15/2656.

104 *See* BAC to MAP, Helldiver, dated 3 October 1943, Briny 11051, AVIA15/2656.

105 *See* Cable BAD to DAMR, Admiralty, dated 2 November 1943, AVIA 15/2656.

106 *See* Supply of Aircraft to the Fleet Air Arm, op cit.

107 *See* BAD to Captain John, Admiralty, dated 26 February 1944. AVIA15/2656.

108 As a matter of interest, the reserve dive bomber strength of the Fleet Air Arm at this time was just over 130 aircraft, figures being (27 November 1943) 25 Vought Chesapeake and 108 Blackburn Skua/Rocs; (25 December 1943) 22 Vought Chesapeake and 107 Blackburn Skua/Rocs. Source; State of the FAA at end of 1943, ADM1/ 13502.

109 *See* Top Secret, A. V. Alexander to Prime Minister, A.5013/44, dated 22 March 1944. ADM1/17095.

110 *See* Admiral of the Fleet Viscount Cunningham of Hyndhope, *A Sailor's Odyssey*, London 1951.

111 *See* R. E. N. Kearney, for C.N.R., dated 21 September 1944. ADP.3369/44, 52/11, AVIA 15/2656.

112 Author's italics.

113 Author's italics.

114 *See* DNAO – Helldiver, No 1820 Squadron's Armament Training Reports, AWD 1908/44, 2145/29, AVIA 15/2656.

115 *See* R. E. N. Kearney for CNR, Helldiver – Modification Procedure, Briny 1345, dated 21 February 1945. ADP. 3846/45, 52/21, AVIA 15/2656.

CHAPTER TWELVE

116 *See* Group Captain C. Clarkson, RAF, Report on the Handling Test of Helldiver (SBW1), April 1944. AVIA 15/2657.

117 *See* BAC representative to chief of Test Branch, SBW-1 Helldiver – General and Operational Assessment, BAC, 15/1/SBW, dated 27 April 1944, AVIA 15/2675.

118 Author's italics.

119 Italics in original.

120 *See* Helldiver I JW.117 – Cockpit Layout; dated 12 May 1944; A & AEE Ref: CTO/AT.164; MAP Ref: Res. Air. 5421/11/RDN5/NEW, AVIA 18/1109.

121 *See* Helldiver I JW.117 – Preliminary Handling Trials, dated July 1944; A & AEE Ref: CTO/AT.164; MAP Ref: Res. Air. 5421/11/RDN5/NEW, AVIA 18/1109.

122 *See* Helldiver I JW.117 – Further Handling Trials and Static Stability measurements, dated 16 May 1944; A & AEE Ref: CTO/AT.164; MAP Ref: Res. Air. 5421/11/RDN5/NEW, AVIA 18/1109.

123 *See* Helldiver I JW.117 – Weights and Loading Data, dated 16 August 1944; A & AEE Ref: CTO/AT.164; MAP Ref: Res. Air. 5421/11/RDN5/NEW, AVIA 18/1109.

124 This chapter owes much to information provided by former 1820 Squadron pilot Don F. Sidnell, of Guildford, Surrey, (at that time), whose detailed notes, log books and photographs were freely made available to the author.

125 *See* F. Denys Walter, *Curtiss Helldiver*, printed in *TAG* magazine, January 1971 edition, edited by Mick Dale, and modified version reprinted in September 1984 edition, edited by Jack Bryant.

126 F. Denys Walter, *Curtiss Helldiver*, op cit.

127 D. F. Sidnell to the author, 27 May 1997.

128 F. Denys Walter, *Curtiss Helldiver*, op cit.

129 D. F. Sidnell, to the author, 5 June 1997.

CHAPTER THIRTEEN

130 *See* Captain Mitsuo Fuchida, IJN, *Shore Based Air in the Marianas*, Interrogation no. (USSBS 448) NAN No. 99, dated 25 November 1945.

131 Captain Akira Sasaki, IJN, *Shore-based aircraft in the Marianas Campaign*, Interrogation no. (USSBS 434) NAV No. 91, dated 23 November 1945.

132 Author's italics.

133 *See* Commander Harold L. Buell, US Navy, Rtd, *Dauntless Helldivers*, New York, 1991.

134 For their story in this battle, *see* Crowood Aviation Series *Douglas SBD Dauntless*, op cit.

CHAPTER FOURTEEN

135 *See* Jack Dean, *Taming the Big-Tailed Beast*, *Wings* Magazine, 1971.

136 *See* Action Report, commander Air Group 18, 3-day strike on Visayas Group, Central Philippines, in support of Stalemate II Operation, 12-14 September 1944, ACA-1 91018, dated 23 September 1944, Office of Naval Records and Library, Washington DC.

137 *See* Action Report – Commander Carrier Air Group 13, dated 6 November 1944, serial 078, NND968133. 3.

138 *See* action report, Bombing Squadron Fourteen, Strikes on Japanese Fleet Units in the Battle for Leyte Gulf, 25, 26 October 1944, dated 1 November 1944; serial 0230-93905, Office of Naval Records & Library, Washington DC.

139 *See* Lieutenant (j.g.) W. C. Harding, VB-14 Radio-Radar Performance during period 25 and 26 October 1944, dated 1 November 1944.

CHAPTER FIFTEEN

140 Rear-Admiral Edwin M. Wilson, USNR (rtd.) letter 'Bombing Eleven,' in *The Hook*, winter 1983.

141 Bill Colleran to Commander J. Alton Chinn, 19 March 1984.

142 *See* Harold L. Buell, *Dauntless Helldivers*, New York 1991.

143 *See* Case History of A-25 Airplane Project, Historical Office, Material Command, Wright Field, microfilm roll number A2061, AF Historical Research Agency, Alabama.

144 *See* Peter C. Smith, Crowood Aviation Series, *Douglas SBD Dauntless*, Ramsbury 1997.

145 *See* Staff Sgt Theron J. Rice, *Helldiving Devildogs*. Curtiss *Flyleaf* magazine.

CHAPTER SIXTEEN

146 *See* commander Carrier Air Group 83, Report of Air Operations conducted by Carrier Air Group Eighty-Three during the period 14 March – 1 June 1945., Vol. 2 of 4 Vols. serial 063-126672, dated 9 June 1945. NDD968133, Office of Naval Records, and library, Washington DC.

147 *See* Bombing Squadron (VB) 9 – ACA Reports 16 Feb – 10 Jun 45, numbered from 1 through 69, NND 968133.

148 *See* Charles M. Crump, air intelligence officer VB-85, *History of Bombing Squadron 85*, privately published, kindly made available for free use by the author, September 1997; copy in author's collection.

149 Ibid.

150 Charles Crump, *History of Bombing Squadron 85*, op cit.

151 Ibid.

CHAPTER SEVENTEEN

152 *See* Lee M. Pearson, *The Attack Bomber*, op cit.

153 Ibid.

154 Charles M. Crump, *History of Bombing Squadron 85*, op cit.

155 For the full story of Mann and Hanna's experiences at the hands of their Japanese captors, *see* The History of Bombing Squadron 85, op cit.

156 *See* Hill Goodspeed, *Unlikely Casualty – A Naval Aviator's path to Hiroshima*, Foundation, Fall, 1995.

157 William B. Pinkerton to the author, 26 September 1997.

158 *See* Charles M. Crump, *History of Bombing Squadron 85*, op cit.

159 Maynard J. Mitchell to the author, 3 September 1997.

160 Maynard J. Mitchell to the author, 18 September 1997.

CHAPTER EIGHTEEN

161 Roger H. Went to the author, 2 October 1997.

162 Edward J. McCarten to Commander J. Alton Chinn, 11 February 1985.

163 Captain John D. Shea Jr. to Commander J. Alton Chinn, 4 January 1982.

164 Charles R. Shuford to the author, 4 April 1997.

165 Benjamin G. Preston to the author, 10 July 1997.

166 CWO John Patrick Piercy to the author, 22 March 1997.

167 Captain Chuck Downey to the author, 27 April 1997.

168 Nos.18780/18774/18884/18830/18891/18828/18751/8825/18828/18836/18784/18888/18870/18836/18809/18781/18859/18948.

169 Nos. 18396/18450/18407/01082/18400/00271/00349/18394/01082/00338/00352/00129/18352/18450/01129/01082/00362/01178/01129/00352/18540/18415/18349/00249.

170 Ron Hinrichs to the author, 8 January 1997.

171 Commander Richard W. Mann, US Navy (rtd), to the author, 11 September 1997.

172 Commander Richard W. Mann, US Navy (rtd), to the Author, 12 August 1997.

173 Ron McMasters to the author, 19 June and 14 July 1997.

174 Colin Shadell to the author, 14 August 1997.

175 Stanley I. Vaughn II to the author, 10 April 1997.

176 *See* the '*In Essence*' column of the *Columbus Dispatch*, issue dated 8 August 1997.

CHAPTER NINETEEN

177 *See Curtiss-Wright Helldiver*, prepared by the West Texas Wing of the Confederate Air Force, Midland International Airport, Midland, Texas, courtesy of Colonel Ted Short.

178 *See Dictionary of American Naval Aviation Squadrons, Volume 1, The History of VA, VAH, VAK, VAL, VAP and VFA Squadrons*, Washington, DC, for full details of these changes.

179 *See* Peter C. Smith, Crowood Aviation Series, *Douglas AD Skyraider.*

180 *See Attack Squadron Six Baker*, dated 1 October 1947, VT-17, VA-6B. op cit.

181 *See Report, Attack Squadron Five Baker*, dated 20 November 1947, VA-5 BAKER/a9-3; 11/JGH/hs. Serial 265. National Archives, Washington DC.

182 Arthur Pearcy to the author, 14 July 1997. *See* also Arthur Pearcy, *U. S. Coast Guard Aviation*, Shrewsbury, 1989.

183 Captain N. Kostakis, Hellenic Navy, to the author, October 1997.

184 *See L'Aviazione di Marina*, by C. De Risio, (Ufficio Storica della Marinea Militare, Roma, 1995; and *Gli Helldiver Italiani*, by Giancarlo Garello, (Aerofan 59 – Ott-Dic 96), Milan.

CHAPTER TWENTY

185 *See* Peter C. Smith, Crowood Aviation Series, *Douglas SBD Dauntless*, op cit.

186 *See* Captain Patou, *Groupe Porte Avions d'Extrême Orient: Rapport d'Opérations, Période du 26 Decembre 1953 au 2 Juin 1954*; detailed combat reports in book form, 250 pages, photocopy in author's collection.

187 *See* Captain Patou, *Groupe Porte Avions d'Extrême Orient: Rapport d'Opérations, Période du 26 Decembre 1953 au 2 Juin 1954*; detailed combat reports in book form, 250 pages, photocopy in author's collection.

188 *See* Jacques Mordal, *Marine Indochine*, Paris, 1953.

189 Op cit.

190 *See* Jules Roy, T*he Battle of Dien Bien Phu*, New York, 1965.

191 *See* General Vo Nguyen Giap, *Dien Bien Phu*, Hanoi, 1964.

192 *See* General Chassin, *Aviation Indochine*, op cit.; et seq.

193 *See* Captain Patou, *Group Porte*, op cit.; et seq.

194 *See* Peter C. Smith, *Jungle Dive Bombers at War*, London, 1987.

195 The names of the various strongpoints that made up the Dien Bien Phu fortress complex were reputedly named after the many mistresses of Colonel de Castries – in which case he certainly had a string of conquests to his name! However, one strongpoint, at least, *Anne-Marie*, was named after the heroine of a lewd Foreign Legion song.

196 *See* Bernard Fall, *Street Without Joy*, Harrisburg, 1966.

197 *See* Charles Favrel, *Le Monde*, 22 April 1954 edition.

198 *See* Bernard Fall, *Street Without Joy*, op cit.

Index